Ontogeny and Phylogeny of the Vertebrate Heart

David Sedmera • Tobias Wang
Editors

Ontogeny and Phylogeny of the Vertebrate Heart

Springer

Editors
David Sedmera
Institute of Anatomy
First Faculty of Medicine
Charles University in Prague
Prague, Czech Republic

Institute of Physiology
Academy of Sciences of the Czech Republic
Prague, Czech Republic

Tobias Wang
Department of Biological Sciences
Zoophysiology, University of Aarhus
Denmark

ISBN 978-1-4614-3386-6 ISBN 978-1-4614-3387-3 (eBook)
DOI 10.1007/978-1-4614-3387-3
Springer New York Heidelberg Dordrecht London

Library of Congress Control Number: 2012937860

Printed on acid-free paper

Springer is part of Springer Science+Business Media (www.springer.com)

Preface

Cardiovascular physiology has always been a central theme in both medicine and biological sciences. While it has long been established that the fundamental mechanisms of cardiovascular function apply to all animals, it is also becoming clear that most animals, i.e., both invertebrates and vertebrates, share the same genetic programming for the development of the heart and major blood vessels. For example, the homeobox gene tinman required in *Drosophila* for the specification of the heart is homologous to a number of *Nkx* genes that are essential for cardiac development in vertebrates. It is also well established that many of the cardiac features of adult fish, amphibians, and reptiles resemble stages in the ontogenetic development of the mammalian heart as well as some common cardiac malformations in humans. As an example, the ventricular septum defects that occur during congenital heart diseases in mammals resemble incomplete ventricular septum in some amphibians and reptiles. Also, the spongious heart of embryonic birds and mammals resembles the adult heart of ectothermic vertebrates.

As the homology of the molecular mechanisms driving cardiac development in different vertebrates and invertebrates has been unraveled, many laboratories, primarily interested in the molecular mechanisms underlying cardiac malformations in humans, are now conducting much of their research on ectothermic vertebrates such as *Xenopus* and zebrafish as well as invertebrates such as *Drosophila*. This is primarily because these animals are much easier to maintain and manipulate than mammals, and hence, does not necessarily reflect an interest per se in the cardiovascular physiology and development of these animals. In parallel to more clinically oriented research on cardiac development, comparative physiologists have studied cardiac function in all vertebrate groups for more than a century and the interest in the ontogenetic development of the cardiovascular function in non-mammalian vertebrates has been a major topic over the past decades. Much of this research has been driven by a desire to understand how the cardiovascular system of these animals functions and how this function can be understood in terms of the evolution of vertebrates and their adaptations to different environments or behaviors.

Hence, as outlined above, two separate research fields—the comparative physiology and the clinically oriented embryology—have studied similar functions in parallel.

Surprisingly, little exchange of information has occurred over the past two decades. The primary objective of this volume is to compile contributions from members of both fields to an in-depth, timely, and comprehensive collection of reviews dealing with the major topics of the ontogeny and phylogeny of cardiac function. The exchange of ideas should be of benefit for both fields, as some techniques originally established in the clinical arena (e.g., the echocardiography or magnetic resonance) recently have become applicable to embryos as well as small adult vertebrates, such as zebrafish. On the other hand, physiological measurements performed in the experimental models have helped to explain some special features to human development such as resistance to ischemia due to its adaptation to physiologically low oxygen tension in the fetal tissues ("Mount Everest in utero") that are not easily measurable in the clinical settings.

Despite the existence of at least two recent excellent textbooks on cardiac development (and regeneration) by Margaret Kirby (Cardiac Development, Oxford University Press, 2007) and Rosenthal and Harvey (Heart Development and Regeneration Vols. 1 and 2, Elsevier, 2010), there is no recent book that integrates the ontogenetic and phylogenetic changes in form and function. Therefore, it is our hope that this collection of reviews will be of considerable interest to biologists and MDs working on any aspect of cardiovascular function. Being state-of-the-art reviews written by competent experts in the field, the content is also of interest for MSc and PhD students in most fields of cardiovascular physiology.

Prague, Czech Republic David Sedmera
Aarhus, Denmark Tobias Wang

Contents

1 **Comparative Evolution and Design in Non-vertebrate Cardiovascular Systems** 1
B.R. McMahon

2 **The Teleost Heart: A Morphological Approach** 35
José M. Icardo

3 **Fish Heart Growth and Function: From Gross Morphology to Cell Signaling and Back** 55
B. Tota and F. Garofalo

4 **A Perspective on the Evolution of the Coronary Circulation in Fishes and the Transition to Terrestrial Life** 75
A.P. Farrell, N.D. Farrell, H. Jourdan, and G.K. Cox

5 **The Sarcoplasmic Reticulum in the Vertebrate Heart** 103
Gina L.J. Galli and Holly A. Shiels

6 **Evolution of the Regulatory Control of the Vertebrate Heart: The Role of the Contractile Proteins** 125
Todd E. Gillis

7 **Ontogenesis of Myocardial Function** 147
David Sedmera and Bohuslav Ostadal

8 **Basic Cardiac Development: The Heart and Its Electrical Components** 177
L.Y.E. Wong, A.F. Moorman, and P. Barnett

9 **The Functional Significance of the Reptilian Heart: New Insights into an Old Question** 207
James W. Hicks and Tobias Wang

Index 229

Contributors

P. Barnett Department of Anatomy, Embryology and Physiology, Academic Medical Center, Amsterdam, The Netherlands

G.K. Cox Department of Zoology, University of British Columbia, Vancouver, BC, Canada

A.P. Farrell Department of Zoology, University of British Columbia, Vancouver, BC, Canada

N.D. Farrell Department of Zoology, University of British Columbia, Vancouver, BC, Canada

Gina L.J. Galli Faculty of Medical and Health Sciences, Core Technology Facility, The University of Manchester, Manchester, UK

F. Garofalo Department of Cell Biology, University of Calabria, Cosenza, Italy

Todd E. Gillis Department of Integrative Biology, University of Guelph, Guelph, ON, Canada

James W. Hicks Department of Ecology and Evolutionary Biology, University of California, Irvine, CA, USA

José M. Icardo Departamento de Anatomía y Biología Celular, Facultad de Medicina, Universidad de Cantabria, Santander, Spain

H. Jourdan Department of Zoology, University of British Columbia, Vancouver, BC, Canada

B.R. McMahon Biological Sciences, The University of Calgary, Calgary, AB, Canada

A.F. Moorman Department of Anatomy, Embryology and Physiology, Academic Medical Center, Amsterdam, The Netherlands

Bohuslav Ostadal Institute of Physiology, Academy of Sciences of the Czech Republic, Prague, Czech Republic

David Sedmera Institute of Anatomy, First Faculty of Medicine, Charles University in Prague, Prague, Czech Republic

Institute of Physiology, Academy of Sciences of the Czech Republic, Prague, Czech Republic

Holly A. Shiels Faculty of Life Sciences, Core Technology Facility, The University of Manchester, Manchester, UK

B. Tota Department of Cell Biology, University of Calabria, Cosenza, Italy

Tobias Wang Department of Biological Sciences, Zoophysiology, University of Aarhus, Denmark

L.Y.E. Wong Department of Anatomy, Embryology and Physiology, Academic Medical Center, Amsterdam, The Netherlands

Chapter 1
Comparative Evolution and Design in Non-vertebrate Cardiovascular Systems

B.R. McMahon

Within the simplest multi-cellular organisms extracellular fluid transport occurs as a result of movements of associated body structures during locomotor and other movements. A similar phenomenon is seen in development within many larval stages of more complex animals where, prior to the initiation of heart beat, fluid movement results from contraction of gut and body muscle systems e.g. in Crustacea (McMahon et al. 2002; Spicer 2006). Later in both evolution and development, specific pumping structures provide the driving force for extracellular fluid movement. Although these vary significantly in anatomy and origin, they are usually termed "hearts". Xavier-Neto et al. (2007) limit the use of the term "heart" to "pumps containing inflow and outflow chambers invested at some point in an animal's lifetime with myocytes". This effectively restricts "hearts" to molluscs and vertebrates. This limitation is not well supported on either anatomical or functional grounds (as discussed later) and this account includes tubular hearts found in several phyla. These structures in combination with the circulating fluid and the distribution and return vessels may be collectively termed the cardiovascular system [effectively the blood vascular system (Ruppert and Carle 1983)].

Origin of Cardiovascular Systems

It seems certain that for most bilateral animal groups the ancestral body plan was organised on a segmental basis. Evidence to support this is found in numerous molecular studies (reviews De Robertis 1997, 2008) which conclude that gene systems involved in the organisation of segmentation and the control over the dorso-ventral and antero-posterior axis during development are essentially similar

B.R. McMahon (✉)
Biological Sciences, The University of Calgary,
Calgary, AB, Canada T2N 1N4
e-mail: mcmahon@ucalgary.ca

D. Sedmera and T. Wang (eds.), *Ontogeny and Phylogeny of the Vertebrate Heart*,
DOI 10.1007/978-1-4614-3387-3_1, © Springer Science+Business Media, LLC 2012

in both insects and vertebrates indicating a common segmented ancestor present before the split between protostome groups (leading to the arthropods and most higher invertebrates) and deuterostome groups (leading to the cephalochordates and vertebrates) i.e. prior to 600 million years ago. Because of the nature of segmentation (serial repetition of homologous parts and separation of the segments by septa) we must conclude that this segmental organisation heavily influenced the evolution of ancestral bilaterian cardiovascular systems (Moorman and Christoffels 2003; Olsen 2006). Some remaining controversial aspects of these conclusions are reviewed by Xavier-Neto et al. (2007). Basically thus invertebrate and vertebrate cardiovascular systems arise from a common segmented ancestor. Thus cardiovascular systems of all vertebrate, and the majority of invertebrate, animals originated as paired tubular vessels, one lying dorsal and one ventral to the alimentary tract (Pass 2000; Hertel and Pass 2002; Burggren and Reiber 2007). Either or both vessels may be contractile. It is likely that this ancestral vascular system was developed for an extended body plan perhaps morphologically similar to those shown in some simpler current invertebrates as shown in Fig. 1.1.

Six hundred million years of evolution has, however, allowed extreme variation from the basic system allowing it to be adapted to suit many diverse body plans and to function in many environments. The resulting cardiovascular systems can be much more complex or much simpler than those illustrated in Fig. 1.1, but their segmental origins are usually clear. This is particularly true in the invertebrate phyla where the cardiovascular system has had to adapt to extreme variability in both size and body plan. Vertebrate cardiovascular systems are much less diverse, as discussed elsewhere in this volume, and this chapter illustrates major trends in the functional evolution of invertebrate cardiovascular systems and where possible compares these to vertebrate systems. The evolutionary diversity of invertebrate cardiovascular systems has been reviewed recently (McMahon 2001; Burggren and Reiber 2007) and for brevity this chapter focuses on the ontogeny and phylogeny of the cardiovascular system in arthropods. Within this group the morphology and physiology of cardiovascular systems of Onycophora and Chilopoda (Hertel et al. 2002), scorpions (Wirkner and Prendini 2007), spiders (Paul et al. 1994) and insects (Pass 2000; Hertel and Pass 2002) have been well reviewed recently as has the morphology of vascular systems in peracarid crustaceans generally (Wirkner and Richter 2010). This review focuses largely on decapod Crustacea where there has also been a surge of recent research interest. Fortunately within this restricted phylogenetic range we can well illustrate the major evolutionary trends which have occurred in the invertebrates generally.

Open and Closed Circulatory Systems

The cardiovascular systems of vertebrates are described as closed systems where, by definition, the circulating fluid (blood) is always separated from the tissue fluid (lymph) by a cellular endothelial layer. Shigei et al. (2001) consider this endothelial

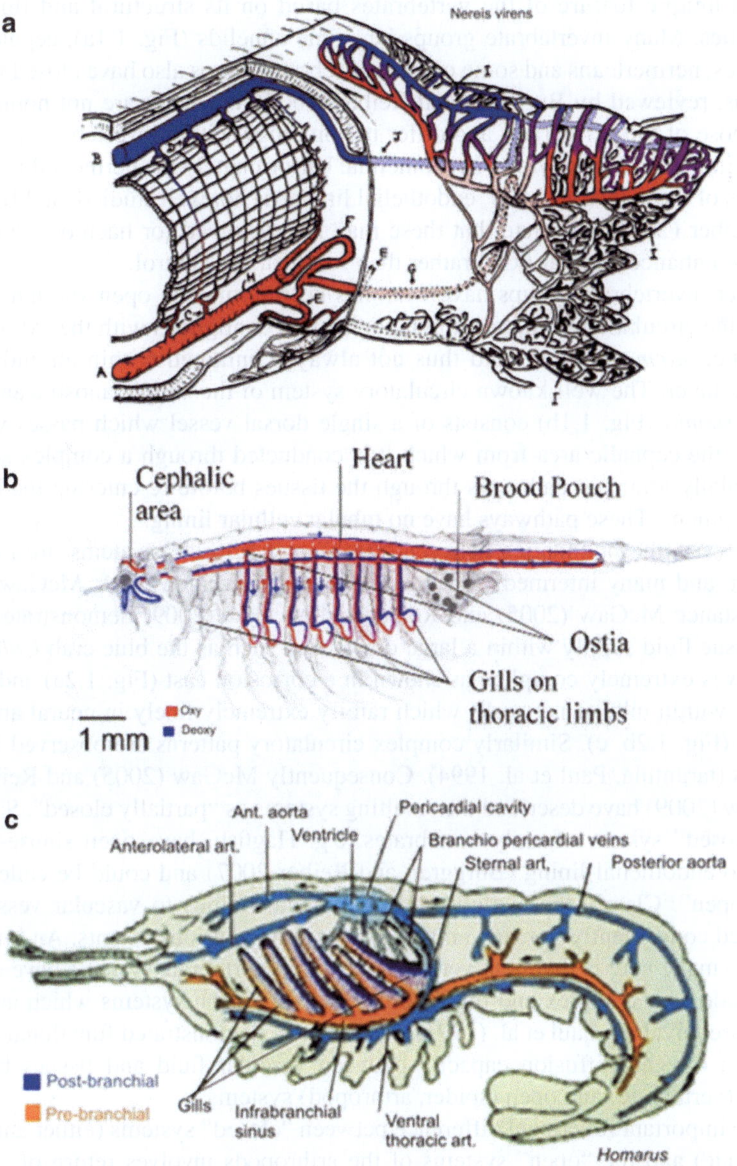

Fig. 1.1 (**a**) Cardiovascular pattern of a typical adult segment of *Nereis virens*. Right parapodium viewed from the posterior surface. A, ventral longitudinal vessel; B, dorsal longitudinal vessel; C, G-I plexus to ventral vessel connectives; D, G–I plexus to dorsal vessel connectives; E, recurrent branch—ventral lateral vessel; F, G-I plexus lateral connective; G, medial branch—ventral lateral vessel; H, stem ventral lateral vessel; I, blind ending "capillaries"; J, dorsal–lateral vessel. Modfied from Nicholl (1954). Oxygenated fluid in *red*, deoxygenated fluid in *blue*. (**b**) Classic open cardiovascular system from the anostracan crustacean *Artemia* showing the extended heart and fluid distribution pathways through each segment of the open haemocoel (From McMahon 2001). (**c**) Condensed heart and extended distribution system in the lobster, a typical macruran decapod crustacean. Oxygenated hemolymph here is *blue* and deoxygenated fluid is *orange*. Chelipeds removed. Modified from McLauchlin (1983)

layer a unique feature of the vertebrates based on its structural and functional properties. Many invertebrate groups i.e. some annelids (Fig. 1.1a), cephalopods, molluscs, nermerteans and some other invertebrate groups also have closed vascular systems, reviewed by Burggren and Reiber (2007), but these are not homologous with those of the vertebrates and differ in homology within invertebrate phyla and even within the same phylum, i.e. Annelida. Unfortunately the permeability characteristics of these invertebrates' endothelial linings is not well studied, and Burggren and Reiber (2007) postulate that these may have evolved for haemodynamic reasons (to enhance laminar flow) rather than for diffusive control.

Other invertebrate groups have what has been termed an open vascular system where the circulating fluid, haemolymph, is often contiguous with the extracellular fluid (i.e. *haemo* + *lymph*) and thus not always contained within an endothelial cellular layer. The well-known circulatory system of the simple anostracan crustacean *Artemia* (Fig. 1.1b) consists of a single dorsal vessel which passes vascular fluid to the cephalic area from which it is conducted through a complex series of segmentally arranged pathways through the tissues before re-entering the heart in each segment. These pathways have no tubular cellular lining.

However, the boundaries between "open" and "closed" systems are rarely so distinct, and many intermediate stages are seen (McMahon 2001; McGaw 2005). For instance McGaw (2005) and Reiber and McGaw (2009) demonstrate clearly that tissue fluid supply within a large crustacean such as the blue crab *Callinectes sapidus* is extremely complex, as shown in a corrosion cast (Fig. 1.2a) and occurs largely within tubular networks which ramify extremely finely in neural and other tissues (Fig. 1.2b, c). Similarly complex circulatory patterns are observed in large spiders (tarantula, Paul et al. 1994). Consequently McGaw (2005) and Reiber and McGaw (2009) have described the resulting systems as "partially closed". Similarly the "closed" system of early vertebrates, e.g. Hagfish, have open sinuses which have no endothelial lining (Burggren and Reiber 2007) and could be called "partially open". Clearly the evolution of endothelial linings to vascular vessels has occurred convergently in many animal groups and to various extents. An important point to make here is that both vertebrate and non-vertebrate systems have evolved to include very complex and highly efficient distribution systems which are functionally equivalent. Paul et al. (1991), for instance, demonstrated functional equivalence in oxygen diffusion capacity between vascular fluid and tissues between closed (vertebrate) and open (spider, arthropod) systems.

One important functional difference between "closed" systems (either annelid or vertebrate) and the "open" systems of the arthropods involves return of vascular fluid to the heart. In a classic "closed" system fluid returning to the heart is conducted in finite tubes lined with cells, which connect directly into the heart. In arthropod "open" vascular systems, however complex, there is no direct tubular connection and fluid returning to the ventricle is collected in a pericardial cavity surrounding the heart. Fluid entry into the ventricle can occur only via paired valved apertures (ostia) which perforate the ventricular wall and filling of the heart must be powered by external mechanisms. This difference is fundamental to the organisation of the heart in arthropods. Without tubular input vessels haemolymph must be

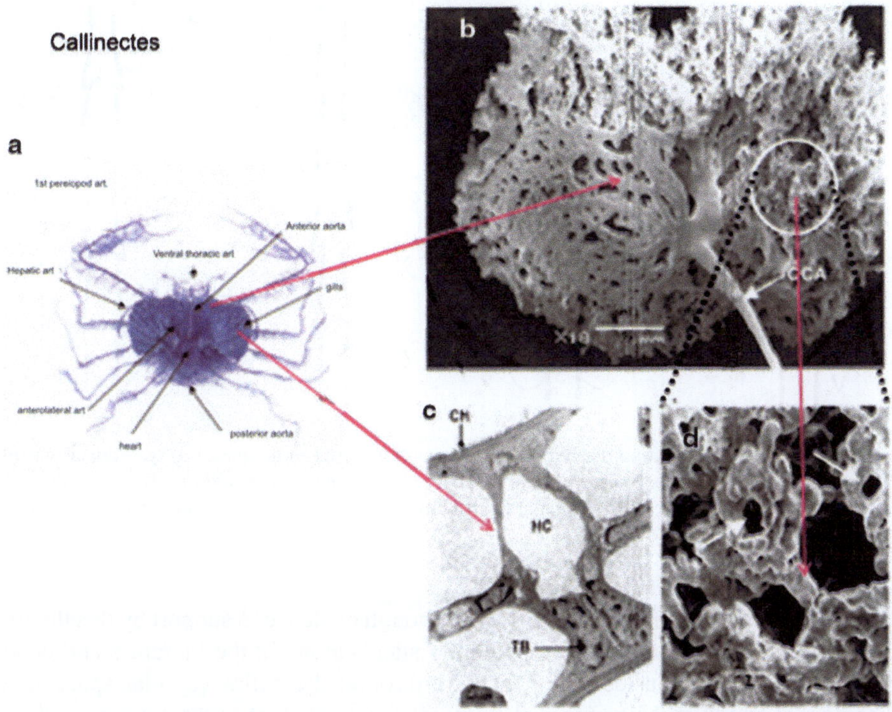

Fig. 1.2 Fine details of the circulatory system of the blue crab *Callinectes* as revealed by resin casting. (**a**) whole cast; (**b**) Fine structure of the antennal gland; (**c**) fine structure of a gill lamella. *ch* chitinous exchange surface, *hc* haemolymph channel, *tb* pillar cell. From Reiber and McGaw (2009)

collected in the pericardial space surrounding the heart and conveyed through the myocardial wall via paired ostial valves. Energy for this inflow is provided for by musculo-elastic ligaments (alary ligaments) connecting the heart to surrounding structures. These ligaments are stretched during systole and thus store some of the energy of myocardial contraction. In diastole this energy acts to return the heart to its presystolic volume as fluid flows in through the ostial valves. In many simple arthropods with the ancestral type extended heart ostial valves and alary ligaments occur in each segment, allowing diastolic filling in each segmental unit.

The arthropod pericardium and pericardial cavity are distinctly different from all other similarly labelled structures in the animal kingdom in that they are derived from the haemocoel and not from the true coelom, as occurs in all other pericardial systems. The exact genesis of this unique structure is not known, but the theory introduced by Lankester (1873, 1887) who postulated that the entire structure of the arthropod heart and pericardial cavity assemblage may have evolved from an ancestral dorsal vessel plus a series of valved segmental vessels perhaps similar to annelids (Fig. 1.1), by swelling and eventual union of the segmental vessels to form a cavity surrounding the heart (phleboedesis) and the formation of segmental ostial

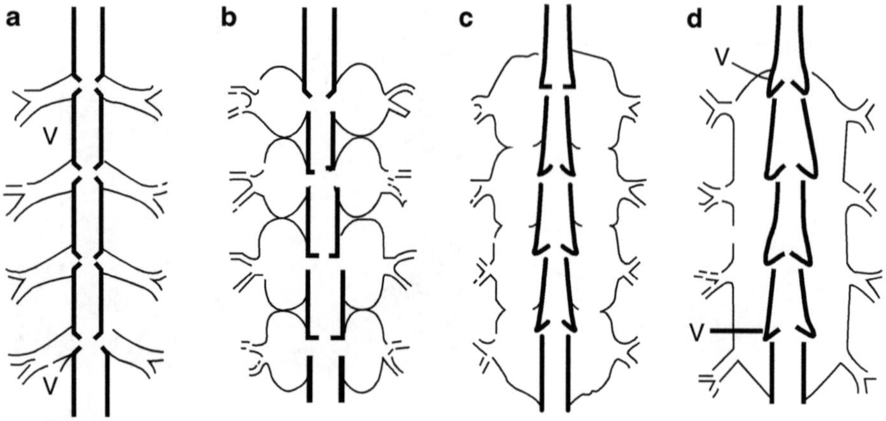

Fig. 1.3 Diagram illustrating the hypothetical transition between an ancestral type dorsal vessel and a generalised arthropod heart with four pairs of ostia; (**a**) ancestral condition; (**b, c**) hypothetical intermediate stages; (**d**) arthropod tubular heart with 4 ostia. *V* valve (from Lankester 1873, 1877)

valves in the wall of the heart (Fig. 1.3). Although evidence to support by this theory is difficult to obtain, the theory apparently stands alone in the literature and is, at least, worthy of reconsideration. In the arthropod the entire vascular space is a haemocoel and thus the pericardial cavity does not connect to the exterior as does the pericardial cavity of the molluscs and (during development) of vertebrates. Thus ultrafiltration which may be a primary influence on the development of renal structures of these latter forms cannot occur in the pericardial cavity of the arthropod open system. Similarly the pericardial cavity of molluscs is of coelomic origin and cannot be compared directly with that of the arthropods.

Phylogeny of the Cardiovascular System in Invertebrates

The postulated extended common ancestral pattern tubular hearts depicted in Fig. 1.1 arose in animals showing extensive metamerism in which many organs (e.g. gills, nephridia etc.) were duplicated segmentally. Contraction in these extended tubular hearts was originally peristaltic (Xavier-Neto et al. 2007 but see Forouhar et al. 2006), and this condition continues in many smaller crustaceans. The process of evolution, however, often leads to the production of larger and more complex forms (e.g. Adamowicz et al. 2008). The increase in tissue mass and complexity requires increased metabolic fuel supply and hence increased fluid flow. Further, because of diffusion limitations, especially of oxygen, this leads to increase in complexity of the distribution systems. Increased fluid supply through a more complex perfusion circuit requires increase in heart power to overcome the substantially increased flow resistance. Several mechanisms have evolved to provide the increased power.

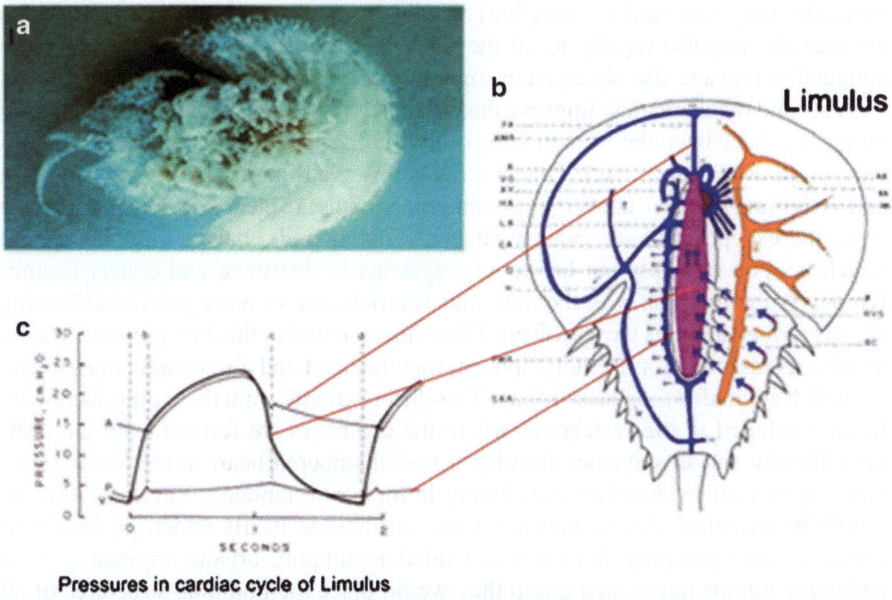

Fig. 1.4 *Limulus* circulatory system (**a**) corrosion cast; (**b**) diagram of heart and major pre- and post-branchial connections; (**c**) hemolymph pressures developed during a single cardiac cycle. Modified from McMahon (2001) and Redmond et al. (1982)

The first is an increase in muscle mass. Again using arthropods as an example, the heart of small anostracan crustaceans such as the brine shrimp *Artemia* consists only of a single-layered myocardium (Økland et al. 1982) which extends over the full length of the animal. Larger animals have developed a more muscular heart. The extended heart of a relatively large crustacean *Squilla* has at least 2 muscle layers whereas the condensed lobster heart may have 50 layers. An early example is illustrated by the heart of the Horseshoe Crab (*Limulus polyphemus:* Xiphosura, Arthropoda). This animal represents an ancient level of organisation, but is relatively large with a correspondingly complex tissue perfusion system (Fig. 1.4a courtesy George Bourne). The *Limulus* heart remains an extended segmentally organised tube but is now heavily muscularised (Fig. 1.4b) and capable of generating reasonably high pressures and flows (Fig. 1.4c; Redmond et al. 1982) despite the complexity of the distribution system.

Increased muscle mass alone is insufficient to fully overcome the increased flow resistance of a complex tissue delivery system. It is unlikely that the original peristaltic hearts could pump large volumes of fluid against the increased tissue resistance found in larger more complex animals. Thus along with the increased muscle mass animals needed to increase the rate of excitation and the rate of conduction within the myocardium so that contraction occurred almost simultaneously over the whole surface. In *Limulus* and in many advanced arthropods these factors are achieved by (a) rapid conduction mechanisms and (b) by the presence of local neuronal networks (e.g. the cardiac ganglion) which set an enhanced beat rate and also conduct the impulse rapidly to all the myocardial cells. Mechanisms to increase conduction rate are also observed in some insects i.e. *Drosophila* (Slama 2010).

A second mechanism to improve the efficiency of the ancestral peristaltic tubular pump may have been the formation (evolution) of separate heart chambers (Xavier-Neto et al. 2007 for review). This process is observed in only two phyla Mollusca and Vertebrata in which convergent evolutionary paths are clearly involved. Mollusca (with the exception of secondarily adapted Cephalopods) have a vascular system which is open centrally but has tubular systems to distribute and collect haemolymph. At the junction with the muscular ventricle one or more pairs of collecting vessels termed auricles have evolved. These are essential in this low-pressure system to serve as reservoir for haemolymph entering the heart and also contain muscle layers which aid in diastolic heart filling. This differs greatly from the multi-chambered heart developed in the vertebrates where the chambers are formed from the heart tube directly and in sequence forming a two-chambered heart in the lower vertebrates with further chambers developing in higher vertebrates. Xavier-Neto et al. (2007) differentiate clearly between these chambered hearts which possess both collecting and pumping chambers and tubular pumping organs originating from peristaltic pumps into which group they would place the pumping structures of all arthropods. This is reasoning is spurious since, as they point out, the initial heart of the lower vertebrates develops as a straight, presumably peristaltic tube. In contrast, the hearts of many advanced arthropods are highly muscularised condensed non-peristaltic structures in which the pericardial cavity clearly functions as a reservoir chamber which clearly aids diastolic filling and may also actively adjust stroke volume.

Regional Specialisation: Especially Gas Exchange Structures

Increased muscularity of the heart has occurred in many larger invertebrates, but other evolutionary trends have also strongly affected cardiac form and efficiency. In arthropods a major trend associated with increase in complexity is the evolution of regional specialisation, particularly in the appendages (Adamowicz et al. 2008; Wirkner 2009). The ancestral biramous arthropod appendages functioned primarily as locomotor and filter feeding devices and, because of their relatively large surface area, also subserved the function of gas exchange. With specialisation for a different function (e.g. walking limbs) the appendages of higher crustaceans have had to increase in size and exo-skeletal thickness, both factors reducing their effectiveness as gas exchange structures. The resultant reduction in gas exchange area, combined with an increased oxygen demand, has necessitated modification of relatively few limb pairs within a specific body region to function as high efficiency gas exchangers, i.e. gills.

This has important consequences for cardiovascular design. In the least complex systems such as *Artemia* (Fig. 1.1b) we note that vascular fluid is oxygenated within each segment and thus all fluids reaching the pericardial cavity and heart, and hence distributed to tissues, are oxygenated. However in an animal in which gas exchange had become restricted to a particular bodily region, oxygenated fluid could only enter the pericardial cavity from that restricted area. Thus in an extended tubular heart system, admixture of deoxy- and oxygenated fluids would occur within the heart and pericardial cavity, reducing the diffusion gradient and thus the efficiency of gas exchange mechanisms. Animals have evolved a variety of mechanisms to prevent this admixture i.e. to separate oxygenated (post-branchial) from deoxygenated (pre-branchial) fluid streams. This account illustrates many of these mechanisms within the Arthropoda and can be compared with those of vertebrates discussed elsewhere in this volume.

First, the active pumping area (heart) needs to be shortened, strengthened and concentrated to lie within the area responsible for gas exchange The location of this varies widely for instance in *Limulus* (Fig. 1.4) and in isopod *crustaceans* and spiders the gill books are located posteriorly in the abdomen, but in most decapod crustaceans gills (Fig. 1.1c) are located more anteriorly. Reference to the ancient arthropod *Limulus* (Fig. 1.4) indicates an early (partial) solution to the problem of potential admixture of oxygenated and deoxygenated fluid. In this large animal the heart is somewhat shortened but remains an elongated tube extending further than the area specialised for gas exchange suggesting that oxy- and deoxy-fluid streams could mix in the pericardium. Oxygenated fluid from the hind end is transported preferentially; however, since (a) the diameter of the heart is greater posteriorly and (b) the propulsive wave propels a bolus of fluid forward into a narrowing lumen producing a surge in pressure, which closes the ostia in the anterior region effectively preventing admixture. Although we may consider this only a partial solution, we should remember that a similar system is retained in scorpions and occurred in a related ancient group, the eurypterids, some of which were more than 2-m-long predators, so it can be presumed to have worked effectively.

In our example, crustaceans, vertebrates and many other invertebrates, the effective pumping (more muscular) region of the heart tube and the surrounding pericardial cavity is further condensed and becomes restricted to those segments involved in gas exchange (Fig. 1.1c). This allows only oxygenated fluid to reach the pericardial cavity and enter the heart. The condensation of the heart has further advantages in that it allows a more rapid and simultaneous contraction of the myocardium allowing more forceful ejection of fluid. This is particularly advantageous in the more highly evolved and complex crustaceans, as it allows simultaneous ejection fluid into the several distribution vessel systems which leave the heart at a range of locations (Fig. 1.1c). Furthermore, as all gill outflow enters the heart and pericardial cavity at this limited location good mixing of the independent branchial streams from each gill occurs.

In both vertebrates (Farrell et al. 2012) and most invertebrates heart coalescence still involves the amalgamation of several segmental regions of the original heart tube. In adult crabs three pairs of ostia and three sets of segmental arteries are seen in the adult heart suggesting that at least three segments are involved. Similarly three pairs of ostia are seen to arise progressively in the development of the sand shrimp *Metapenaeus ensis* (McMahon et al. 2002) but Mayrat et al. (2006) suggest that as many as six segments may have been be involved during evolution. Wilkens (1999) suggested that bending of the heart tube into an S shape could be involved, as seen in vertebrates (Wong et al. 2012), but this was not visible during the development of the sand shrimp *M. ensis* (McMahon et al. 2002).

Functioning of the Heart

The functioning of invertebrate hearts generally has been reviewed by McMahon et al. (1995a, b) and Burggren and Reiber (2007); that of the hearts of several other arthropod groups is reviewed by Paul et al. (1994), Hertel and Pass (2002), Hertel et al. (2002) and Wirkner and Prendini (2007); and of crustaceans by Reiber and McGaw (2005) and Mayrat et al. (2006). As previously, this review briefly outlines cardiac function in several decapod crustaceans and reviews new material from recent studies.

The Pericardium and Pericardial Cavity

Functionally the arthropod pericardium is unlike that of the mammalian heart in that it is not closely applied to the heart (compliant), but is variable in volume (non-compliant) and supported by ligaments which connect to the exoskeleton and with a variety of connections to muscular and other abdominal systems. Non-compliant pericardial systems occur commonly in, but are not restricted to, invertebrates, occurring also in elasmobranchs and Dipnoi. As in the elasmobranchs, the pericardial cavity of arthropods is essential to the functioning of the heart since it acts as a reservoir for heart filling (McMahon 2001; Mayrat et al. 2006).

The arthropod pericardial cavity acts as a collecting reservoir for oxygenated fluid arriving from the gas exchange areas. In higher crustaceans, at least, innervated muscle systems are found in the pericardial floor and in the alary ligaments (Volk 1988; Mayrat et al. 2006) while the pericardial floor is connected to other muscle systems. Contraction of these muscles may allow variation of pericardial volume and pressure and the tension of the alary ligaments, affecting diastolic filling and hence stroke volume. Functionally it has the same role as the auricles of molluscs or vertebrates. As noted above the pericardial sinus is a functional heart chamber, and the complete structure (i.e. ventricle plus the pericardial septum and cavity) should be regarded as equivalent to the hearts of vertebrate and other invertebrate phyla.

The Ventricle

Ultra structure of the myocardial cells of a range of arthropods has been well described (Sanger and McCann 1968; Økland et al. 1982; Nylund and Tjonneland 1989; Wirkner 2009).

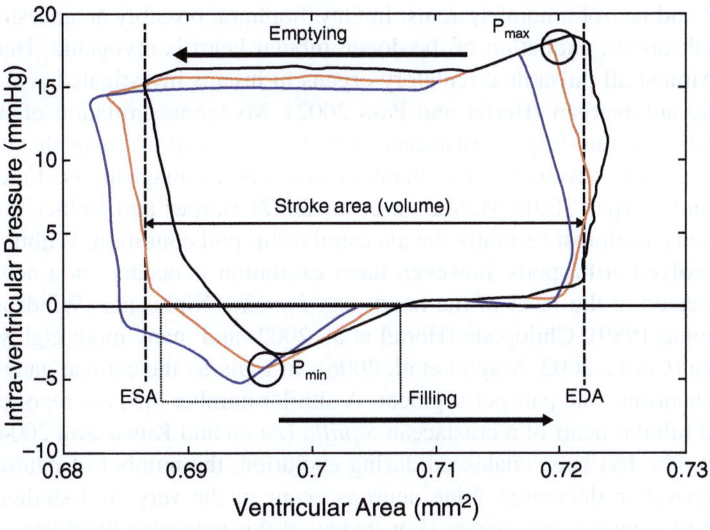

Fig. 1.5 Pressure/volume (calculated) relationships recorded from the heart ventricle of a small decapod crustacean *Palaemonetes pugio* (Guadagnoli et al. 2007)

In systole in the crustacean heart contraction of the ventricular myocardium causes closure of the ostial valves and ejection of fluid through several distribution systems as discussed below. Part of the contractile force is stored in musculo-elastic elements in the alary ligaments which suspend the heart within the pericardial cavity. In diastole the ostial valves open and the energy stored in the alary ligaments allows the ventricle to refill as fluid flows through the ostia. A somewhat similar process occurs in the heart of bivalve molluscs (e.g. *Anadonta*; Brand 1972 and see review by McMahon et al. 1997) and in elasmobranchs (Eckert et al. 1997) except that in these latter animals the circulatory system is closed, the heart is multi-chambered and fluid flows first into the auricle(s) before entering the ventricle.

Pressures developed within the heart and arterial systems, pressure profiles, stroke volumes and cardiac output measured within a range of intact invertebrate animals have been reviewed by McMahon and Wilkens (1983), McMahon and Burnett (1990) and McMahon et al. (1997). Most recently intra-cardiac pressures, stroke volumes and the pressure/volume relationship have been calculated for the heart of a smaller decapod crustacean *Palaemonetes pugio* by Guadagnoli et al. (2007) (Fig. 1.5). Interestingly, despite the striking differences in anatomy between these examples the pressure/volume loops from the ventricle of this decapod are essentially similar to those recorded from vertebrate hearts.

Excitation Mechanisms

Excitation of ancestral tubular dorso-ventral vessels was probably myogenic and in many invertebrate groups, in protochordates, and in vertebrates heart-beat is still essentially myogenic with overriding cholinergic and adrenergic influences from the CNS and neurohumoral systems. In Onychophora, possibly an ancestral cousin of the arthropods, excitation of the dorsal tubular heart is myogenic (Hertel et al. 2002). Almost all pulsatile circulatory organs in insects investigated also exhibit a myogenic automatism (Hertel and Pass 2002). Myogenic initiation of heart-beat also occurs in some simpler crustaceans e.g. *Artemia triops* (Yamagishi et al. 1997) and during early development in many crustaceans (Yamagishi and Euichi 1997; Reiber and Harper 2001; McMahon et al. 2002; Harper and Reiber 2004), and myogenicity is almost certainly the ancestral arthropod condition. Within the more highly evolved arthropods, however, heart excitation is occurs via a neural pacemaker located at the level of the heart, e.g. *Limulus* Xiphosura (Windsor-Watson and Groome 1989), Chilopoda (Hertel et al. 2002) and many more highly evolved Crustacea (Cooke 2002; Mayrat et al. 2006). In *Limulus* the cardiac ganglion contains 14 neurons, one pair per segment. A similar number (15) are reported for the extended tubular heart of a crustacean *Squilla* (Ando and Kuwasawa 2004). But as the heart tube has been coalesced during evolution, the number of neurons in the cardiac ganglion decreases. Nine neurons occur in the very well-studied cardiac ganglion of decapod crustaceans (For review of this extensive literature see Cooke 2002). An ultimate reduction is seen in the tiny copepod *Vargula* where single neuron apparently initiates the heart beat (Ishii and Yamagishi 2002).

It is probable that in early ontogeny all decapod hearts are myogenic, and the transition between myogenic and neurogenic control mechanisms during development in the hearts of decapod and isopod crustaceans (McMahon et al. 2002; Yamagishi and Euichi 1997) is discussed in the ontogeny section below.

Control of Cardiac Performance

As is well known in vertebrates, stroke volume and cardiac output are regulated by a combination of local (stretch), central nervous and peripheral mechanisms as discussed elsewhere in this chapter. It is less well known that a similar suite of mechanisms are involved in crustaceans (McMahon 1995) and many other invertebrates (McMahon et al. 1995a, b; Hertel and Pass 2002; Hertel et al. 2002). This review concentrates on recent work on decapod crustaceans which have received significant recent attention.

Autoregulation

In decapod crustaceans heart rate has been reported as responsive to stretch (Kuramoto and Ebara (1984) review earlier data), but Wilkens and McMahon (1992), Wilkens (1993) and Wilkens et al. (1996) studied several decapod crustaceans and concluded that the heart rate of these species is only minimally responsive to stretch, and thus these hearts do not show classic auto-regulation as seen in vertebrates. These differences may be methodological. Earlier studies reporting stretch-related tachycardia were performed with isolated ventricles disconnected from their alary suspensory ligaments thus unable to fill or beat properly. Such hearts rapidly become hypoxic and responded only to reperfusion with aerated media rather than stretch per se. Wilkens and McMahon (1992) developed a new isolated heart preparation with intact ligaments and perfused this heart continually to maintain natural beating. Perfusion of these hearts with aerated saline did allow increased heart rate but equivalent perfusion with deoxygenated saline did not, indicating that oxygen availability rather than stretch was responsible for the increased beat rate seen in the earlier studies. Similarly increase in pre- and post-loading of the hearts was without major effect. Wilkens (1993) concluded that because of the anatomy of the crustacean (all arthropod) open systems the only natural stretch that can occur results from the elastic return of the alary ligaments during diastole. However, in contrast to most vertebrates the pericardial cavity volume can vary (see above) increasing or decreasing the tension of the suspensory system and thus diastolic filling. A combination of mechanisms may allow the ventricular muscle mass to respond to these changes by increase or decrease in stroke volume and thus cardiac output. Despite the reduced sensitivity to stretch isolated heart preparations showed great sensitivity to a range of cardio-active substances including amines (e.g. 5-HT), and peptides [e.g. Proctolin, FMRFamids; Wilkens and McMahon (1992)] and the circulating haemolymph oxygen level (Wilkens et al. 1996). More recently Sakurai and Wilkens (2003) reported sensitivity to stretch in the extended cardiac ganglion of an isopod crustacean *Ligia*, but in this preparation stretch decreased rather than increased the ganglion firing rate.

Neural Regulation

In all of the higher arthropods studied, even in hearts which display only myogenic rhythms, heart beat and heart performance are affected by neural control and/or neurohormonal mechanisms which effectively control moment to moment heart performance. Physiology of these heart control mechanisms has been reviewed recently for a number of groups [e.g. Hertel and Pass (2002) for insects; Hertel et al. (2002) for onycophorans and chilopods; Harper and Reiber (2001) and Mayrat et al. (2006) for crustaceans].

In decapod crustaceans neural control is affected by neurons from the ventral nerve cord which innervate the pacemaker cells of the cardiac ganglion as well as innervating cardiac muscle directly (Wilkens and Walker 1992; Cooke 1988, 2002). In decapod crustaceans three pairs (two pairs of accelerator and one pair of inhibitory axons) innervate the heart (Alexandrowicz 1932; Cooke 1988, 2002). Wilkens and Walker (1992) stimulated these nerves *en passant* in Crayfish (*Procambarus*) and concluded that cardio-regulatory nerves from the CNS can act as cardiac regulators principally via adjustment in heart rate. These authors also reported that 5-HT (injection) had independent cardio-accelerator effects on the crayfish heart.

Several recent studies have described similar neural control mechanisms in other crustacean groups including the relatively primitive stomatopod *Squilla* (Ando and Kuwasawa 2004) and the isopods *Bathynomus* (Kihara and Kuwasawa 1984; Tanaka and Kuwaswa 1990, 1991) and *Ligia* (Sakurai and Yamagishi 1998a, b).

Hormonal and Neurohormonal Regulation

A large number of hormones are released into the circulating haemolymph of arthropods, most of which have either direct or indirect effects on heart function. The effects of these substances on cardiac function have been well reviewed (Windsor-Watson and Groome (1989) for Limulus; Hertel and Pass (2002) for insects; Hertel et al. (2002) for onycophorans and chilopods; Mayrat et al. (2006), Wilkens and Taylor (2003) and Harper and Reiber (2001) for crustaceans) while new cardio-active hormones continue to be reported (e.g. Christie et al. 2010). The complexity resulting from this multiplicity of hormonal effectors is made more complicated since the reported effects of specific substances may vary between species and even within individuals. Some of the variability seen may stem from the variety of techniques used as follows:

Of the plethora of studies conducted many have used isolated or semi-isolated hearts. Although apparently simple in concept, isolation techniques can vary widely and the results observed depend on many factors:

(a) Isolation from the alary ligaments i.e. the normal heart suspensory system without which the heart cannot function properly (Wilkens et al. 1996).
(b) Isolation from (or presence of) other effector systems (i.e. the pericardial organs) which could be stimulated independently by the test preparations and cause secondary effects on the heart.
(c) Isolation from the regulating influence of the nervous system.
(d) Route of administration of the test substance i.e. whether released into the perfusate or injected into the heart directly: these differing in effect by their initial action on the cardiac ganglion or on the myocardium directly.
(e) The suitability of the perfusing medium e.g. rate of flow, respiratory gas levels, ionic balance, temperature, etc.

Without equivalence in all of these factors, it is very difficult to directly compare the results for any test substance. McGaw Wilkens and McMahon (1995) show that a particular hormone may have dramatically different (opposite) results on heart function when tested on isolated vs intact preparations as well illustrated for the Frmrfamid-like peptides F1 and F2 shown in Fig. 1.6.

Many other studies have been conducted by injection of the test substance into the circulating haemolymph, and while this may best mimic the natural case, care must still be taken in the interpretation of the results. Particular care should be taken when considering the administration route. Many effector organs at a range of locations can be expected to respond to a single dose of a cardio-active agent. For instance in a decapod crustacean these include *at least:*

(a) CNS which may have direct effects on the heart via the cardio-arterial nerves.
(b) The pericardial organs which could be stimulated to release a blend of cardio-active agents.
(c) The cardiac ganglion which could be stimulated directly by the test agents or indirectly by stretch caused by change in cardiac volume.

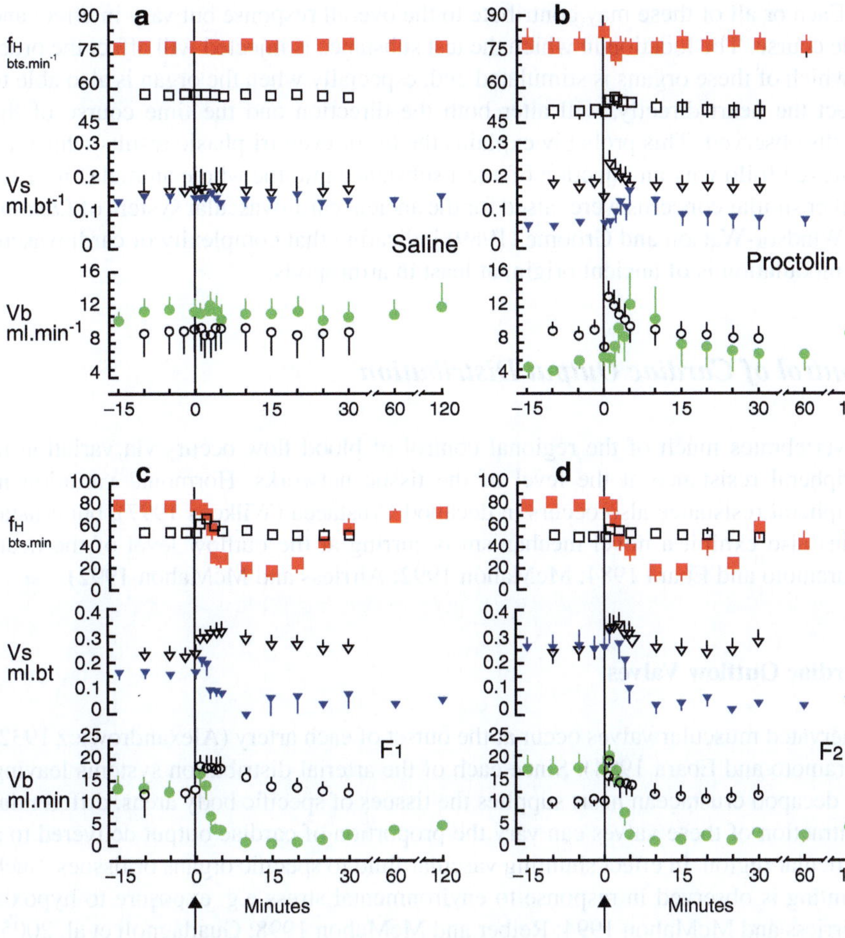

Fig. 1.6 Differential effects of peptide cardio-active substances on the performance of the heart in isolated and in vivo heart preparations of *Cancer magister* modified from McGaw et al. (1994). *Filled symbols* indicate responses of hearts in whole animal in vitro *open symbols* indicate the responses of semi-isolated hearts

(d) The myocardium as in (c) above.
(e) The cardio-arterial valves in direct response to the test agent or via neural or neurohormonal action.
(f) The pericardial septum by action on contained or attached muscles.
(g) Arterial or peripheral resistance (Wilkens 1997; Wilkens et al. 2008), and even
(h) The oxygen binding state of the respiratory carrier haemocyanin which can respond to cardio-active substances by additional release of oxygen (Morris and McMahon 1989) which strongly affects cardiac performance (Wilkens et al. 1996).

Each or all of these may contribute to the overall response but vary in effect and time course. The location in which the test substance is injected will affect the order in which of these organs is stimulated and, especially when the organ is also able to affect the heart directly, will alter both the direction and the time course of the results observed. This probably explains the bi- or even tri-phasic results which are observed following an injection of a test substance into the whole animal (Fig. 1.6). Rather similar concerns were raised for the ancient cardiovascular system of *Limulus* by Windsor-Watson and Groome (1989) indicating that complexity of cardiovascular modulation is of ancient origin, at least in arthropods.

Control of Cardiac Output Distribution

In vertebrates much of the regional control of blood flow occurs via variation in peripheral resistance at the level of the tissue networks. Hormonal variation in peripheral resistance also occurs in decapod Crustacea (Wilkens 1997), but crustaceans also exhibit a novel mechanism occurring at the outflow level of the heart (Kuramoto and Ebara 1984; McMahon 1992; Airriess and McMahon 1992).

Cardiac Outflow Valves

Innervated muscular valves occur at the outset of each artery (Alexandrowicz 1932; Kuramoto and Ebara 1984). Since each of the arterial distribution systems leaving the decapod crustacean heart supplies the tissues of specific body areas, differential contraction of these valves can vary the proportion of cardiac output delivered to a particular region, in effect shunting vascular fluid to specific organs or tissues. Such shunting is observed in response to environmental stress e.g. exposure to hypoxia (Airriess and McMahon 1994; Reiber and McMahon 1998; Guadagnoli et al. 2005) hyposalinity (McGaw and McMahon 1996) and during enforced activity (Dewachter and McMahon 1996).

These cardiac valves also respond differentially to stimulation with a variety of cardio-active agents e.g. amines (Airriess and McMahon 1992) and particularly peptides such as proctolin and Frmfamides (McGaw et al. 1994, 1995; Wilkens et al. 1996). It is also interesting to note that vascular shunting also occurs during startle responses (McMahon et al. 2002) indicating that neural networks may also be involved.

Arterial Resistance

Control of regional vascular flow distribution can also occur by peripheral mechanisms. Wilkens (1997) and Wilkens and Taylor (2003) show that a variety of cardio-active substances can affect the resistance to flow in several decapod crustacean distribution systems. Amongst possible mechanisms responsible for these resistance

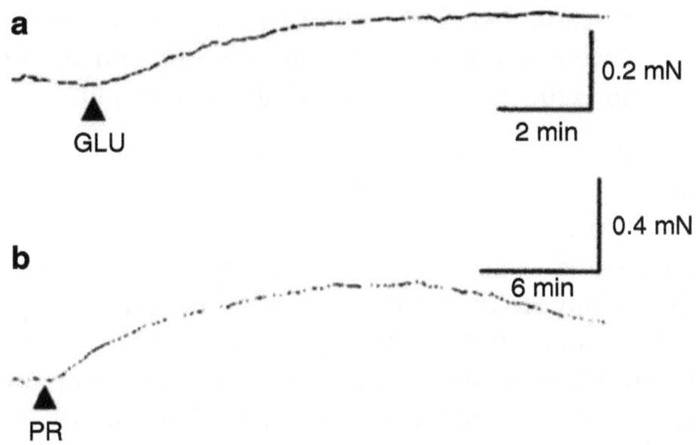

Fig. 1.7 Circumferential contractility in tissue rings from the hepatic artery in *Homarus americanus* in response to Glutamate (GLU) and Proctolin (PR) administration. From Wilkens et al. (2008)

changes Wilkens (1997) suggested that resistance changes in posterior dorsal artery (posterior aorta) may be because of the action of valves occurring at the base of the segmental arteries. Wilkens and Taylor (2003) pointed out that changes in flow resistance could occur both in the gills and in the "venous return" circulatory beds, but ruled out these as effectors in the responses seen in their study on the lobster *Panulirus*. Most recently Wilkens and co workers (2008, for review) have shed new light on the structure and functioning of the distribution (arterial) vessels of the lobster *Homarus americanus*. Previously considered as largely passive elastic structures (McMahon and Burnett 1990; Shadwick et al. 1990), all major arterial trunks are now shown to contain contractile elements such as actin, myosin and tropomyosin and to also to show circumferential contractility in response to the neurotransmitter Glutamate and the neurohomone Proctolin (Fig. 1.7). Furthermore, the dorsal abdominal artery (posterior aorta) contains striated muscle cells and, additionally, contracts in response to electrical stimulation. This artery is of course a remnant of the original dorsal heart tube and also possesses other cardiac structures giving rise to paired valved segmental arteries.

Clearly a wide range of cardio-active substances may participate in the control of flow resistance in crustacean circulatory systems and must play a role in the regional control of blood flow and the shunting mechanisms outlined above.

Control of Heart Rate Versus Stroke Volume

The heart beat is initiated by the ganglion firing rate, but heart rate in vivo is adjusted to suit current conditions either via the CNS (Guirguis and Wilkens 1995) or via circulating hormones (Wilkens and Taylor 2003). Heart stroke volume is also variable in crustaceans via a number of mechanisms:

(a) By increased systolic contraction causing additional stretch on the alary ligaments and increased subsequent diastolic filling.
(b) By movements of the pericardial septum either by its intrinsic muscle fibres or by movement of attached or adjacent structures.

(c) Increased "venous" return as a result of bodily movements, e.g. the escape movements of the abdomen of lobster and shrimps which displace fluid back along the return pathways to the pericardium.

Cardiac beat rate and cardiac stroke volume may vary not only in concert but may also vary independently, e.g. in response to hypoxia in crabs and lobsters where a decrease heart rate is compensated for by an increase in stoke volume which restores or even enhances blood flow to tissues (Airriess and McMahon 1994; Reiber and McMahon 1998) or in response to activity where heart rate is increased <20% while stroke volume increases >70% (Dewachter and McMahon 1996). Independent variation in heart rate vs stroke volume also occurs in response to acute change in temperature (Dewachter and Wilkens 1996; Worden et al. 2006).

As heart rate responds negatively to oxygen deficiency (Airriess and McMahon 1994; Wilkens et al. 1996; Wilkens 1993) and inadequately in response to acute temperature change (Dewachter and Wilkens 1996) and increased activity (Dewachter and McMahon 1996), adjustments in stroke volume provide an essential regulatory component increasing cardiac output and hence oxygen supply in response to increased metabolic demand.

The fine control of circulation in crustaceans is not fully understood as yet but appears rival that seen in lower vertebrates.

Auxiliary Pumps

Auxiliary pumps occurring in addition to, or instead of, the heart drive flow of vascular fluid occur widely in invertebrates particularly in arthropods and especially insects (Pass 2000), in the cephalochordate *Amphioxus/Branchiostoma* (Randall and Davy 1980; Ruppert 1997) and in an agnathan fish (Burggren and Reiber 2007). These pumps usually function to boost vascular flow through areas of high demand or low resistance. In the highly active cephalopod molluscs the gills are of very high area (high resistance), and a branchial (auxiliary) heart has evolved to boost vascular fluid flow through the branchial circuit (McMahon et al. 1997). In malacostracan crustaceans the cor-frontale may operate as a booster pump to supply highly complex anterior nervous tissue networks (Steinacker 1987; Mayrat et al. 2006). Windsor-Watson and Groome (1989) described the accessory pumping action of the gill books in *Limulus* which boost fluid flow back to the heart. Auxiliary pumps occur very commonly in insects at many locations where flow boost is required (Pass 1998, 2000). In insects the evolution of the tracheal system has freed the dorsal vessel from a primary function in gas exchange and in consequence the heart retains little power necessitating boosted flow to such structures as the legs, antennae and wings (Hertel and Pass 2002). It is possible that the radically different architecture of most invertebrate systems where the gills are located at the low pressure end of the circuit (as opposed the high pressure input to the gills in aquatic vertebrates) has encouraged the development of the many accessory pumping structures seen.

Ontogeny of the Heart

Considering the abundance and economic importance of invertebrates generally and arthropods in particular, ontogenetic aspects have been poorly studied. With the exception of the insect *Drosophila* and protochordates which have been well studied (see Xavier-Neto et al. 2007 for integrated review) there is little available information on either development or molecular genetics for the majority of higher invertebrates and even less on the development of the heart. This paucity of information is a pity when one considers that the majority of invertebrate larvae are semitransparent and organogenesis can be readily observed in the living animal.

This situation is only slightly better for the crustaceans (Harper and Reiber 2001, 2004; Mayrat et al. 2006; Burggren and Reiber 2007). In the simple branchiopod crustacean *Artemia* heart beat is first observed in thoracic stage 4 following which the heart was incremented posteriorly segment by segment (Spicer 1994, 2006). As each segment is added, pairs of ostia and alary ligaments are added sequentially ending when development of the last segment is completed.

Similarly in more highly evolved crustaceans such as the decapod *M. ensis* the heart is not functional in the first six (naupliar) larval stages (McMahon et al. 2002; Harper and Reiber 2001, 2004), although fluid moment can be seen to accompany body and limb movements in earlier stages (cf Spicer 2006 above). In *M. ensis* heart beat starts in the first protozoea larva where a short, globular shaped, contractile area ("heart") is present in the posterior–dorsal cephalopthorax (Fig. 1.8a; McMahon et al. 2002). Fluid enters via a single pair of ostial valves and, although only a single artery leading to the anterior appears open at this stage, nonetheless vascular fluid flow can be seen throughout both the thorax and abdomen (Table 1.1, McMahon et al. 2008). As in *Artemia* (Spicer 2006) contractions in the gut commence at the same time as heart beat and as these are retrograde, they potentially also serve to complement flow through the abdominal region. By the third protozoeal stage flow through the dorsal vessel commences and supplements vascular fluid flow posteriorly along the gut and throughout the abdomen (Fig. 1.8c). After the transition to the first mysid stage a pair of lateral arteries arising from the anterior of the heart is seen to open (these are equated to the anterolateral arteries of the adult). Meanwhile a second, seemingly separate, contractile area starts to develop in the dorsal vessel immediately posterior to the first (Fig. 1.8c). The seemingly separate origin of these two concentrations of contractile tissue may lead some support to Mayrat et al.'s suggestion (2006) that the crustacean heart might contain a contribution from the original ventral arterial system which had been dragged dorsally during early development.

By the third mysid stage (Fig. 1.8d) this "new" muscular area has grown to greatly exceed the size of the original contractile site and two new pairs of ostia are seen dorsally. The general shape of this second contractile area now begins to resemble the heart ventricle of adult decapod crustaceans. A second pair of lateral arteries can now be observed leaving the dorsal vessel just posterior to the ventricle. A branch (visceral?) of these arteries heads ventrally and will form the adult sternal artery which descends to lie below the nerve cord and splits to provide the ventral thoracic and ventral abdominal arteries.

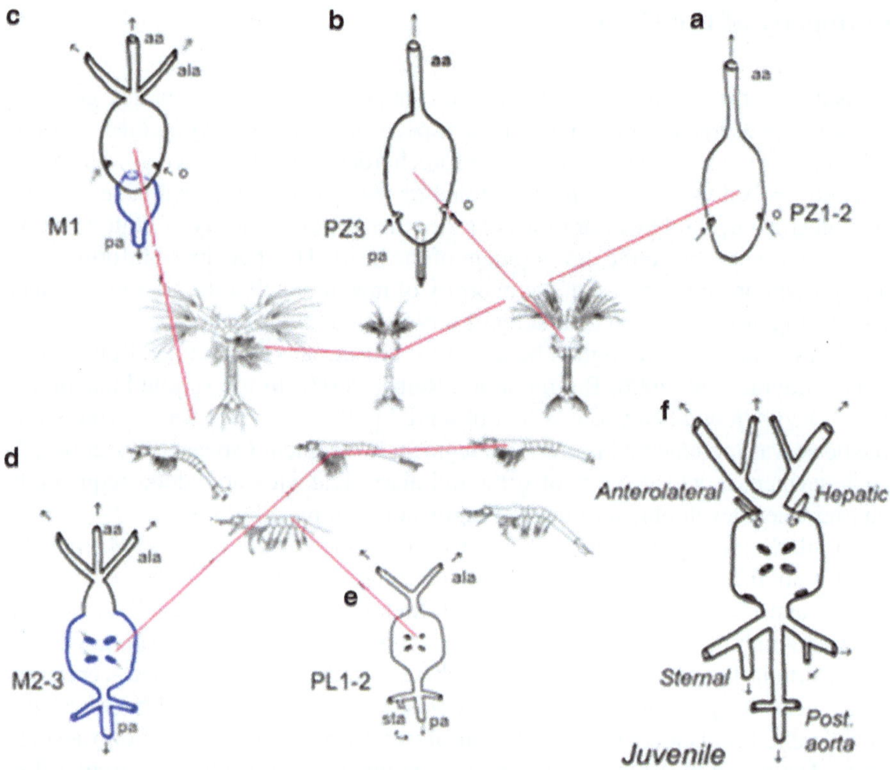

Fig. 1.8 Ontogeny of the heart in *Metapenaeus ensis*. *aa* anterior aorta, *ala* anterior lateral artery, *M1* Mysid stage 1, *PZ 1,2,3* Protozoeal stages1,2,3, *PL 1-2* Post-larval stages 1,2, *o* ostia, *pa* posterior aorta, *sta* sternal artery. Naupliar stages 1–6 not shown since no heart activity is seen. Area drawn in *blue* in M1 and M2–M3 is the later developing contractile area. Central background figures show developmental stages PZ1 to Pl of peneaid shrimps. From Reiber and Harper (2001). Surrounding figures are drawings of the developing heart in *Metapenaeus ensis* (from McMahon et al. 2002)

In the early post-larval stages of *M. ensis* (Fig. 1.8e) the anterior aorta closes leaving the anterolateral arteries to supply the entire cephalic area. The heart continues to develop in subsequent juvenile stages adding the last pair of arteries to develop from the heart area. These are the hepatic arteries which leave the heart ventrally to supply the viscera. Full development of the adult heart is not seen until after several juvenile stages (Fig. 1.8f).

Thus full development of the heart extends beyond larval development. Despite this extended developmental period, good circulation throughout the body is seen from the first stages of heart development (McMahon et al. 2008). Flow velocity in the abdomen, for instance, increases only slightly after opening of the posterior aorta (Table 1.1).

Together with the continued structural development of the heart, changes in heart excitation mechanism also progress into later development (Yamagishi and Euichi 1997; McMahon et al. 2002; Mayrat et al. 2006). McMahon et al. (2002) injected tetrodotoxin (TTX) into the pericardial cavity of a range of larval stages of the

Table 1.1 Variation in flow velocity (±SD) with development in *Metapenaeus ensis*

Instar	Outward cephalothorax	Return cephalothorax	Outward abdomen	Return abdomen	Outward cephalothorax	Return abdomen	Outward combined	Return combined
PZ1								
Lateral + dorso-ventral combined	29±5	6±3	20±17	7±3			25±12	7±3
PZ2								
Lateral	25±9	12±7	412±9	14±7	33±21	14±7	34±17	13±7
Dorsoventral	32±17	11±8	37±17	13±5	34±17	12±5		
PZ3								
Lateral	24±4	13±7	34±20	10±8	33±3	10±3	38±17	9±6
Dorsoventral	51±13	11±3	34±20	12±6	38±17	11±4		
Mysid								
Lateral	35±17	11±6	36±15	10±8	35±16	10±7	33±3	12±8
Dorsoventral	30±9	14±7	31±11	11±8	31±9	13±8		
Postlarva								
Lateral	49	11±8	46		47±22	14±7	48	14±7
Dorsoventral		12±5		16±7				

Flow velocity estimated from movements of haemolymph cells through vascular channels as recorded by video microscopy
Velocities measured are shown both for the cephalopthorax and for the abdomen (outward = leaving, inward = returning to the heart)

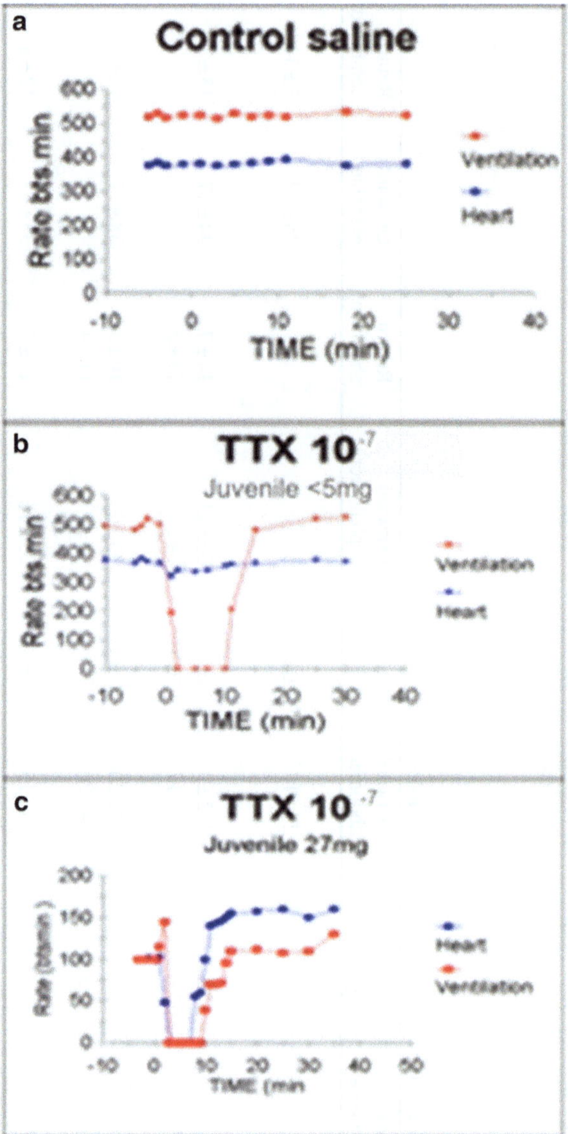

Fig. 1.9 Change in the effects of TTX injection on heart and ventilatory rates during development in the shrimp *Metapenaeus ensis* to show acquisition of neurogenic drive. From Doyle and McMahon 1997, McMahon et al. 2002

shrimp *M. ensis* while recording heart and respiratory movements (Fig. 1.9). In larvae and early juvenile stages (<5 mg mass), TTX injection stopped the respiratory movements but not the heart beat (Fig. 1.9b). In later juvenile stages >25 mg mass TTX (Fig. 1.9c) stopped both, indicating that neural overdrive from the cardiac ganglion was present. While it is difficult to stage these events exactly, as they occur in juvenile stages, it seems likely that the dominance of the neurogenic pacemaker can only occur when the cardiac ganglion is complete which apparently does not occur until after several juvenile stages.

Cardiovascular Adaptation and Evolution of Aerial Breathing

The transition from an aquatic to a terrestrial mode of life has occurred many times and in many phyla during animal evolution. It is a complex process involving changes at all levels within the animal and occurring over long periods of evolutionary time. During the evolution of air breathing competence the original gas exchange organ which had become highly specialised for aquatic exchange must undergo massive redesign or more usually be replaced by a second gas exchanger more suitable to exchange gases with air. Since the animal must necessarily depend at least originally on a partially aquatic existence the animal remains dependent on aquatic exchange and clearly both exchangers must occur and be functional at the same time during evolution. These animals are necessarily amphibious.

The problems faced by either vertebrate or invertebrate animals in this transition involve major changes in all aspects of bodily function, but, in terms of the cardiovascular system, the most important are those involving aquatic gas exchangers (e.g. gills). In the aquatic milieu the gills are responsible for many functions including gas exchange, pH and ionic regulation, and water balance. In fully terrestrial animals these functions must be transferred to other structures often requiring extensive modification of the cardiovascular system as the new exchange systems are "plugged into" the existing circulation.

In the vertebrates these changes have been in process for more than 500 million years and the eventual development of multi-chambered hearts and separated circulatory systems is relatively well known (see elsewhere in this symposium), but the equivalent changes which have occurred in many invertebrate groups have been less well studied.

As before, adaptations to terrestrial life have occurred many times in invertebrate evolution, occur in many invertebrate phyla and involve an extreme diversity of gas exchange and cardiovascular systems. Thus we continue to limit our focus to two arthropod groups, the decapod crustaceans and pterygote insects which will suffice to illustrate the mechanisms of adaptation used. The capability to utilise aerial gas exchange has evolved largely in transitional environments where aquatic oxygen availability has been limited and modification of and eventual replacement of the aquatic gas and ion exchangers has occurred. In the decapod crustaceans adaptation to the terrestrial habitat is an apparently ongoing process with many species living in intertidal and transitional habitats being variously adapted to utilise aerial oxygen. For instance most intertidal crabs have good tolerance to air exposure, but many show reduced gas exchange and are not active in air (e.g. *Cancer productus*, Defur et al. 1983). Animals which can maintain activity during air exposure vary from burrowing crabs such as, e.g. *Holthuisana transvera* (Taylor and Greenaway 1984) and burrowing crayfish, e.g. *Procambarus clarkii* (McMahon and Hankinson 1993), *Coenobita* and other terrestrial hermit crabs, which require periodic exposure to water stores (Burggren and McMahon 1988; Wheatly et al. 1986; Farrelly and Greenaway 2005) to several species of land crab such as the Christmas island Red crab *Gecarcoidea natalis* (Adamczewska and Morris 2001) *Discoplax* (*Cardisoma*) *hirtipes* (Morris 2005) and the Coconut crab *Birgus latro* (Greenaway et al. 1988; Morris et al. 1988;

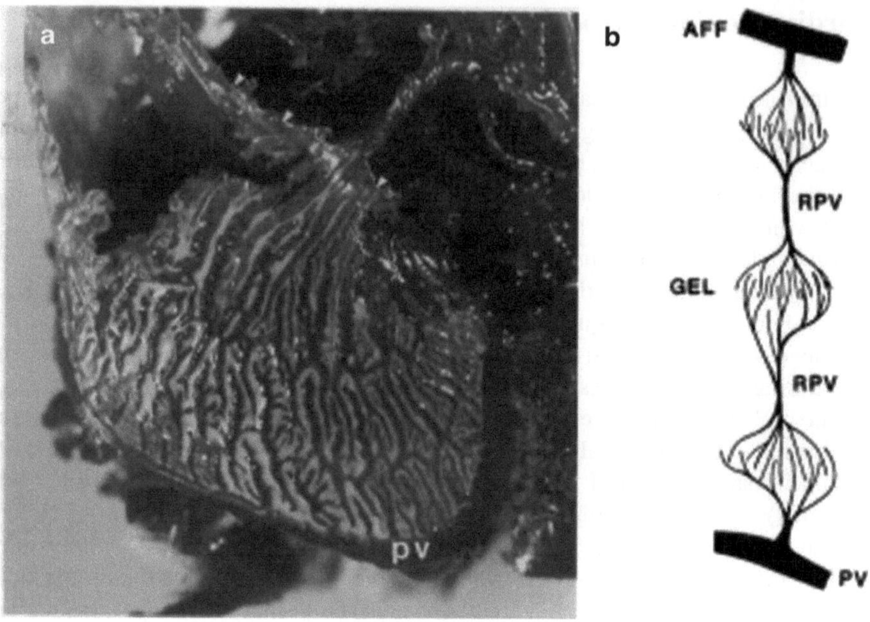

Fig. 1.10 (**a**) A circulatory cast of the surface of the lung in the terrestrial crab *Geograpsus grayii*. *Arrows* indicate the large afferent vessel; pv indicates the pulmonary venous collecting vessel. (**b**) Diagram of the portal blood system present in the "lung" of many land crabs. *AFF* afferent vessel, *RPV* portal vien, *GEL* gas exchange lacunae, *PV* pulmonary vein. From Farrelly and Greenaway (1993)

Farrelly and Greenaway 2005). These latter animals are essentially permanently terrestrial requiring water immersion only for the purposes of reproduction.

Gecarcoidea, *Birgus* and many other land crabs have developed a secondary air breathing organ (lung) within the branchial cavity but with a blood vascular system separate from that supplying the gills. Land crab lungs develop from the wall of the branchiostegal cavity and can become very complex gas exchange organs (Greenaway and Farrelly 1984; Innes and Taylor 1986; Farrelly and Greenaway 1993, 2005). The walls of the branchiostegal cavity become excessively infolded or fenestrated producing a very large surface area. They are perfused with fluid originating from the anterior arteries of the heart which divide into several consecutive capillary systems within the pulmonary structure (Fig. 1.10, Greenaway and Farrelly 1990) before entering a collecting sinus which reaches the pericardial cavity separately from fluid returning from the branchial system.

Gills and lungs are both functional even in fully terrestrial crustaceans, but there is a tendency for a reduction in gill area and a strengthening of the lamellae with increasing exposure to air (Farrelly and Greenaway 1994, 2005). With increasing terrestrial competence the lungs come to play a larger role in oxygen uptake while the gills remain important for carbon dioxide elimination. In the more fully terrestrial crabs tested in their natural environment both post-pulmonary and post-branchial fluid are relatively well oxygenated (Red Crab—Adamczewska and

Morris 2001; Coconut crab *Birgus latro*—Morris et al. 1988; *Pseudothelphusa*—Innes and Taylor 1986). While this avoids the perfusion of the heart with deoxygenated fluid as occurs in fish, such admixtures could reduce the efficiency of gas exchange, e.g. during immersion. Mechanisms to minimise the effects of such admixture have, however, been described.

Taylor and Greenaway (1984) reported change in the distribution of fluid between lung and gill pathways in between emersed and immersed specimens of the amphibious burrowing crab *Hothuisana transversa*, and it seems probable that haemolymph can be diverted into either the branchial or pulmonary circuits. The mechanisms involved are not clearly understood but could involve changes to the resistance to fluid flow between the two circuits. These include:

1. In amphibious crabs the gills tend to collapse in air increasing their resistance to flow and diverting fluid flow to the alternative circuit.
2. When the branchial cavity is water filled the respiratory pumps generate subambient pressure in the branchial chambers which could facilitate fluid flow through the bronchial circuit.
3. Thirdly, Taylor and Taylor (1986) and Wilkens and Taylor (2003) report the presence of valves in the branchial perfusion circuit which could control resistance to gill blood flow.
4. It is also probable that shunting mechanisms utilising the arterial valves (see above) may act to alter resistance to flow through the branchiostegal circuit. Haemolymph entering the pulmonary circuit leaves the heart by arteries at the anterior of the heart. The arterial valves at the bases of these are highly sensitive to octopamine which allows shunting of fluid into these arteries and thus through the alternate circuit. Direct evidence for such shunting is not established for the fully terrestrial land crabs, but supportive evidence has been provided by several studies on amphibious crabs, e.g. flow through the anterior arterial circuit is promoted when intertidal amphibious crabs (*Cancer magister*) are air exposed (Airriess and McMahon 1994, 1996) and injection of octopamine into the pericardial cavity promotes flow through these arteries (Airriess and McMahon 1992). Release of octopamine thus causes increased flow through the anterior arteries and thus through the branchiostegal (putative pulmonary) circuit. The evidence presented suggests that mechanisms observed in these less well-adapted amphibious crustaceans may allow selective distribution of flow in the more fully adapted land crabs, but the work to support this is urgently needed in this area.

Decapod crustaceans are thus better pre-adapted for the integration of a second branchial air-breathing organ in that the circulation system which supplies the wall of the branchiostegal cavity is separate from that supplying the gills. This part of the vascular system is important early in development in decapod crustaceans where it is responsible for gas exchange prior to the development of the gills. The presence of an alternate vascular system bypassing the gills much facilitates the adaptation of the vascular system to perfuse the new branchiostegal pulmonary gas exchange organ. Furthermore this arrangement allows the heart to remain essentially unmodified avoiding the duplication of pumps required to operate the pulmonary circuit of most air breathing vertebrates (Sedmera and Ostadal 2012; Wong et al. 2012).

Both gills and lungs of the terrestrial decapod crustaceans are external structures and although protected to same degree by the branchiostegites must remain moist to carry out important respiratory and ion regulatory functions and thus pose a potential route for water loss. For this reason even the most terrestrial of the land crabs remain limited to moist tropical environments.

Several other arthropod groups have, however, become fully terrestrial and occupy the driest of environments. Several mechanisms have evolved to internalise the gas exchange surfaces but certainly the most efficient is the tracheal system which is found in several arthropod groups but is best developed in the pterygote insects. Tracheal gas exchange is an elegant solution where an expanded internal micro-tubular air sac system allows gas exchange directly between air and tissues at the cellular or even sub-cellular level. This requires little or no involvement of the vascular system in gas transport and under these circumstances the vascular system is usually limited to a single dorsal vessel with segmentally arranged ostia and alary structures (Pass 2000; Hertel and Pass 2002).

The presence of segmentally arranged arteries in some more primitive insects e.g. cockroaches and in the majority of arthropod groups indicates that ancestral insect vascular systems may have been more complex but have been reduced in insect evolution as a result of (a) the uncoupling of the vascular system from gas exchange and (b) a specialisation in small size.

Evolutionary Effects of Reduction in Size

There has been a tendency to equate the simple systems of the non-malacostracan crustaceans with early evolutionary stage that is to say that these represent an early stage of arthropod evolution. This however seems unlikely. It is more likely that these systems simply exhibit secondary evolutionary response to a combination of minimal size which can supply tissue exchange needs by simple circulation of fluid throughout the animal and gas exchange through the tissues such as is seen today in the larval stages of animals with elaborate adult vascular systems.

Invertebrate/Vertebrate Transitions

To this point this account has been restricted to discussion of protostome invertebrate species but I would like to close by commenting on the similarities between and linkages with vertebrate systems. In terms of phylogeny the Protochordates and Cephalochordates are deuterostome invertebrates occupying an evolutionary position somewhere between invertebrates and vertebrates. Cardiovascular systems within this group have been reviewed by Randall and Davy (1980), Burggren et al. (1997) and Burggren and Reiber (2007) but the functioning of their vascular systems remains poorly understood. The Urochordates have been considered ancestral to the vertebrates via the ascidian larval stages which display both a notochord and segmental myotome blocks. Metamorphosis to the adult form results in radically changed anatomy but a large brachial basket is still perfused by vascular fluid from the single (ventral?) tubular heart. A close relationship between Urochordates and vertebrates remains controversial and is challenged on molecular genetic and other grounds (e.g. Bassham et al. 2008).

Several reviews (Randall and Davy 1980; Ruppert 1997; Burggren et al. 1997) describe the best known cephalochordate amphioxus (*Branchiostoma*) as having no heart but also describe contractile vessels at a variety of sites largely within the ventral area. One of these, the endostylar artery, has been described as being homologous with the heart of Urochordates. Rahr (1981) described a ventral heart in this species, raising the question that a progressive array (segmental?) of contractile areas might exist in this animal and form the basis for a segmented heart tube similar that observed in the embryos of lower vertebrates. Fluid then passes though a branchial basket after which fluid is collected by paired dorsal aortae which supply the tissues segmentally. Although this system has clear similarities to the ancestral vertebrate system neither cephalochordate (*Amphioxus*) nor urochordate vascular systems are lined with epithelium (Randall and Davy 1980; Ruppert 1997; Shigei et al. 2001), thus the latter authors consider the protochordate systems as not equivalent to extant vertebrate closed systems and consider the vertebrate epithelium a unique development occurring later in vertebrate evolution.

Although this account has focused on invertebrate hearts and circulatory systems it may be useful in conclusion to focus on the basic similarities between invertebrate and vertebrate systems. Considerable similarities between the two are seen in function with both groups demonstrating the considerable powers of adaptation in the evolving of highly efficient and complex cardiovascular systems adapted to all environments.

Basic similarities are evident in the segmental organisation of the cardiovascular systems which arose very early in animal evolution. Over evolutionary time huge differences in body plans have evolved within animal phyla and cardiovascular systems have adapted to these in both vertebrates and invertebrate groups. The most apparent morphological difference i.e. that the heart and nervous system of the deuterostome chordates (including vertebrates) has come to lie dorsal to the gut whereas that of protostome invertebrates is ventral, and this has become a defining difference separating the two groups.

This apparently major difference, however, may have resulted simply from a mutational inversion of the Chordin/BMP gene network which controls dorsoventral organisation in chordates vs its counterpart in the protostome invertebrate groups (De Robertis 2008; Saudemont et al. 2008). A result of inversion in molecular organising systems is that the heart of protostome animals comes to be located dorsally rather than ventrally as in the chordates. An examination of the simplest heart and vascular plans when this inversion is removed reveals essentially similar basic systems. The following quote from Saudemont et al. (2008) “expression of the homologue of NK4/tinman in the pulsatile dorsal vessel of *Platynereis* strongly supports the hypothesis that the vertebrate heart derived from a dorsal vessel relocated to a ventral position by D/V axis inversion in a chordate ancestor" clearly expresses the ease with which so major an anatomical separation criterion could have been achieved. Evidence contrary to this view is provided by Xavier-Neto et al. (2007), but seems unconvincing to this author.

Thus there are few important differences between invertebrate and vertebrate cardiovascular systems. Both evolved from a common ancestor 600 million years ago. The vertebrate cardiovascular system is thus just one part of an incredible diversity of similar systems resulting from 600 million years of post-Cambrian evolution.

References

Adamczewska A, Morris S (2001) Metabolic status and respiratory physiology of *Gecarcoidea natalis*, the Christmas Island Red Crab, during the annual breeding migration. Biol Bull 200:321–335

Adamowicz SJ, Purvis A, Wills MA (2008) Increasing morphological complexity in multiple parallel lineages of the Crustacea. Proc Natl Acad Sci USA 105:4786–4791

Airriess CN, McMahon BR (1992) Aminergic modulation of circulatory performance in the crab Cancer magister. In Hill RB, Kuwasawa K, McMahon BR, Kuramoto T (eds) Comparative physiology, vol 11, Phylogenetic models in functional coupling of the CNS and the cardiovascular system. Karger, Basel, pp 123–131

Airriess CN, McMahon BR (1994) Cardiovascular adaptations enhance tolerance of environmental hypoxia in the crab *Cancer magister*. J Exp Biol 190:23–41

Airriess CN, McMahon BR (1996) Short-term emersion affects cardiac function and regional hemolymph distribution in the crab *Cancer magister*. J Exp Biol 199:569–578

Alexandrowicz JS (1932) The innervations of the heart of Crustacea 1. Decapods. Q J Microsc 75:181–249

Ando H, Kuwasawa K (2004) Neuronal and neurohormonal control of the heart in the stomatopod crustacean, *Squilla oratoria*. J Exp Biol 207:4663–4677

Bassham S, Cañestro C, Postlethwait JH (2008) Evolution of developmental roles of *Pax2/5/8* paralogs after independent duplication in urochordate and vertebrate lineages. BMC Biol 6:35

Brand AR (1972) The mechanism of blood circulation in *Anodonta anatine* L. (Bivalvia Unionidae). J Exp Biol 56(362):379

Burggren WW, McMahon BR (1988) Circulation. In: Biology of the land crabs, Chap 1. Cambridge University Press, Cambridge

Burggren WW, Reiber CL (2007) Evolution of cardiovascular systems. In: Aird WC (ed) Endothelial biomedicine. Cambridge University Press, Cambridge, pp 29–49

Burggren WW, Farrell A, Lillywhite H (1997) Vertebrate cardiovascular systems. In: Danzler WH (ed) Handbook of physiology, Sect 13. Comparative physiology, vol 1. Oxford University Press, Oxford, pp 215–308

Christie AE, Stevens JS, Bowers MR et al (2010) Identification of a calcitonin-like diuretic hormone that functions as an intrinsic modulator of the American lobster, *Homarus americanus*, cardiac neuromuscular system. J Exp Biol 213:118–127

Cooke IM (1988) Studies on the crustacean cardiac ganglion. Comp Biochem Physiol 91C: 205–218

Cooke IM (2002) Reliable, responsive pacemaking and pattern generation with minimal cell numbers: the crustacean cardiac ganglion. Biol Bull 202:108–136

De Robertis EM (1997) Evolutionary biology: the ancestry of segmentation. Nature 387:25–26

De Robertis EM (2008) Evo-devo: variations on ancestral themes. Cell 132:185–195. doi:10.1016/j.cell.2008.01.003

Defur PL, McMahon BR, Booth CE (1983) Analysis of hemolymph oxygen levels and acid-base status during emersion 'insitu' in the red rock crab, *Cancer productus.* Biol Bull 165:582–590

Dewachter B, McMahon BR (1996) Haemolymph flow distribution, cardiac performance and ventilation in *Cancer magister* (Decapoda, Crustacea) during moderate activity. J Exp Biol 199:627–633

Dewachter B, Wilkens JL (1996) Comparison of temperature effects on heart performance of the dungeness crab, *Cancer magister, in vitro* and *in viva.* Biol Bull 190:385–395

Eckert R, Randall D, Burggren WW, French K (1997) Animal physiology, 4th edn. Freeman, New York, 479

Farrell AP, Farrell ND, Jourdan H, Cox G (2012) A perspective on the evolution of the coronary circulation in fishes and the transition to terrestrial life. In: Sedmera D, Wang T (eds) Ontogeny and phylogeny of the vertebrate heart. Springer, New York

Farrelly CA, Greenaway P (1993) Land crabs with smooth lungs: Grapsidae, Gecarcinidae, and Sundathelphusidae: ultrastructure and vasculature. J Morphol 215:245–260

Farrelly CA, Greenaway P (1994) Gas exchange in air-breathing crabs: lungs versus gills. J Exp Biol 187:113–130

Farrelly CA, Greenaway P (2005) The morphology and vasculature of the respiratory organs of terrestrial hermit crabs (*Coenobita* and *Birgus*): gills, branchiostegal lungs and abdominal lungs. Arthropod Struct Dev 34:65–89

Forouhar AS, Liebling M, Hickerson A, Nasiraei-Moghaddam A, Tsai HJ, Hove JR, Fraser SE, Dickinson M, Gharib M (2006) The embryonic vertebrate heart tube is a dynamic suction pump. Science 312:751–753

Greenaway P, Farrelly CA (1984) The venous system of the terrestrial crab *Ocypode cordimanus* (Desmarest 1825) with particular relevance to the vasculature of the lungs. J Morphol 181:133–142

Greenaway P, Farrelly CA (1990) Vasculature of the gas exchange organs in air breathing brachyurans. Physiol Zool 63:117–139

Greenaway P, Morris S, McMahon BR (1988) Adaptations to a terrestrial existence by the Robber Crab *Birgus latro.* II. In Vivo respiratory gas exchange and transport. J Exp Biol 140:493–509

Guadagnoli JA, Jones LA, Reiber CL (2005) The influence of reproductive state on cardiac parameters and hypoxia tolerance in the Grass Shrimp, *Palaemonetes pugio*. Funct Ecol 19:976–981

Guadagnoli JA, Tobita K, Reiber CL (2007) Assessment of the pressure–volume relationship of the single ventricle of the grass shrimp, *Palaemonetes pugio*. J Exp Biol 210:2192–2198

Guirguis MS, Wilkens JL (1995) The role of the cardioregulatory nerves in mediating heart rate responses to locomotion, reduced stroke volume and neurohormones in *Homarus americanus*. Biol Bull 188:179–185

Harper SL, Reiber CL (2001) Ontogeny of neurohormonal regulation of the cardiovascular system in the crayfish Procambarus clarkii. J Comp Physiol 7B:577–583

Harper SL, Reiber CL (2004) Physiological development of the embryonic and larval crayfish heart. Biol Bull 206(2):78–86

Hertel W, Pass G (2002) An evolutionary treatment of the morphology and physiology of circulatory organs in insects. Comp Biochem Physiol 133A:555–575

Hertel W, Wirkner CS, Pass G (2002) Studies on the cardiac physiology of Onychophora and Chilopoda. Comp Biochem Physiol 133A:605–609

Innes AJ, Taylor E (1986) The evolution of air-breathing in crustaceans: a functional analysis of branchial, cutaneous and pulmonary gas exchange. Comp Biochem Physiol 85A:621–637

Ishii Y, Yamagishi H (2002) Cardiac pacemaker mechanisms in the ostracod crustacean *Vargula hilgendorfii*. Comp Biochem Physiol 133A:589–594

Kirara A, Kuasawa K (1984) A neuroanatomical and electrophysiological analysis of nervous regulation in the heart of an isopod crustacean *Bathyonomus doerderleini*. J Comp Physiol 154A:88–94

Kuramoto T, Ebara A (1984) Neurohormonal modulation of the cardiac outflow through the cardioarterial valve in the lobster. J Exp Biol 11:123–130

Lankester ER (1873) On the primitive cell-layers of the embryo as the basis of genealogical classification of animals, and on the origin of vascular and lymph systems. Ann Mag Nat Hist 11(4):321–338

Lankester ER (1887) Note on the coelom and vascular system of Mollusca and Arthropoda. Nature 37:498

Mayrat A, McMahon BR, Tanaka K (2006) The circulatory system. In: Treatise on zoology—anatomy, taxonomy, biology—The Crustacea. Revised and updated from the Traite de Zoologie 2:3–84

McGaw IJ, McMahon BR (1996) Cardiovascular responses resulting from variation in external salinity in the Dungeness Crab *Cancer magister*. Physiol Zool 69:1384–1401

McGaw IJ (2005) The decapod crustacean circulatory system: a case that is neither open nor closed. Microsc Microanal 11:18–36

McGaw IJ, Airriess CN, McMahon BR (1994) Peptidergic modulation of cardiovascular dynamics in the Dungeness crab *Cancer magister*. J Comp Physiol 164B:103–111

McGaw IJ, Stillman JH (2010) Cardiovascular system of the Majidae (Crustacea: Decapoda). Arthropod Struct Dev 39:340–349

McGaw IJ, Wilkens JL, McMahon BR (1995) Crustacean cardio-excitatory peptides may be inhibitory in vivo. J Exp Biol 204:923–932

McLaughlin PA (1983) Internal anatomy, Chap 1. In: Mantel L (ed) Internal anatomy and physiological regulation, vol 5. In: Bliss DE (Series ed) Biology of Crustacea. Academic Press, New York, pp 1–54

McMahon BR, Butler PJ, Taylor EW (1978) Changes in pH and carbon dioxide levels during recovery from disturbance and during long term exposure to hypoxia in the lobster *Homarus vulgaris*. J Exptl Zool 205(3):361–370

McMahon BR, Wilkens JL (1983) Ventilation perfusion and oxygen uptake. In: Mantel L (ed) vol 5. Bliss D (Series ed) The biology of Crustacea. Academic Press, New York, pp 289–372

McMahon BR, Burnett L (1990) The crustacean open circulatory system: a re-examination. Physiol Zool 63:35–71

McMahon BR, Hankinson JJ (1993) Respiratory adaptations in burrowing crayfish. Freshw Crayfish 9:174–182

McMahon BR (1992) Factors controlling the distribution of cardiac output in crustaceans. In Hill RB, Kuwasawa K, McMahon BR, Kuramoto T (eds) Comparative physiology, vol 11. Phylogenetic models in functional coupling of the CNS and the cardiovascular system. Karger, Basel, pp 51–61

McMahon BR (1995) Integrated neural and neurohormonal control of respiratory and circulatory function in crustaceans: is their evidence for an "autonomic" control system. Verb Deustch Zool Ges 88:87–101

McMahon BR, Chu KH, Mak EMT (1995a) Development of the heart in the Shrimp *Metapenaeus ensis*. Fish Shell Fish Culture Symp 34:470–473

McMahon BR, Smith PJS, Wilkens JL (1997) Invertebrate circulatory systems. In: Danzler WH (ed) Handbook of physiology, comparative physiology, Sect 13. American Physiological Society, New York, pp 931–1008

McMahon BR (2001) Control of cardiovascular function and its evolution in Crustacea. J Exp Biol 204:923–932

McMahon BR, Tanaka K, Doyle JE, Chu KH (2002) A change of heart: cardiovascular development n the shrimp *Metapenaeus ensis*. Comp Biochem Physiol 133A:577–587

McMahon BR, Chu KH, Tanaka K, Doyle JE, Mak E, Yu C, Nelson M (2008). Functional development of the vascular system in larval shrimp *Metapenaeus ensis*. In: Abstracts of Proceedings of the Society for Experimental Biology, Barcelona

Moorman AF, Christoffels VM (2003) Cardiac chamber formation: development, genes, and evolution. Physiol Rev 83:1223–1267

Morris S, Greenaway P, McMahon BR (1988) Adaptations to a terrestrial existence by the Robber Crab *Birgus latro*. I. An in vitro investigation of haemolymph gas transport. J Exp Biol 140:477–491

Morris S, McMahon BR (1989) Neurohumor effects on crustacean haemocyanin oxygen affinity. J Exp Zool 249:334–337

Morris S (2005) Respiratory and acid–base responses during migration and to exercise by the terrestrial crab *Discoplax* (*Cardisoma*) *hirtipes*, with regard to season, humidity and behaviour. J Exp Biol 208:4333–4343

Nylund A, Tjonneland A (1989) Crustacean heart ultrastructure and its phylogenetic implications, with special reference to the position of the isopods within the eumalacostracan phylogeny. Monitore Zoologico Italiano Monogr Suppl 4:29–42

Økland S, Tjønneland A, Larsen L, Nylund A (1982) Heart ultrastructure in *Branchinecta paludosa*, *Artemia salina*, *Branchipus schaefferi*, and *Streptocephalus* sp. (Crustacea, Anostraca). Zoomorphology 101:71–81

Olsen EN (2006) Gene regulatory networks in the evolution and development of the heart. Science 29(313):1922–1927

Pass G (1998) Accessory pulsatile organs: microscopic anatomy of invertebrates, vol 11B: Insecta. Wiley-Liss, New York, pp 621–640

Pass G (2000) Accessory pulsatile organs: evolutionary innovations in insects. Annu Rev Entomol 45:495–518

Paul RJ, Zahler S, Werner R (1991) Adaptation of an open circulatory system to the oxidative capacity of different muscle types. Naturwissenshaften 78:134–135

Paul RJ, Bihlmayer S, Colmorgen M, Zahler S (1994) The open circulatory system of spiders (*Eurypelma californicum*, *Pholcus phalangioides*): a survey of functional morphology and physiology. Physiol Zool 67(6):1360–1382

Rahr H (1981) The ultrastructure of the blood vessels of *Branchiostoma lanceolatum* (Pallas) (Cephalochordata). Zoomorphology 97:53–74

Randall DJ, Davy PS (1980) The hearts of urochordates and cephalochordates. In: Bourne GH (ed) Hearts and heart-like organs, vol 1. Comparative anatomy and development. Academic Press, New York, pp 41–59

Redmond JR, Jorgenson DD, Bourne GB (1982) Circulatory physiology of *Limulus*. In: Bonaventura JC (ed) Physiology and biology of horseshoe crabs: studies on normal and environmentally stressed animals. Alan Liss, New York, pp 135–146

Reiber CL, McMahon BR (1998) The effects of progressive hypoxia on the crustacean circulatory system: a comparison of the freshwater crayfish *Procambarus clarkii* and the lobster *Homarus americanus*. J Comp Physiol B 168:168–176

Reiber CL, Harper SL (2001) Perspectives on cardiac physiological ontogeny in crustaceans. Zoology (Jena) 104:103–113

Reiber CL, McGaw IJ (2009) A review of the "open" and "closed" circulatory systems: new terminology for complex invertebrate circulatory systems in light of current findings. Int J Zool Suppl 1:1–8 (Article ID 301284)

Ruppert E (1997) Cephalochordates, in Microscopic Anatomy of Invertebrates Hemichordata, Chaetognatha and the invertebrate chordates. New York Wiley Liss 15:349–504

Ruppert EE, Carle KJ (1983) Morphology of metazoan circulatory systems. Zoomorphology 103:193–208

Sakurai A, Yamagishi H (1998a) Identification of two cardioacceleratory neurons in the isopod crustacean *Ligia exotica* and their synaptic effects on cardiac ganglion cells. J Comp Physiol 182A:145–152

Sakurai A, Yamagishi H (1998b) Cardioacceleratory neurons in the isopod crustacean *Ligia exotica*: visualization of peripheral projection onto the heart muscle. Zool Sci 15:19–25

Sakurai A, Wilkens JL (2003) Tension sensitivity of the heart pacemaker neurons in the isopod crustacean *Ligia pallasii*. J Exp Biol 206:105–115

Sanger JW, McCann FV (1968) Observations on the ultrastructure of the myocardium of the moth *Hyalophora cecropia*. J Insect Physiol 14(8):1105–1106

Saudemont A, Dray N, Hudry B, Le Gouar M, Vervoort M, Balavoine G (2008) Complementary striped expression patterns of NK homeobox genes during segment formation in the annelid *Platynereis*. Dev Biol 317:430–443
Sedmera D, Ostadal B (2012) Ontogenesis of myocardial function. In: Sedmera D, Wang T (eds) Ontogeny and phylogeny of the vertebrate heart. Springer, New York
Shadwick RE, Pollock CM, Striker SA (1990) Structure and biomechanical pathways of the crustacean blood vessels. Physiol Zool 63:90–101
Shigei T, Tsuru H, Ishikawa N, Yoshioka (2001) Absence of endothelium in invertebrate blood vessels: significance of endothelium and sympathetic nerve/medial smooth muscle in the vertebrate vascular system. Jpn J Pharmacol 87:253–260
Slama K (2010) Physiology of heartbeat reversal in adult *Drosophila melanogaster* (Diptera: Drosophilidae). Eur J Entomol 107(1):13–31
Spicer JI (1994) Ontogeny of cardiac function in the brine shrimp *Artemia franciscana* Kellogg 1906 (Branchiopoda: Anostraca). J Exp Zool 270:508–516
Spicer JI (2006) Gut reaction by heartless shrimps: experimental evidence for the role of the gut in generating circulation before cardiac ontogeny. Biol Lett 22:580–582
Tanaka K, Kuwasawa K (1991) Central outputs for extrinsic control of the heart in an isopod crustacean *Bathynomus doederleini* Neuroanatomy and electrophysiology. Comp Biochem Physiol 98C:79–86
Tanaka K, Kuwasawa K (1990) Bradycardia and tachycardia induced by extrinsic cardiac nerves in an isopod *Bathynomus doederleini.* Wiese K, Krenz K-D, Tautz J, Riechert H, Mulloney B, Editors In Frontiers in Crustacean Neurobiology Basel: Birkhauser 492–497
Taylor HH, Greenaway P (1984) The role of the gills and branchiostegites in gas exchange in a bimodally breathing crab, *Holthuisana transversa* Von Martens. Evidence for a facultative change in the distribution of the respiratory circulation. J Exp Biol 111:103–121
Taylor HH, Taylor E (1986) Observations of valve-like structures and evidence for rectification of flow within the gill lamellae of the crab *Carcinus maenas* (Crustacea, Decapoda). Zoomorphology (Berlin) 106:1–11
Volk EL (1988). The role of suspensory ligaments in modifying cardiac output in crustaceans. MSc Dissertation, The University of Calgary
Wheatly MG, McMahon BR, Burggren WW, Pinder AW (1986) Haemolymph acid-base, electrolyte and gas status during sustained voluntary activity in the land hermit crab (*Coenobita compressus*). J Exp Biol 125:225–243
Wilkens JL, McMahon BR (1992) Intrinsic properties and extrinsic neurohormonal control of crab hemodynamics. Experientia 48:827–834
Wilkens JL, Walker RL (1992) Nervous control of crayfish hemodynamics. In: Hill RB, Kuwasawa K, McMahon BR, Kuramoto T (eds). Phylogenetic models in functional coupling of the CNS and the cardiovascular system. Comprehensive Physiology, vol 11. Karger, Basel, pp 115–122
Wilkens JL (1993) Re-evaluation of the stretch sensitivity hypothesis of crustacean hearts: hypoxia, not lack of stretch, causes reduction in heart rate of isolated hearts. J Exp Biol 176:223–232
Wilkens JL, Kuramoto T, McMahon BR (1996) The effects of six pericardial hormones and hypoxia on the semi-isolated heart and sternal artery of the lobster *Homarus americanus*. Comp Biochem Physiol C114:57–65
Wilkens JL (1997) Possible mechanisms of control of vascular resistance in the lobster *Homarus americanus*. J Exp Biol 200:487–493
Wilkens JL (1999) Evolution of the cardiovascular system of Crustacea. Am Zool 39:199–204
Wilkens JL, Taylor HH (2003) The control of vascular resistance in the southern rock lobster, *Jasus edwardsii* (Decapoda: Palinuridae). Comp Biochem Physiol 135A:369–376
Wilkens JL, Cavey MJ, Shovkivska I, Zhang ML, Ter Kuers H (2008) Elasticity, unexpected contractility and the identification of actin and myosin in lobster arteries. J Exp Biol 211:766–772
Windsor-Watson H III, Groome JR (1989) Modulation of the *Limulus* heart. Am Zool 29:1287–1303

Wirkner CS (2009) The circulatory system in Malacostraca—evaluating character evolution on the basis of differing phylogenetic hypotheses. Arthrop Syst Phyl 67:57–70

Wirkner CS, Pass G (2002) The circulatory system in Chilopoda: evolutionary and phylogenetic aspects. Acta Zool 83:197–202

Wirkner CS, Prendini L (2007) Comparative morphology of the hemolymph vascular system in scorpions–a survey using corrosion casting, MicroCT, and 3D-reconstruction. J Morphol 268(5):401–13

Wirkner CS, Richter S (2010) Evolutionary morphology of the circulatory system in Peracarida (Malacostraca; Crustacea). Cladistics 26:143–167. doi:10.1111/j.1096-0031.2009.00278.x

Wong LYE, Moorman AF, Barnett P (2012) Basic cardiac development: the heart and its electrical components. In: Sedmera D, Wang T (eds) Ontogeny and phylogeny of the vertebrate heart. Springer, New York

Worden MK, Clark CM, Conaway M, Qadri AM (2006) Temperature dependence of cardiac performance in the lobster *Homarus americanus*. J Exp Biol 209:1024–1034

Xavier-Neto J, Castro RA, Sampaio AC, Azambuja AP, Castillo HA, Cravo RM, Simoes-Costa MS (2007) Parallel avenues in the evolution of hearts and pumping organs Cell. Mol Life Sci 64:719–734

Yamagishi H, Euichi H (1997) Transfer of the heart pacemaker during juvenile development in the isopod crustacean Ligia exotica. J Exp Biol 200:2393–2404

Yamagishi H, Ando H, Makioka T (1997) Myogenic heartbeats in the primitive crustacean *Triops longicaudatus*. Biol Bull 193:350–358

Chapter 2
The Teleost Heart: A Morphological Approach*

José M. Icardo

Introduction

Numerous review articles have reported on the morphology and physiology of the fish heart (see for instance Santer 1985; Satchell 1991; Farrell and Jones 1992; Burggren et al. 1997) through the years. The recent use of the teleost heart as an organ model has focused a lot of attention on this fish group. Genetic (see Harvey and Rosenthal 1998; Chen and Fishman 2000; Lohr and Yost 2000; Yelon 2001, and references herein) and epigenetic (Taneda et al. 2010) studies carried out in teleosts have provided deep insights into several mechanisms of great developmental significance for the heart of vertebrates. Furthermore, teleosts are being used to study several other biological mechanisms ranging from the establishment of the left–right axis (Chen et al. 1997; Nagai et al. 2010), to heart regeneration (Poss et al. 2002; Lepilina et al. 2006), to organ development in the absence of gravity (Niihori et al. 2004). Full understanding of all these data requires a thorough knowledge of the morphological aspects of the teleost heart.

The heart of modern teleosts has classically been described as being formed by four segments arranged in series: sinus venosus, atrium, ventricle, and bulbus arteriosus. Recent studies indicate that, in addition to those four chambers, all teleosts contain a conus arteriosus (Schib et al. 2002; Grimes et al. 2006; Icardo 2006) and a distinct atrioventricular segment (Icardo and Colvee 2011). Accordingly, the morphology of the heart is reviewed to include these two segments. Attention is also focused on the various structural patterns of the conus and bulbus. The architecture of the heart ventricle has been divided into four main types (Tota et al. 1983) depending on the absence, or on the presence and the extent of a compact layer.

*Contract grant number: CGL2008-04559/BOS from the "Ministerio de Educación y Ciencia," Spain.

J.M. Icardo (✉)
Departamento de Anatomia y Biología Celular, Facultad de Medicina,
Universidad de Cantabria, c/Cardenal Herrera Oria, s/n, 39011 Santander, Spain
e-mail: icardojm@unican.es

D. Sedmera and T. Wang (eds.), *Ontogeny and Phylogeny of the Vertebrate Heart*,
DOI 10.1007/978-1-4614-3387-3_2, © Springer Science+Business Media, LLC 2012

Ventricle classifications are not merely academic. Different modes of heart performance have been attributed to the several ventricle types. For instance, completely trabeculated ventricles are thought to be unable to cope with increasing amounts of afterload. Evidence is reviewed here supporting the notion that the teleost heart can maintain high ventricular work in the absence of a compacta. In higher vertebrates, the collagenous skeleton of the heart plays an important role in myocardial mechanics (Weber 1989). However, this subject has received very little attention in the fish heart. The possible role of the heart collagenous skeleton in the maintenance of the ventricular shape and in the performance of the teleost heart is also discussed. Finally, the structure of the venous pole of the heart is reviewed. Despite the great biological importance of several species such as the zebrafish and the medaka as model organisms, this review is not focused on any particular species. Rather, it intends to reflect the enormous morphological diversity of the teleost heart, put the focus on controversial subjects, and addresses several issues of general morpho-functional significance.

The Outflow Tract: The Bulbus, the Conus, and the Conus Valves

The outflow tract (OFT) is the morphological division of the heart located between the ventricle and the beginning of the dorsal aorta. In most primitive fish, the OFT is formed by two segments: a proximal, muscular, conus arteriosus, and a distal, arterial-like, bulbus arteriosus (Icardo et al. 2002, 2005a; Durán et al. 2008; Grimes and Kirby 2009). The anatomical composition of the OFT in several other ancient fish, like hagfishes and lampreys, is unclear, but most uncertainties appear to derive from partial observations (Parsons 1930; Yamauchi 1980). In fact, the overall picture that emerges from the different studies is that the OFT of all primitive fish abides to the same general rule. Like primitive fish (Parsons 1930; Icardo 2006; Grimes and Kirby 2009), ancient teleosts show a conus and a bulbus. In all genera studied (Albula, Pterothrissus, Megalops, Elops, Tarpon), the heart shows a muscular conus arteriosus (Parsons 1930; Santer 1985; Satchell 1991). The conus is of variable length and contains up to two valve rows, with a total of four to six valves. These are the conus valves, which regulate the ventricular flow dynamics. In addition to the conus, the heart of ancient teleosts presents a distal outflow segment, the bulbus arteriosus, which opens in the ventral aorta. This segment contains connective tissue and elastic fibers (Parsons 1930). By contrast, the OFT of modern teleosts has classically been described as being formed solely by the elastic bulbus arteriosus. Consequently, the outflow valves were named bulboventricular valves.

In modern teleosts, the prominent bulbus arteriosus dominates the morphology of the heart outflow (Fig. 2.1). The bulbus is an elastic chamber with a “windkessel” function. It expands during ventricular ejection to store a large part of the cardiac stroke volume. Gradual elastic recovery allows a steady flow of blood toward the gills, preventing damage of the delicate gill vasculature (Priede 1976; Satchell 1991;

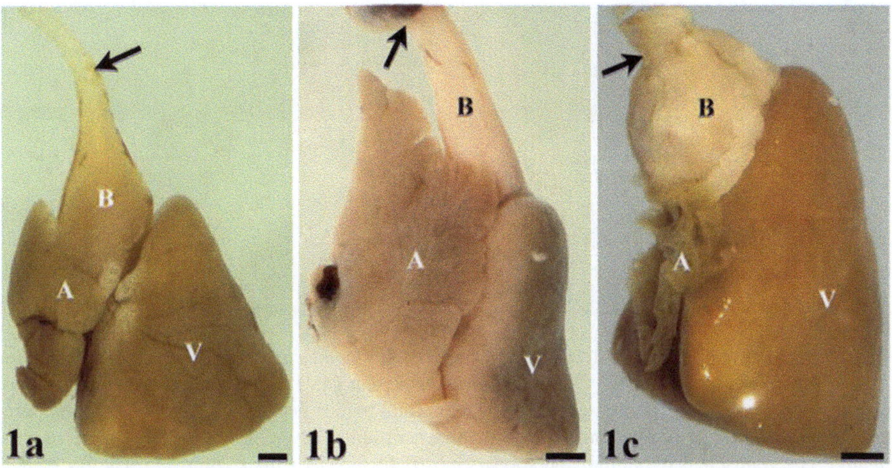

Fig. 2.1 Right lateral views of *Trachurus trachurus* (**a**), *Trigla lucerna* (**b**), and *Sparus pagrus* (**c**) hearts. The ventricle (V) may be pyramidal (**a, c**) or sac-like (**b**). The atrium (A) may be very large (**b**) or much smaller (**a**). The outflow tract (OFT) is dominated by the prominent bulbus arteriosus (B). No other OFT component is apparent. The bulbus is pear-shaped in (**a**), cylindrical and elongated in (**b**), and short and thick in (**c**). Note the relationship between the length of the bulbus and that of the ventricle: the ratio is close to a 1:1 in (**a**) and (**b**), and approximately 1:3 in (**c**). The *arrows* mark the upper insertion of the pericardium and the boundary between the bulbus and the ventral aorta. Scale bars: 0.1 cm (From Icardo 2006, Anat. Rec. A, 288:900–908)

Farrell and Jones 1992; Jones et al. 1993). To this end, the wall of the bulbus is endowed with high amounts of elastin material and an external (subepicardial) collagen layer, which probably controls bulbus compliance by limiting circumferential deformation (Icardo et al. 1999a).

From the morphological point of view, the external shape of most bulbus ranges from pear-shaped, to elongated, to thick and robust (Fig. 2.1). The wall of the bulbus is organized into layers: the endocardium, the endocardial ridges, the middle layer, the subepicardium, and the epicardium (Fig. 2.2a, b). Given the high level of radiation in fish, the existence of various shape patterns is not surprising. The surprising fact is the high variability in bulbus structure. A series of studies carried out in our laboratory (Icardo et al. 1999a, b, 2000a, b) has shown that the structure of the bulbus is close to being species-specific. The inner surface of the bulbus is characterized by the presence of ridges (Fig. 2.2a, b). These are longitudinal columns, which occupy the full length of the bulbus (Fig. 2.2a). On the whole, they are thicker at the base of the bulbus and attenuate toward the ventral aorta. Depending on the species, the ridges may be very prominent (Fig. 2.2a, b) or much more discrete (Icardo et al. 2000b). However, the real significance of these variations is currently unknown. The inner surface of the ridges is covered by endocardium (Fig. 2.2b). The ridge endocardium shows histological differences across species, ranging from squamous to columnar (Icardo et al. 2000b). In addition, endocardial cells in many species contain moderately dense bodies (Benjamin et al. 1984) of (mostly) unknown nature. The presence of the dense bodies indicates a secretory function, which may

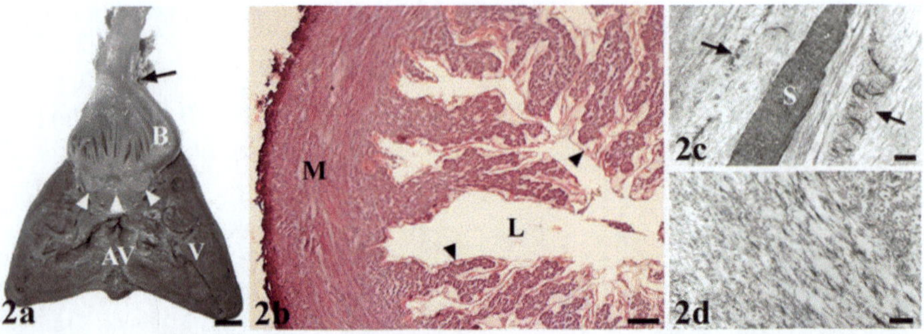

Fig. 2.2 Bulbus and conus in different teleost species. (**a**) *Thunnus alalunga*. Internal heart structure. The outflow tract, the ventricle (V), and the atrioventricular (AV) regions are exposed. The bulbus (B) shows well-marked longitudinal ridges. The cranial bulbus boundary corresponds to the pericardial insertion (*arrow*). The conus supports three semilunar valves of roughly equal size (*arrowheads*). The ventricle is pyramidal and appears mostly formed by a thick compacta. The AV orifice is guarded by the AV valves. (**b**) *Trigla lucerna*. Cross-section of the bulbus. Hematoxylin–eosin. The ridges protrude into the lumen (L) and are covered by endocardium (*arrowheads*). The middle layer (M) contains large amounts of extracellular material. The subepicardial layer (*dark in the picture*) is rich in collagen. (**c**) *Anguilla anguilla*. TEM micrograph of the middle layer of bulbus. A smooth muscle cell (S) is surrounded by elastin and fibrillar material interspersed with collagen fibers (*arrows*). (**d**) *Trematomus bernacchii*. TEM micrograph of the middle layer of bulbus. The extracellular matrix is formed by a filamentous network. Note the absence of collagen and elastin. Scale bars: (**a**) 0.5 cm; (**b**) 100 μm; (**c**) 1 μm; (**d**) 0.5 μm; (**c**) from Icardo et al. 2000a, Cells Tissues Organs, 167:184–198; (**d**) from Icardo et al. 1999b, Anat. Rec. 256:116–126

be different between older and more modern teleost families (Leknes 2009). This secretory function appears to be enhanced in the bulbus of the Antartic teleosts (Icardo et al. 1999a, b), where endocardial cells may be implicated in the production of anti-freeze mucins. Of note, endocardial cells in other heart chambers show scavenger functions (Seternes et al. 2002), bind natriuretic peptides (Cerra et al. 1997), are a source of endogenous nitric oxide (Imbrogno et al. 2003), or may be involved in autocrine/paracrine regulation of the subjacent tissue (for a recent review, see Icardo 2007). The middle layer of the bulbus contains smooth muscle cells and variable amounts of elastin. It may also contain, as in the common eel (Fig. 2.2c), a few collagen layers interspersed with the elastin material (Icardo et al. 2000a), or, as in tuna, collagen bundles, blood vessels and nerves (unpublished observations). However, it may lack elastin, as in the Antarctic teleosts (Icardo et al. 1999a, b). In these species, the elastin material is replaced by a fibrillar network (Fig. 2.2d), which is probably made up of glycosaminoglycans. The subepicardium is a thin layer rich in collagen and elastin, fibroblasts, vessels, and nerves (Icardo et al. 2000b). Although this constitutes a general description, the subepicardium of the bulbus can be a more complex layer. For instance, it may contain lymphocytes, plasma cells, and dendrite-like cells, and has been implied to be involved in the development of the humoral immune response (Icardo et al. 1999b). This is surprising, but it is not an isolated feature in fish. The subepicardium of the sturgeon contains thymus-like tissue (Icardo et al. 2002) which has been implicated in the

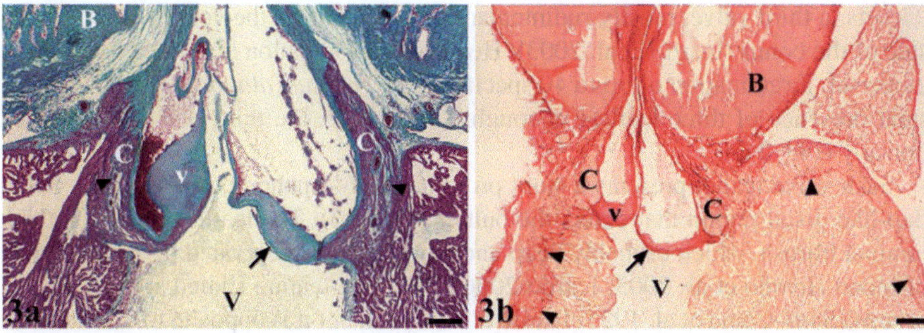

Fig. 2.3 Conus arteriosus in two hearts with different ventricular structure. B, bulbus; C, conus; V, ventricle; v, conus valves. (**a**) *Spondylosoma cantharus*. Martin's trichrome. The compact musculature of the conus differentiates easily from the completely trabeculated ventricle. Conus vessels (*arrowheads*) are apparent. (**b**) *Echiicthys vipera*. Sirius red. A collagenous ring isolates the conus myocardium from the ventricular myocardium. The conus valves are anchored to the conus and show a proximal stout body and a thick, luminal fibrosa (*arrow*). In (**b**), a thin rim of collagen (*arrowheads*) locates at the boundary between the compact and spongy layers of the ventricle. Scale bars: 200 μm

establishment and the maintenance of the cellular immune responses. The absence of elastin in stenothermal teleosts may be considered a sort of adaptation to subzero temperatures. However, the rationale for the structural variety in temperate teleosts is unknown. A number of factors such as cardiovascular dynamics, lifestyle, ecophysiology, range of diversification, may have specifically adapted the fine bulbus structure to comply with its "windkessel" function.

The conus arteriosus, present in ancient teleosts (see above), was thought to have disappeared in more derived species. The loss of the conus was attributed to intussusception into the ventricle (Smith 1918), or it was considered a direct consequence of heart evolution (Santer 1985). However, evidence raised in the last decade indicates that the conus is not lost at all in the heart of modern teleosts (Schib et al. 2002; Icardo et al. 2003; Grimes et al. 2006; Icardo 2006). Examination of up to 28 species belonging to different families and orders (Icardo 2006) has revealed that the conus arteriosus is a distinct anatomical segment interposed between the ventricle and the bulbus arteriosus. The Macrouridae (Greer Walker et al. 1985) could also be added to that list (discussed in Icardo et al. 2003). However, species differences do exist (Fig. 2.3). The conus is easily recognizable in hearts whose ventricles lack a compact layer (Fig. 2.3a), and may be more difficult to discern in hearts possessing a compacta (Fig. 2.3b). It is formed by compact myocardium (very evident in completely trabeculated ventricles), and contains more collagen, elastin, and laminin than the ventricular muscle. With very few exceptions, the conus contains vessels (Fig. 2.3a) even when the neighboring myocardium is not vascularized (Icardo 2006). As occurs with other fish hearts having conus, the teleost conus arteriosus supports the outflow valves, which should more properly be named conus valves (Schib et al. 2002).

Most teleosts possess a single row of conus valves formed of two (left and right) pocket-like leaflets and the supporting sinus (Fig. 2.3). Only a few species pertaining to the order of the Elopiformes show two valve rows (Parsons 1930). In modern teleosts, a third valve, mostly rudimentary, may appear on the ventral or the dorsal side of the conus (Icardo et al. 2003). However, this situation does not appear to be universal. Examination of up to 40 specimens of *Thunnus alalunga* shows the constant presence of three leaflets of roughly equal size, the third one being located dorsally (Fig. 2.2a).

Each valve leaflet presents a stout proximal body, anchored to the conus, and a flap-like distal region that enters the bulbus (Fig. 2.3). The leaflets present a thick luminal fibrosa, which probably bears most of the stress generated by the backflow of blood (Icardo et al. 2003; Icardo 2006). This is a feature shared with other fish groups (Sans-Coma et al. 1995; Icardo et al. 2002), but it is opposite to the situation observed in mammals, where the fibrosa is located on the parietal side of the leaflet. Differences in extracellular matrix composition have been described in the conus valves of several teleost species (Greer Walker et al. 1985; Raso 1993; Icardo et al. 2003). These differences in composition may be implicated in the mechanical function of the valve. At least in several species (Schib et al. 2002; Icardo et al. 2003; Icardo 2006), the conus myocardium has a distinct structural organization, which has also been implicated in valve physiology.

The Ventricle

The teleost heart ventricle is a chamber that shows (Fig. 2.1) considerable species variability (Santer 1985; Farrell and Jones 1992). Indeed, this assertion can be applied to all fish heart ventricles (Santer 1985). The external shape, the architectural organization, the histology, the coronary distribution, the relative mass, the work dynamics, etc., vary widely between species. In an attempt to classify this variability, several authors have grouped the ventricles into broad categories (Greer Walker et al. 1985; Santer 1985; Tota 1989). The problem is that the ventricles do not quite fit into any particular category very often. This is especially true when data between different categories are crossed. Nonetheless, divisions based on the external shape of the ventricle (Santer 1985), or on the degree of "muscularization" and vascularization of the ventricular wall (Tota 1989), have proven useful.

The external ventricular shape has been grouped (Santer 1985) into three main categories: tubular, sac-like, and pyramidal (Fig. 2.1). This division has several functional implications. For instance, pyramidal ventricles have been related to an active lifestyle, a robust ventricular wall, and a high output work. The salmonid and scombrid families present this type of ventricle (Farrell and Jones 1992). The very active tuna also shows a pyramidal ventricle (Fig. 2.2a). However, the presence of a pyramidal ventricle does not correlate with either a robust ventricle or a very active lifestyle in many other cases such as in the Antarctic teleosts and in members of the sparid and serranid families. The significance of the two other morphologies is unclear (Farrell and Jones 1992). Furthermore, the relation between the external ventricular shape and the inner architecture is not constant (Simoes et al. 2002). Sac-like ventricles are observed in many marine teleosts, and tubular ventricles are frequently observed in fish which, like the eel, present and elongated body shape.

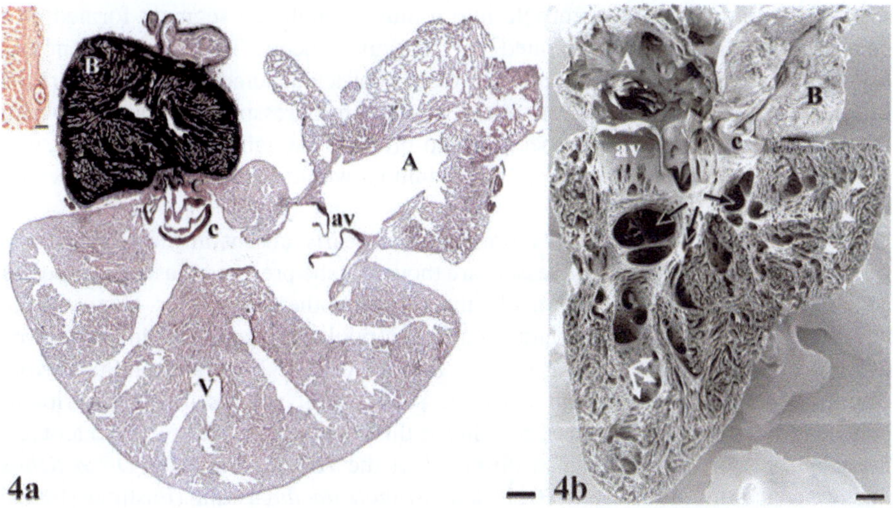

Fig. 2.4 Hearts with completely trabeculated ventricles. A, atrium; av, atrioventricular valves; B, bulbus; C, conus; c, conus valves; V, ventricle. (**a**) *Balistes carolinensis*. Orcein. The elastic bulbus is intensely stained. The ventricle is saccular and entirely trabeculated. The AV orifice is delimited by a ring of compact myocardium. The atrium shows a complex network of thin trabeculae which originates from the AV orifice. *Inset*: Detail of coronary vessels in the ventricular subepicardium. (**b**) *Sparus auratus*. SEM composite shows the architectural organization of the heart, sagital section. The left side of the heart is shown from the right. In the ventricle, an interconnected system of trabecular sheets and lumina (*black* and *white arrows*) radiate outward from the main ventricular lumen. The lumina become smaller (*white arrows*) toward the periphery. At the ventricular periphery, the trabecular sheets give origin to a system of single trabeculae that reach the outer myocardial layer. Note the formation of arch systems (*arrowheads*). Scale bars: (**a**) 100 μm; *inset*, 20 μm; (**b**) 50 μm. (**b**, from Icardo et al. 2005b. J Exp Zool., 303A:665–675)

Another heart classification (Tota et al. 1983; Tota 1989; Tota and Garofalo 2012) relies on whether the ventricle presents a compact layer, on the relative thickness of the compacta, and on the extent of myocardial vascularization. Type-I hearts show entirely trabeculated ventricles and lack a compacta. The ventricles of the rest of the heart types present both an external compacta and an inner spongiosa. Type-II hearts show vessels in the compacta but not in the spongiosa, and type-III hearts have vessels in both the compacta and the spongiosa. Type-IV hearts are different from type-III hearts in that a large proportion of their ventricular mass is formed by a compacta.

Most teleost ventricles (close to 80%, Santer 1985) are entirely trabeculated and thus belong to type-I hearts (Fig. 2.4a). The trabecular network has been described as a highly organized system of small lumina and trabecular sheets which radiate outward from a central lumen (Fig. 2.4b) (Munshi et al. 2001; Icardo et al. 2005b). The size of the lumina decreases progressively toward the ventricular periphery. Of note, a similar pattern is observed in the spongy component of ventricles having a compacta (Pieperhoff et al. 2009). Although observations have been made in just a few species, this architectural arrangement may be more common than previously realized. It transforms the ventricle into a multi-chambered segment formed of a small number of lumina separated by trabecular sheets. This arrangement has important functional implications. The main ventricular lumen would support the highest amount of stress, and the stress would be progressively attenuated toward the periphery. The trabecular sheets would produce enough contractile force, and the communication between the different lumina would facilitate blood squeezing (Icardo et al. 2005b).

Coronary vessels have been reported to be nearly absent in the ventricles of type-I hearts. In fact, coronary vessels are thought to be present in just a few species of temperate teleosts and to be mostly absent in stenothermal species. However, the conus arteriosus (Icardo 2006) and the atrioventricular segment (Icardo and Colvee 2011, see below) show vascular profiles in most of type-I hearts examined. More importantly, this is accompanied by the presence of vessels in the ventricular subepicardium (Fig. 2.4a, inset). Strikingly, this feature is not restricted to temperate teleosts. Coronaries are also observed in the Antarctic species *Dissostichus mawsoni* (Icardo and Colvee 2011) and *Nothotenia angustata* (Eastman 1993). Overall, ventricular subepicardial vessels are not very numerous, and they do not appear to form a very rich plexus. Consequently, their presence may have been underestimated. On the other hand, their exact role in this type of hearts is unclear. It is assumed that myocardial cells in the entirely trabeculated ventricles are supplied by the blood flowing through the ventricular chamber.

The ventricle of many other teleost species shows both a compacta and a spongiosa (Fig. 2.5), belonging to type-II hearts. The compacta is formed by myocardial cells arranged into bundle layers which appear more complex and thicker in more athletic fish (Sánchez-Quintana and Hurle 1987; Farrell and Jones 1992). The bundles are oriented in different directions (Fig. 2.5b) forming interrelated loops and coils, which provide the structural basis for developing high blood pressures (Farrell and Jones 1992). The compact layer is always vascularized, while the spongiosa is not. However, several species do not fit well within this classification. For instance, a large part of the ventricular wall thickness in tuna is formed by a very thick, extensively vascularized compacta (Fig. 2.5b). Curiously, an apparently extensive vascular network is also observed in the spongy layer (Fig. 2.5c). Thus, the overall ventricular architecture of tuna is closer to that of type-IV hearts.

The existence of the two muscular components does not mean that all ventricles have a similar structure. The thickness of the teleost compacta may range from a mere two- to three-cell-thick layer (as in *Echiicthys vipera*) to occupy a large part of the ventricular thickness (as in tuna). Estimations of the thickness relation between the spongiosa and the compacta in type-II hearts indicate that it ranges from about 30 (as in *E. vipera*), through 10 (as in *Oncorhynchus mikiss* and *Salvelinus alpinus*), to 3 (as in *Trachurus trachurus*). It should be stressed that these values are rough estimates and have been obtained at the mid-ventricle level. The thickness of the compacta is not completely regular, undergoing variations from apex to base. Percentage values between the relative mass of the ventricle and that of the compacta have been reported earlier (Farrell and Jones 1992) for several teleost species. However, the relation between the two muscle components, albeit specific,

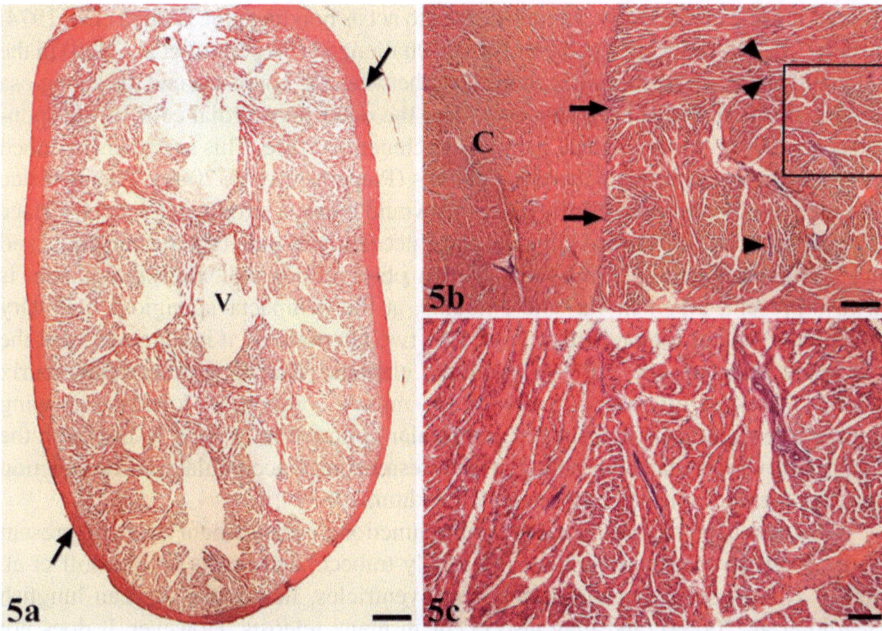

Fig. 2.5 Ventricular organization in hearts with compacta and spongiosa. (**a**) *Oncorhynchus mikiss*. Hematoxylin–eosin. Detail of the ventricular chamber (V). The outer compacta (*arrows*) limits the ventricle. Note the unequal thickness of the compacta and spongiosa. (**b**) *Thunnus alalunga*. Detail of the ventricular structure. The compacta (C) is organized into bundles showing different orientations. *Arrows* indicate the compacta–spongiosa boundary. The spongiosa shows numerous vascular profiles (*arrowheads*). (**c**) Detail of the *boxed area* in (**a**). Coronaries of different sizes are clear. Scale bars: (**a**) 1 mm; (**b**) 300 μm; (**c**) 100 μm

is not an invariable value. The proportion of the compacta has been reported to vary with changing seasons (Farrell and Jones 1992) and growth (Farrell et al. 1988; Cerra et al. 2004), but not with physical activity (Farrell et al. 1990). It should be mentioned that both the compacta and the spongiosa are formed by typical cardiomyocytes (Yamauchi 1980; Santer 1985; Cerra et al. 2004; Icardo et al. 2005b) whose structural, functional, and metabolic characteristics have been reviewed a number of times through the years (Yamauchi 1980; Santer 1985; Farrell and Jones 1992; Burggren et al. 1997). The different morphological arrangements do not depend on distinct characteristics of the myocardial cells, but on the specifics of the heart design.

Much less interest has been placed on the way the compacta and the spongiosa become structurally connected to form a functional unit. The basic problem derives from the overall circumferential arrangement of the compacta, to which the overall perpendicularly arranged spongiosa must be attached. If the two components are not connected tightly, they would tend to separate from each other. In fact, the spongiosa can be peeled off from the compacta quite easily after fixation (Farrell et al. 2007). The existence of a layer of connective tissue located at the boundary between the two muscle components was considered to act as bonding glue (Poupa et al. 1974; Tota 1978). This concept has recently been challenged (Pieperhoff et al. 2009) in the salmonid heart, where the collagenous layer between the compacta and the spongiosa is quite discrete. It has been suggested that the outer myocardial cells of the spongiosa bend their tips to create a parallel attachment surface. This surface is enriched in desmosome and fascia adherens elements (Pieperhoff et al. 2009). The presence of a high number of intercellular junctions would create a strong attachment surface (Pieperhoff et al. 2009), providing the force necessary to maintain the compacta and the spongiosa together. However, our own observations indicate that collagen is always present, albeit in variable amounts, at the compacta–spongiosa boundary (Fig. 2.3b). Rather than rejecting any of the two alternatives, it is suggested that the combined effect of the cellular junctional elements and the extracellular matrix establish the synergy needed to sew the two muscle components together, allowing at the same time for a coordinated ventricular contraction. This is in line with the early observation that both collagen and desmosomes accumulate at the junction between the two muscle compartments (Midttun 1983).

It has also been suggested that a similar junctional arrangement could be present at the outer myocardial boundary in entirely trabeculated hearts (Pieperhoff et al. 2009). This occurs in hearts with spongy ventricles, like in the African lungfish (Icardo et al. 2005a), and may also occur in many teleosts. However, it does not appear to be a universal feature. In the teleost *Sparus auratus*, the outer myocardium forms a continuous single-cell layer, like a shell, to which the trabecular muscle cells become attached in a mostly perpendicular direction (Icardo et al. 2005b).

Another subject which has received little attention in fish is the possible role of the connective tissue in the maintenance of the architectural design of the ventricle and in the mechanical performance of the heart. The presence of a collagenous scaffold in the avian and mammalian hearts provides structural support for the myocardium and appears to play an important role in myocardial mechanics (Caulfield and Borg 1979; Weber 1989; Icardo and Colvee 1998). Collagen is an important component of the subepicardial tissue in fish. In teleosts, it may increase ventricular resilience and limit ventricle deformation (Icardo et al. 2005b). Subendocardial collagen and coiled collagen fibers running along the main trabecular axis have been described in the teleost heart ventricle (Sánchez-Quintana et al. 1995). However, the exact role of this collagen, or even the existence of a collagen scaffold, is still unclear.

A simple method to visualize the collagen network is to digest the tissue with NaOH. This preserves the collagenous component, which can then be observed with the scanning microscope (Ohtani 1987). When this procedure is applied to entirely trabeculated ventricles, the pieces of tissue are reduced to threads during processing. This indicates the absence of a collagenous scaffold which could have strong implications in either the maintenance of shape or the ventricular performance. These negative findings reinforce the role of the trabecular architecture in heart dynamics. In heart ventricles having a compacta the situation appears to be quite different. Unpublished observations in the common eel show the presence of an extensive collagen network extending between the subepicardium and the spongiosa (Fig. 2.6a). This network mimics the distribution of the muscular bundles in the compacta and the architecture of the spongiosa. Collagen connections between the

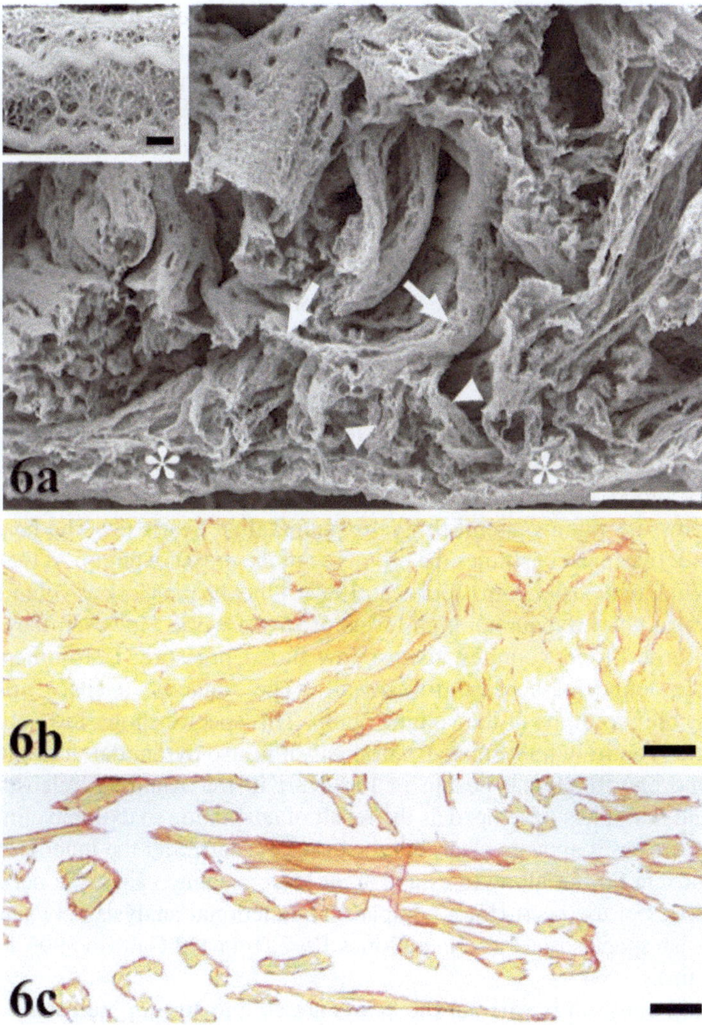

Fig. 2.6 Collagen arrangement in the ventricular chamber. (**a**) *Anguilla anguilla*. SEM micrograph depicting a portion of the ventricle digested with NaOH. The collagenous skeleton reproduces the ventricular architecture. Trabecular sheets (*arrows*) reach the compacta (*asterisks*). *Arrowheads*, arch system. *Inset*: Trabecular surface. Wavy collagen bundles run superficially and are joined by thin collagen fibrils. (**b**) *Anguilla anguilla*. Sirius red. Wavy collagen bundles (in *red*) run along the surface of the ventricular trabeculae. Collagen is nearly absent at the inner side of the trabeculae. (**c**) *Monopterus albus*. Sirius red. The trabecular collagen shows the same distribution as in (**b**), despite that the ventricle lacks a compacta. Scale bars: (**a**) 150 μm; *inset*, 5 μm; (**b**) 100 μm; (**c**) 100 μm

compacta and the spongiosa are numerous (Fig. 2.6a). Furthermore, collagen in the trabeculae occupies a subendocardial location (Fig. 2.6a, inset), the muscle cells occupying a central position (Fig. 2.6b). Curiously, the distribution of collagen in the trabeculae is similar in type II (Fig. 2.6b) and type I (Fig. 2.6c) hearts.

These observations are not meant to infer the existence of a common pattern for all teleost species having a compacta. However, the collagenous skeleton observed in the eel ventricle is similar to that obtained in sturgeons (Icardo et al. 1996). At least in the eel, the collagen network should play an important role in the maintenance of the ventricular architecture. In addition, its presence raises several questions of biological importance. For instance, does this network play a functional role similar to that described in the mammalian heart? The answer is unclear as yet, but the improvement of ventricular performance observed in the eel heart during growth occurs concomitantly with an increase in the amount of interstitial collagen (Cerra et al. 2004). On the other hand, the collagen connections between the compacta and the spongiosa should play an important role in the bonding of the two muscular components of the ventricle (see above).

From a functional point of view, the presence of pyramidal ventricles having a compacta has been associated with species showing active lifestyles (Santer et al. 1983). These hearts are able to sustain high levels of stroke work by pumping small volumes of blood at high heart rates against relatively high blood pressures. They work as pressure pumps, as much as the mammalian hearts do. The heart of the extremely active tuna constitutes the prototype of the pressure pump. At the opposite end of the functional spectrum (Tota et al. 1997), other hearts work as volume pumps. They are also able to maintain high levels of stroke work. However, they do it by pumping large blood volumes against low blood pressure. Cardiomegalia and bradycardia define these hearts functionally. The Antarctic teleost *Chionodraco hamatus* is the prototype of the volume pump. This species shows low activity, and its ventricle is entirely trabeculated. Curiously, it is also pyramidal. It can be argued that the morpho-functional design of the heart in the Antarctic teleosts is very specific and that it has developed as the result of adaptation to extreme climate conditions. However, it appears clear that the external shape and the inner architecture of the ventricle do not allow to establish, at least in many cases, the performance characteristics of the heart (for a comparative functional analysis between teleosts and other fish species, see Farrell and Jones 1992; Tota and Gattuso 1996, and references herein).

A related question is whether the hearts are able to sustain increasing levels of afterload. For instance, the extreme morpho-functional adaptation of the hearts of the Antarctic teleosts makes them fail when afterload is increased. This is very patent in the icefish *C. hamatus* (Tota and Gattuso 1996), and less remarkable in other Antarctic species such as *Trematomus bernacchii* (Farrell and Jones 1992). In fact, only hearts with a pyramidal ventricle and a compacta were thought to be able to cope with significant increases in afterload (Farrell and Jones 1992). A recent study has challenged this view. The teleost *S. auratus* that has a pyramidal ventricle (Fig. 2.4b) is able to increase ventricular work significantly, and to maintain cardiac output, when the output pressure is increased (Icardo et al. 2005b). This occurs without significant variations in heart rate. Thus, the heart of *S. auratus* works like a pressure pump, in a similar way as the hearts of very active species such as salmonids and tuna (Farrell and Jones 1992). The remarkable thing is that the ventricle of *S. auratus* lacks a compacta (Fig. 2.4b). It is still unknown whether other hearts with a similar morphological pattern may perform similarly. What appears clear is that the functional capabilities of the teleost heart cannot be directly inferred from examination of the external heart shape or the myoarchitectural design.

The Atrioventricular Region

The atrioventricular (AV) region is formed by a ring of cardiac tissue which supports the AV valves (Santer and Cobb 1972; Farrell and Jones 1992). This succinct description, together with several references to the presence of a delay in the electrical conduction in the heart (Satchell 1991; Sedmera et al. 2003), sums up most of our knowledge of this part of the heart. However, recent morphological analyses show a more complex picture (Icardo and Colvee 2011). When the heart of teleosts with completely trabeculated ventricles is examined, the AV region appears formed by a distinct ring of myocardium (Fig. 2.7a, b). This myocardium is compacted, shows vascular profiles in most of the species, and contains variable amounts of collagen and elastin (Icardo and Colvee 2011). These three characteristics differentiate clearly the AV area from the ventricular and atrial chambers. A ring of connective tissue contributes to delineating the AV muscle from that of atrium and ventricle. In hearts possessing a compacta (Fig. 2.7c, d), the histological differences with the neighboring musculature are maintained, and the ring of connective tissue also contributes to differentiating the AV segment (Sedmera et al. 2003, for observations in the zebrafish). It should be stressed that the isolation of the AV muscle ring from the surrounding musculature is by no means complete. Areas of continuity with the atrial and ventricular muscle are always observed (Fig. 2.7) (Icardo and Colvee 2011). From a morphological point of view, the AV region constitutes a distinct segment of the adult teleost heart. Furthermore, the morphological appearance of the AV segment parallels that of the conus in all hearts examined (Icardo 2006; Icardo and Colvee 2011). This includes the presence of vessels.

Regarding the vascular supply of these two segments, there are several relevant features which should be mentioned since they reflect both the diversity of the teleost heart and the difficulty to establish categories of general significance. As stated above, the conus arteriosus and the AV segment show vessels in most of the species examined. However, vascular profiles could not be demonstrated in several species such as *Mullus surmuletus*, *Coris julis*, and most of the Antarctic species (Icardo 2006; Icardo and Colvee 2011). In these cases, endocardial extensions (endocardial sinusoids) penetrate the compact muscle and appear to substitute the coronary vessels. That is, the compact myocardium takes the blood supply directly from the heart lumen. In another species, *Periophthalmodon schlosseri*, the vascular endothelium present in the AV muscle is continuous with the atrial endocardium. These vascular profiles, instead of representing a true coronary circulation, may correspond to some kind of endothelial sinusoids. (The term sinusoid is applied here in a general sense. The existence of endothelial fenestrations is currently unknown.) The observations made in *P. schlosseri* cast some doubts on the real nature of the

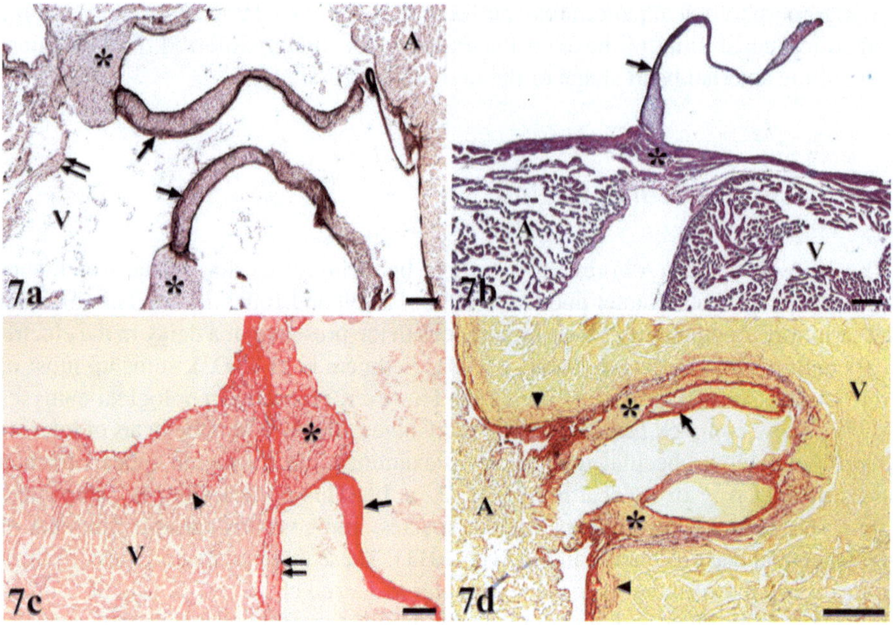

Fig. 2.7 Composite showing the atrioventricular (AV) region of type I (**a**, **b**) and type II (**c**, **d**) hearts. A, atrium. V, ventricle. In all cases, the AV valves (*arrows*) anchor to a ring of compact, vascularized myocardium (*asterisks*). The *arrows* also indicate the thick atrial fibrosa of the leaflets. The myocardial AV ring is partially isolated from the atrial and ventricular musculature by a layer of connective tissue rich in collagen. The collagen appears *red* in (**c**) and (**d**). *Double arrows* in (**a**) and (**c**) indicate continuity between the AV muscle and the ventricular trabeculae. *Arrowheads* in (**c**) and (**d**) indicate the presence of collagen at the compacta–spongiosa boundary. (**a**) *Serranus cabrilla*. Orcein. Note the robustness of the AV ring. The dense cellular core of the valve leaflets is apparent. (**b**) *Balistes carolinensis*. Hematoxylin–eosin. The AV ring is thin but the compactness of the myocardium contrasts with the delicate ventricular and atrial musculature. (**c**) *Echiicthys vipera*. Sirius red. The connective tissue ring separates the AV myocardium from the ventricular compacta. (**d**) *Anguilla anguilla*. Sirius red. The entire AV ring is exposed. Note the continuity between the AV and the atrial muscle. Scale bars: 100 μm

myocardial vessels found both in the conus and in the AV segment in several teleosts. Yet in other cases, like in *E. vipera*, coronary vessels co-exist with endocardial sinusoids in continuity with the heart lumen. It appears that several species have developed a dual mode of blood supply for the myocardium, or that part of the vascular profiles may represent a primitive form of the mammalian Thebesian system.

Irrespective of the mode of vascular supply, the morphological evidence that the conus arteriosus and the AV region are distinct segments of the teleost heart is important from the phylogenetic point of view. Interestingly, the segmental division of the adult fish heart is similar to that found in the heart of higher vertebrates during embryogenesis (Moorman and Christoffels 2003; Wong et al. 2012). In higher vertebrates, the two heart segments exhibit specific patterns of gene expression (He and Burch 1997; Franco et al. 1999; Horsthuis et al. 2009) during development. These genetic patterns appear to be conserved across the evolutionary scale (Chang et al. 2004;

Beis et al. 2005; Rutenberg et al. 2006; Scherz et al. 2008; Shimada et al. 2009). The two segments also exhibit specific morphogenetic properties such as the induction of cushion tissue and the formation of valves (Eisenberg and Markwald 1995). Thus, despite that the conus and the AV segment are not septated in fish, they are conserved across the evolution of the vertebrate heart and appear to share many molecular and functional characteristics.

The AV valves are generally formed by two leaflets (Figs. 2.4b and 2.7) that contain numerous cells grouped into a dense core, and large amounts of connective tissue. The leaflets exhibit a strong atrial fibrosa rich in collagen (Fig. 2.7) and elastin. Within the leaflets, the cell number, the cell morphology, and the amount of extracellular material vary widely between species (Icardo and Colvee 2011). A system of chordae, similar to that observed in mammals, is always absent (also, see Hu et al. 2000). However, ventricular trabecular sheets can often be seen anchored in the AV muscle ring (Sedmera et al. 2003; Icardo and Colvee 2011). These sheets should bear some of the stress generated by the ventricular contraction and could represent a primitive form of the papillary system.

The Atrium and the Sinus Venosus

The teleost atrium is a single chamber which shows considerable variability in size and shape between species (Fig. 2.1) (Santer 1985; Farrell and Jones 1992). It is formed of an external rim of myocardium and of a complex network of thin trabeculae (pectinate muscles) (Fig. 2.4a). The presence of two arcuate systems of pectinate muscles, fanning out from the atrioventricular aperture, has been described in several teleosts (Santer 1985). The atrial myocardium is surrounded by a subepicardial, thick layer of collagen (Fig. 2.3b). Collagen also encircles the atrial trabeculae. In general, the trabecular collagen is more abundant in the atrium than in the ventricle (Figs. 2.3b and 2.7d). It probably helps to support the atrial architecture. However, the significance of this feature in terms of chamber contraction and distension is unclear.

The sinus venosus is a thin-walled chamber whose composition varies between species. It is generally described as being formed by muscle and connective tissue. However, the proportion of the two components appears to vary widely. The sinus venosus wall may be mostly made up of connective tissue (as in *Danio rerio*), of connective tissue with sparse myocardial bundles (as in *Pleuronectes platessa*), or mostly of myocardium (as in *Anguilla anguilla*) (Santer and Cobb 1972; Yamauchi 1980; Farrell and Jones 1992). To add more variation, the myocardium may be replaced by smooth muscle cells in other species such as *Cyprinus carpio* (Yamauchi 1980). The sinus venosus conveys the blood into the atrium from which it is separated by the sinus valve (Yamauchi 1980).

An important characteristic of the sinus is that it contains the heart pacemaker. In most teleosts, the presence of a specialized ring of tissue located at the sinoatrial region has been identified as the primary pacemaker region. This area is also densely innervated (Yamauchi 1980). Other components of the cardiac conduction system, similar to those present in mammals, have not been identified in the teleost heart. Despite that an electrocardiogram with P, QRS, and T waves can be recorded (Satchell 1991), the teleost heart lacks a morphologically defined conduction system (Nair 1973; Sedmera et al. 2003). The absence of a regionalized pattern of connexin expression in the zebrafish (Christie et al. 2004) also argues against the presence of a conduction system in teleosts. It has been suggested that the geometry of the muscle trabeculae allows for the preferential spread of electrical excitation (Sedmera et al. 2003), thus being the functional correlate of the His-Purkinje system. The trabeculae anchored in the AV muscle ring (Fig. 2.7) may constitute that preferential pathway (Sedmera et al. 2003; Icardo and Colvee 2011).

Acknowledgments The author wishes to thank L. González and B. Gallardo for technical assistance.

References

Beis D, Bartman T, Jin S-W, Scott IC, D'Amico LA, Ober EA, Verkade H, Frantsve J, Field HA, Wehman A, Baier H, Tallafuss A, Bally-Cuif L, Chen J-N, Stainier DYR, Jungblut B (2005) Genetic and cellular analyses of zebrafish atrioventricular cushion and valve development. Development 132:4193–4204

Benjamin M, Norman D, Scarborough D, Santer RM (1984) Carbohydrate-containing endothelial cells lining the bulbus arteriosus of teleosts and the conus arteriosus of elasmobranchs (Pisces). J Zool (London) 202:383–392

Burggren WW, Farrell A, Lillywhite H (1997) Vertebrate cardiovascular systems. In: Dantzler WH (ed) Handbook of physiology, sect 13, Comparative physiology, vol 1. Oxford University, New York, pp 215–308

Caulfield JB, Borg TK (1979) The collagen network of the heart. Lab Invest 40:364–372

Cerra MC, Canonaco M, Acierno R, Tota B (1997) Different binding activities of A- and B-type natriuretic hormones in the heart of two Antarctic teleosts, the red-blooded *Trematomus bernacchii* and the hemoglobinless *Chionodraco hamatus*. Comp Biochem Physiol 118A:993–999

Cerra MC, Imbrogno S, Amelio D, Garofalo F, Colvee E, Tota B, Icardo JM (2004) Cardiac morphodynamic remodelling in the growing eel (*Anguilla Anguilla* L.). J Exp Biol 207:2867–2875

Chang C-P, Neilson JR, Bayle JH, Gestwicki JE, Kuo A, Stankunas K, Graef IA, Crabtree GR (2004) A field of myocardial-endocardial NFAT signalling underlies heart valve morphogenesis. Cell 118:649–663

Chen J-N, Fishman MC (2000) Genetics of heart development. Trends Genet 16:383–388

Chen JN, van Eeden JM, Warren KS, Chin A, Nússlein-Volhard C, Haffter P, Fishman MC (1997) Left-right pattern of cardiac *BMP4* may drive asymmetry of the heart in zebrafish. Development 124:4373–4382

Christie TL, Muir R, White TW, Valdimarsson G (2004) Molecular cloning, functional analysis, and RNA expression analysis of connexin 45.6: a zebrafish cardiovascular connexin. Am J Physiol 286:H1623–H1632

Durán AC, Fernández B, Grimes AC, Rodríguez C, Arqué JM, Sans-Coma V (2008) Chondrichthyans have a bulbus arteriosus at the arterial pole of the heart: morphological and evolutionary implications. J Anat 213:597–606

Eastman JT (1993) Antarctic fish biology. Evolution in a unique environment. Academic, New York

Eisenberg LM, Markwald RR (1995) Molecular regulation of atrioventricular valvuloseptal morphogenesis. Circ Res 77:1–6

Farrell AP, Jones DR (1992) The heart. In: Hoar WS, Randall DJ, Farrell AP (eds) Fish physiology, Vol XII, The cardiovascular system, Part A. Academic, San Diego, pp 1–87

Farrell AP, Hammons AM, Graham MS, Tibbits GF (1988) Cardiac growth in rainbow trout, *Salmo gairdneri*. Can J Zool 66:2368–2373

Farrell AP, Johansen JA, Steffensen JF, Moyes CD, West TG, Suarez TK (1990) Effects of exercise training and coronary ablation on swimming performance, heart size and cardiac enzymes in rainbow trout, *Oncorhynchus mikiss*. Can J Zool 68:1174–1179

Farrell AP, Simonot DL, Seymour RS, Clark TD (2007) A novel technique for estimating the compact myocardium in fish reveals surprising results for an athletic air-breathing fish, the Pacific tarpon. J Fish Biol 71:389–398

Franco D, Markman MMW, Wagenaar GTM, Ya Y, Lamers WH, Moorman AFM (1999) Myosin light chain 2a and 2v identifies the embryonic outflow tract myocardium in the developing rodent heart. Anat Rec 254:135–146

Greer Walker M, Santer M, Benjamin M, Norman D (1985) Heart structure of some deepsea fish (Teleostei: Macrouridae). J Zool (London) 205:75–89

Grimes AC, Kirby ML (2009) The outflow tract of the heart in fishes: anatomy, genes and evolution. J Fish Biol 74:963–1036

Grimes AC, Stadt HA, Sheperd IT, Kirby ML (2006) Solving an enigma: arterial pole development in the zebrafish heart. Dev Biol 290:265–276

Harvey RP, Rosenthal N (1998) Heart development. Academic, San Diego

He C-Z, Burch JBE (1997) The chicken GATA-6 locus contains multiple control regions that confer distinct patterns of heart region-specific expression in transgenic mouse embryos. J Biol Chem 272:28550–28556

Horsthuis T, Christoffels VM, Anderson RH, Moorman AFM (2009) Can recent insights into cardiac development improve our understanding of congenitally malformed hearts? Clin Anat 22:4–20

Hu N, Sedmera D, Yost HJ, Clark EB (2000) Structure and function of the developing zebrafish. Anat Rec 260:148–157

Icardo JM (2006) Conus arteriosus of the teleost heart: dismissed, but not missed. Anat Rec A 288:900–908

Icardo JM (2007) The fish endocardium. A review on the teleost heart. In: Aird WC (ed) Endothelial biomedicine. Cambridge University, Cambridge, pp 79–84

Icardo JM, Colvee E (1998) Collagenous skeleton of the human mitral papillary muscle. Anat Rec 252:509–518

Icardo JM, Colvee E (2011) The atrioventricular region of the teleost heart. A distinct heart segment. Anat Rec 294:236–242

Icardo JM, Colvee E, Tota B (1996) Morphological organization of the sturgeon (*Acipenser naccarii*) heart with special reference to the collagenous architecture. In: VII International symposium on fish physiology, Oslo, Norway, 3–6 Aug 1996, p 93

Icardo JM, Colvee E, Cerra MC, Tota B (1999a) Bulbus arteriosus of Antarctic teleosts. I. The white-blooded Chionodraco hamatus. Anat Rec 254:396–407

Icardo JM, Colvee E, Cerra MC, Tota B (1999b) Bulbus arteriosus of Antarctic teleosts. II. The red-blooded Trematomus bernacchii. Anat Rec 256:116–126

Icardo JM, Colvee E, Cerra MC, Tota B (2000a) Light and electron microscopy of the bulbus arteriosus of the European eel (*Anguilla anguilla*). Cells Tissues Org 167:184–198

Icardo JM, Colvee E, Cerra MC, Tota B (2000b) The bulbus arteriosus of stenothermal and temperate teleosts: a morphological approach. J Fish Biol 57(suppl A):121–135

Icardo JM, Colvee E, Cerra MC, Tota B (2002) The estructure of the conus arteriosus of the sturgeon (Acipenser naccarii) heart: II. The myocardium, the subepicardium, and the conus-aorta transition. Anat Rec 268:388–398

Icardo JM, Schib JL, Ojeda JL, Durán AC, Guerrero A, Colvee E, Amelio D, Sans-Coma V (2003) The conus valves of the adult gilthead seabream (*Sparus auratus*). J Anat 202:537–550

Icardo JM, Brunelli E, Perrotta I, Colvee E, Wong WP, Ip YK (2005a) Ventricle and outflow tract of the African lungfish *Protopterus dolloi*. J Morphol 265:43–51

Icardo JM, Imbrogno S, Gattuso A, Colvee E, Tota B (2005b) The heart of *Sparus auratus*: a reappraisal of cardiac functional morphology in teleosts. J Exp Zool 303A:665–675

Imbrogno S, Cerra MC, Tota B (2003) Angiotensin II-induced inotropism requires an endocardial endothelium-nitric oxide mechanism in the in-vitro heart of *Anguilla anguilla*. J Exp Biol 206:2675–2684

Jones DR, Brill RW, Bushnell PG (1993) Ventricular and arterial dynamics of anaesthetised and swimming tuna. J Exp Biol 182:97–112

Leknes IL (2009) Structural and histochemical studies on the teleostean bulbus arteriosus. Anat Histol Embryol 38:424–428

Lepilina A, Coon AN, Kikuchi K, Holdway JE, Roberts RW, Burns CG, Poss KD (2006) A dynamic epicardial injury response supports progenitor cell activity during zebrafish heart regeneration. Cell 127:607–619

Lohr JL, Yost HJ (2000) Vertebrate model systems in the study of early heart development: Xenopus and zebrafish. Am J Med Genet 97:248–257

Midttun B (1983) Ultrastructure of the junctional region of the fish heart ventricle. Comp Biochem Physiol 76A:471–474

Moorman AF, Christoffels VM (2003) Cardiac chamber formation: development, genes, and evolution. Physiol Rev 83:1223–1267

Munshi JSD, Olson KR, Roy PK, Ghosh U (2001) Scanning electron microscopy of the heart of the climbing perch. J Fish Biol 59:1170–1180

Nagai Y, Asaoka Y, Namae M, Saito K, Momose H, Mitani H, Furutani-Seiki M, Katada T, Nishina H (2010) The LIM protein Ajuba is required for ciliogenesis and left-right axis determination in medaka. Biochem Biophys Res Commun 396:887–893

Nair MG (1973) The development of the nervous system in the heart of the Chinese carp, *Cyprinus carpio* (Linnaeus), with a special reference to its conduction system. Mikroskopie 29:1–7

Niihori M, Mogami Y, Naruse K, Baba SA (2004) Development and swimming behaviour of medaka fry in a spaceflight aboard the space shuttle Columbia (STS-107). Zool Sci 21: 923–931

Ohtani O (1987) Three-dimensional organization of the connective tissue fibers of the human pancreas. A scanning electron microscopic study of NaOH treated tissues. Arch Histol Jpn 50:557–566

Parsons CW (1930) The conus arteriosus in fishes. Q J Microsc Sci 73:145–176

Pieperhoff S, Bennett W, Farrell AP (2009) The intercellular organization of the two muscular systems in the adult salmonid heart, the compact and the spongy myocardium. J Anat 215:536–547

Poss KD, Wilson LG, Keating MT (2002) Heart regeneration in zebrafish. Science 298:2188–2190

Poupa O, Gesser H, Jonsson S, Sullivan L (1974) Coronary-supplied compact shell of ventricular myocardium in salmonids: growth and enzyme pattern. Comp Biochem Physiol A 48:85–95

Priede IG (1976) Functional morphology of the bulbus arteriosus of rainbow trout (*Salmo gairdneri* Richardson). J Fish Biol 9:209–216

Raso DS (1993) Functional morphology of laminin, collagen type IV, collagen bundles, elastin, and proteoglycans in the bulbus arteriosus of the white bass, *Morone chrysops* (Rafinesque). Can J Zool 71:947–952

Rutenberg JB, Fischer A, Jia H, Gessler M, Zhong TP, Mercola M (2006) Developmental patterning of the cardiac atrioventricular canal by Notch and hairy-related transcription factors. Development 133:4381–4390

Sánchez-Quintana D, Hurle JM (1987) Ventricular myocardial architecture in marine fishes. Anat Rec 217:263–273

Sánchez-Quintana D, Garcia-Martinez V, Climent V, Hurle JM (1995) Morphological analysis of the fish heart ventricle: myocardial and connective tissue architecture in teleost species. Ann Anat 177:267–274

Sans-Coma V, Gallego A, Muñoz-Chápuli R, de Andrés AV, Durán AC, Fernández B (1995) Anatomy and histology of the cardiac conal valves of the adult dogfish (*Scyliorhinus canicula*). Anat Rec 241:496–504
Santer RM (1985) Morphology and innervation of the fish heart. Adv Anat Embryol 89:1–102
Santer RM, Cobb JLS (1972) The fine structure of the heart of the teleost, *Pleuronectes platessa* L. Z Zellforsch 131:1–14
Santer RM, Greer Walker M, Emerson L, Witthames PR (1983) On the morphology of the heart ventricle in marine teleost fish (teleostei). Comp Biochem Physiol 76A:453–457
Satchell GH (1991) Physiology and form of fish circulation. Cambridge University, Cambridge
Scherz PJ, Huisken J, Shai-Hernandez P, Stainier DYR (2008) High-speed imaging of developing heart valves reveals interplay of morphogenesis and function. Development 135:1179–1187
Schib JL, Icardo JM, Durán AC, Guerrero A, López D, Colvee E, de Andrés AV, Sans-Coma V (2002) The conus arterious of the adult gilthead seabream (*Sparus auratus*). J Anat 201:395–404
Sedmera D, Reckova M, deAlmeida A, Sedmerova M, Biermann M, Volejnik J, Sarre A, Raddatz E, McCarthy RA, Gourdie RG, Thompson RP (2003) Functional and morphological evidence for a ventricular conduction system in zebrafish and Xenopus hearts. Am J Physiol 284:H1152–H1160
Seternes T, Sorensen K, Smedsrod B (2002) Scavenger endothelial cells of vertebrates: a nonperipheral leukocyte system for high-capacity elimination of waste macromolecules. Proc Natl Acad Sci USA 99:7594–7597
Shimada E, Kinoshita M, Murata K (2009) Expression of cardiac myosin light chain 2 during embryonic heart development in medaka fish, *Oryzias latipes*, and phylogenetic relationship with other myosin light chains. Dev Growth Differ 51:1–16
Simoes K, Vicentini CA, Orsi AM, Cruz C (2002) Myoarchitecture and vasculature of the heart ventricle in some freshwater teleosts. J Anat 200:467–475
Smith WC (1918) On the process of disappearance of the conus arteriosus in teleosts. Anat Rec 15:65–71
Taneda Y, Konno S, Makino S, Morioka M, Fukuda K, Imai Y, Kudo A, Kawakami A (2010) Epigenetic control of cardiomyocyte production in response to a stress during the medaka heart development. Dev Biol 340:30–40
Tota B (1978) Functional cardiac morphology and biochemistry in Atlantic bluefin tuna. In: Sharp G, Dizon A (eds) The physiological ecology of tuna. Academic, New York, pp 89–112
Tota B (1989) Myoarchitecture and vascularisation of the elasmobranch heart ventricle. J Exp Zool (Suppl) 2:122–135
Tota B, Garofalo F (2012) Fish heart growth and function: from gross morphology to cell signalling and back. In: Sedmera D, Wang T (eds) Ontogeny and phylogeny of the vertebrate heart. Springer, New York
Tota B, Gattuso A (1996) Heart ventricle pump in teleosts and elasmobranchs: a morpho-dynamic approach. J Exp Zool 275:162–171
Tota B, Cimini V, Salvatore G, Zummo G (1983) Comparative study of the arterial and lacunary systems of the ventricular myocardium of elasmobranch and teleost fishes. Am J Anat 167:15–32
Tota B, Cerra MC, Mazza R, Pellegrino D, Icardo JM (1997) The heart of the antarctic icefish as paradigm of cold adaptation. J Thermal Biol 22:409–417
Weber KT (1989) Cardiac interstitium in health and disease: the fibrillar collagen network. J Am Coll Cardiol 13:1637–1652
Wong LYE, Moorman AF, Barnett P (2012) Basic cardiac development: the heart and its electrical components. In: Sedmera D, Wang T (eds) Ontogeny and phylogeny of the vertebrate heart. Springer, New York
Yamauchi A (1980) Fine structure of the fish heart. In: Bourne G (ed) Heart and heart-like organs, vol 1. Academic, New York, pp 119–148
Yelon D (2001) Cardiac patterning and morphogenesis in the zebrafish. Dev Dyn 222:552–563

Chapter 3
Fish Heart Growth and Function: From Gross Morphology to Cell Signaling and Back

B. Tota and F. Garofalo

Cardiac Nonuniformity: From Genes to Shape

Complex structures evolve from simple module components that, like the origami structures, can be rearranged many times in various ways to make complicated shapes and forms. However, origami does not grow, being only generated by repeated folding without growth. On the contrary, biological growth is a self-assembling autonomous process, which obeys the internal stimuli, so that genetic and epigenetic events allow for the one-dimensional genetic code to specify for a three-dimensional animal in time. Research during the past two decades shows that both the *phylotypic* body plan, i.e. the organization shared by all members of a phylum, and the organ plan can be described not only anatomically, but also in terms of compartments of expression of selector genes and signaling systems. Accordingly, a "second anatomy" has emerged, which is only visible when the embryo is treated with color reagents revealing the location of the RNAs and proteins encoded by these genes (Gerhart and Kirschner 1997). The spatially organized selector genes and signaling systems orchestrate the topological organization of compartmentalized phylotypic processes by activating and orienting subsequent time-dependent more local developmental events.

As best documented in zebrafish, the heart develops as an assembly of distinct genetic modules added to the primitive ancestor chordate heart tube during evolution. According to both the long-held "serial" model (Fishman and Olson 1997) and the novel "ballooning" model (Moorman and Christoffels 2003) of chamber formation, the modules are supposed to be common morpho-functional components of the heart that can be selectively affected by single gene mutations. Conceivably, continuous adaptation of the genomic regulatory programs that control cardiac development

B. Tota (✉) • F. Garofalo
Department of Cell Biology, University of Calabria,
Arcavacata di Rende 87030, Cosenza, Italy
e-mail: tota@unical.it

D. Sedmera and T. Wang (eds.), *Ontogeny and Phylogeny of the Vertebrate Heart*,
DOI 10.1007/978-1-4614-3387-3_3, © Springer Science+Business Media, LLC 2012

resulted from modification of existing elements and addition of new elements (Olson 2006). Many studies also show that various traits considered unique to vertebrates were already present in the early chordate ancestors, while new functions became associated with modified duplicated sets of genes (Fishman and Stainier 1994; Olson 2006). As a result, this new knowledge uncovers the deep evolutionary conservation of molecular pathways in the hearts from fruitfly to humans and, at the same time, the diversity related to novel modules of the building plan of the present-day vertebrate (and human) heart (Olson 2006).

The final adult heart configuration is characterized by remarkable morpho-functional nonuniformity (or heterogeneity) encompassing all levels of cardiac organization from gross morphology to molecular biology. This may result from the modular morphogenesis of the heart driven by distinct transcriptional regulatory programs that control the regional specification of structures of each anatomical region. Examples of such modular morphogenesis include cardiac chamber formation, trabeculation pattern and the localized pacemaker activity. Recently, Icardo (see also the present volume) has characterized the atrio-ventricular region (Icardo and Colvee 2001) and the outflow tract (the arterial pole) (Icardo 2006) of the teleost heart in terms of distinct species-specific morphological segments with high degree of evolutionary diversification. Several genes, including transcription factors, have been localized in these regions during the period of cushion formation. Knowledge on their cell and tissue organization adds new information regarding the nonuniformity of the lower vertebrate heart. At the histological level, an universal trait of diversity is represented by the three principle cardiac tissues, i.e. the endocardium (EE), the myocardium (MYO) and the epicardium (EPI) that not only play a key morphogenetic role, like in the process of chamber formation and compartmentalization, but also control the morpho-functional integrity of the adult heart pump through their close morphological and functional integration. This multilevel spatio-temporal integration, therefore, seems to be a prerequisite to compensate for the nonuniformity of the vertebrate heart, which results from its ontogenetic and phylogenetic history.

Here we use heart ventricle nonuniformity in fish to illustrate and review aspects of the two-way interaction (cross-talk) between different regions, as well as between EPI, MYO and EE through which the organ integration is accomplished.

Cardiac Growth, Size and Life History-Related Allometry

There is a surprising uniformity of scaling relationship in living animals (Hochachka and Somero 2002). Namely, during growth of an individual organism or its organs and tissues, the component parts typically increase in allometric fashion, i.e. out of proportion to one another (Vogel 1988). Shape is therefore a function of size. This universal rule, valid for both the ontogenetic growth and for lineage evolution, also drives cardiac growth.

A property which seems partially intrinsic to the organ is that larger animals have larger hearts. This means that cell number and cellular hypertrophy are under the

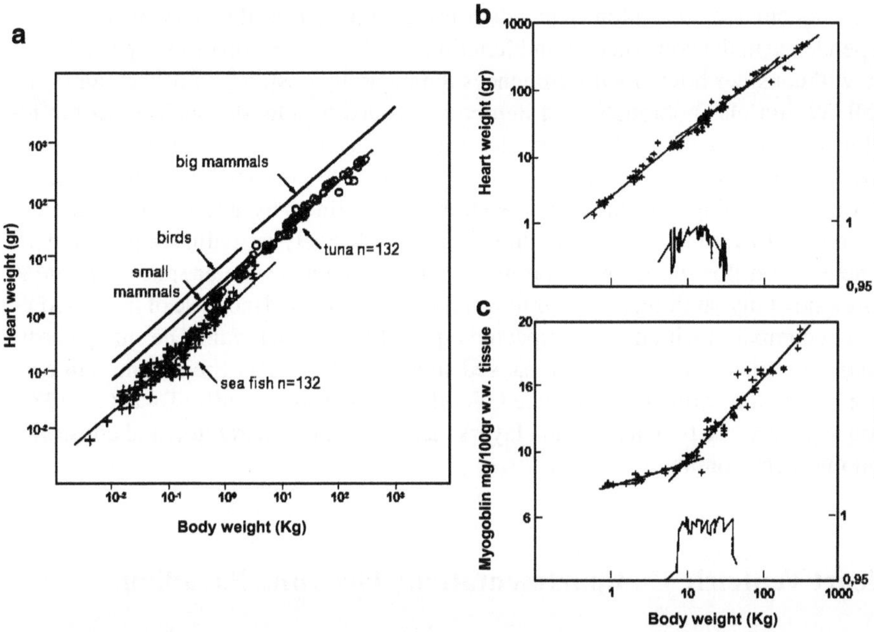

Fig. 3.1 (**a**) The relation between weight and body weight for different groups of animals. The regression lines [with $y = {}^{10}\log$ (heart weight in g) and $x = {}^{10}\log$ (body weight in kg)] are as follows: (Poupa and Ostadal 1969). Birds: $y = 0.89x + 0.89$; large mammals: $y = 0.87x + 1.02$; small mammals: $y = 0.85x + 0.58$. Tuna fish: $y = 0.90x + 0.47$; sea fish: $y = 0.89x + 0.08$. (**b**) *Upper panel*: logarithmic plot of heart weight vs. body weight for tuna (*T. Thynnus thynnus*) with regression lines for younger and older animals. The plot at the *bottom* is the probability of having different slopes of the regression lines for the two groups as a function of the point of separation between them. *Lower panel*: semi-logarithmic plot of concentration of cardiac myoglobin vs. body weight with regression lines for younger and older animals. The plot at the *bottom* is the probability of having different slopes of the regression lines for the two groups as a function of the point of separation between them (reproduced from: Poupa et al. 1981)

control of inherited programs, in addition to the epigenetic factors (hemodynamic forces) that influence the degree of heart growth. As outlined by Poupa and Ostadal (1969), heart mass of vertebrates scales with body mass ($a = 2.94$, where a is the slope of regression line), while heart rate (HR) declines with increased mass ($a = -0.75$). However, beyond this universal trend, comparative allometric analyses reveal distinct growth patterns (regression lines) that are of eco-physiological or bio-constructural significance: (1) smaller animals have relatively larger hearts than big animals; (2) parallel regression lines and "jumps" in relative heart weight (*RHW*, i.e. heart mass times 100 divided by body mass, *BW*) can be identified, so that fishes, mammals and birds are usually characterized by shifts from a lower to an upper regression line, respectively (Fig. 3.1). Specific scaling factors include size-dependent influence of Earth's gravity, body surface, skeletal muscle mass, heat production and metabolic rate (Poupa and Lindström 1983).

In conclusion, the allometric relationship emphasizes the importance of size dependence in the construction and functioning of the organism or its organs. Below, we will consider how growth influences the internal construction of the ventricular wall. Within elasmobranchs and numerous groups of teleosts, such as scombrids, salmonids and anguillids, the ventricular wall consists of two layers, an inner *spongiosa* and an outer *compacta*. The former consists of highly organized sheets of cardio-myocytes forming muscular ridges (trabeculae) lined by endocardial cells and protruding toward the ventricular lumen; it is exclusively or mainly supplied by the venous blood flowing from the periphery into the intertrabecular spaces (*lacunae*). This kind of myoarchitecture permits a surprisingly elevated increase of the EE-MYO area as compared with more compacted types of ventricular wall. The subepicardial *compacta*, made up by densely packed myocardial bundles (hence the name), is supplied by the coronary oxygenated blood flowing from the gills (Tota et al. 1983). Consequently, the two myocardial layers face different myoarchitectural constraints and microenvironments.

Heart Ventricle Compartmentation: The Tuna Paradigm

Bluefin tuna (*Thunnus thynnus thynnus*) are unique athletic pelagic teleosts that due to their retial heat exchangers conserve heat in their locomotory muscle, brain, eyes and viscera (Shiels et al. 2011). As shown in Fig. 3.1, tuna heart growth shifts from the fish to the small mammals regression line (Poupa et al. 1981), due to the elevated metabolic rate and extraordinary locomotory, respiratory and cardio-circulatory features of the animal. However, beyond this allometric continuity, the detailed analysis of regression coefficients reveals a physiologically significant discontinuity between the heart growth of younger and older animals with a maximum probability (0.998) of different slopes at around 17 kg BW. Of note, this BW approximately coincides with the attainment of the sexual maturity. Interestingly, the regression line of myoglobin (Mb) concentration values in the heart ventricle also shows different slopes with the breaking point position coincident with that of the HRW/BW relationship (Poupa et al. 1981). However, while the heart is relatively larger in juvenile compared to adult tunas, cardiac Mb concentration increases with body mass. This may be important to preserve aerobic performance of the thicker myocardial wall occurring in large animals.

The tuna ventricle also represents a hallmark of nonuniformity in myoarchitecture and blood supply. In fact, the thickness and the complexity of the *compacta* reach the highest level in teleosts, while the coronary arteries also vascularize the *spongiosa* ending in the lumen via artero-lacunary and thebesian vessels (Tota et al. 1983). Interestingly, the *spongiosa* shows higher aerobic potential than the *compacta*, as documented by its higher mitochondrial content and metabolism (Tota et al. 1983; Di Maio and Block 2008). An aspect of metabolic cooperativity between the two myocardial compartments is depicted in Fig. 3.2. The *spongiosa*, exposed to particularly high plasma levels of lactate generated in large amounts by the working

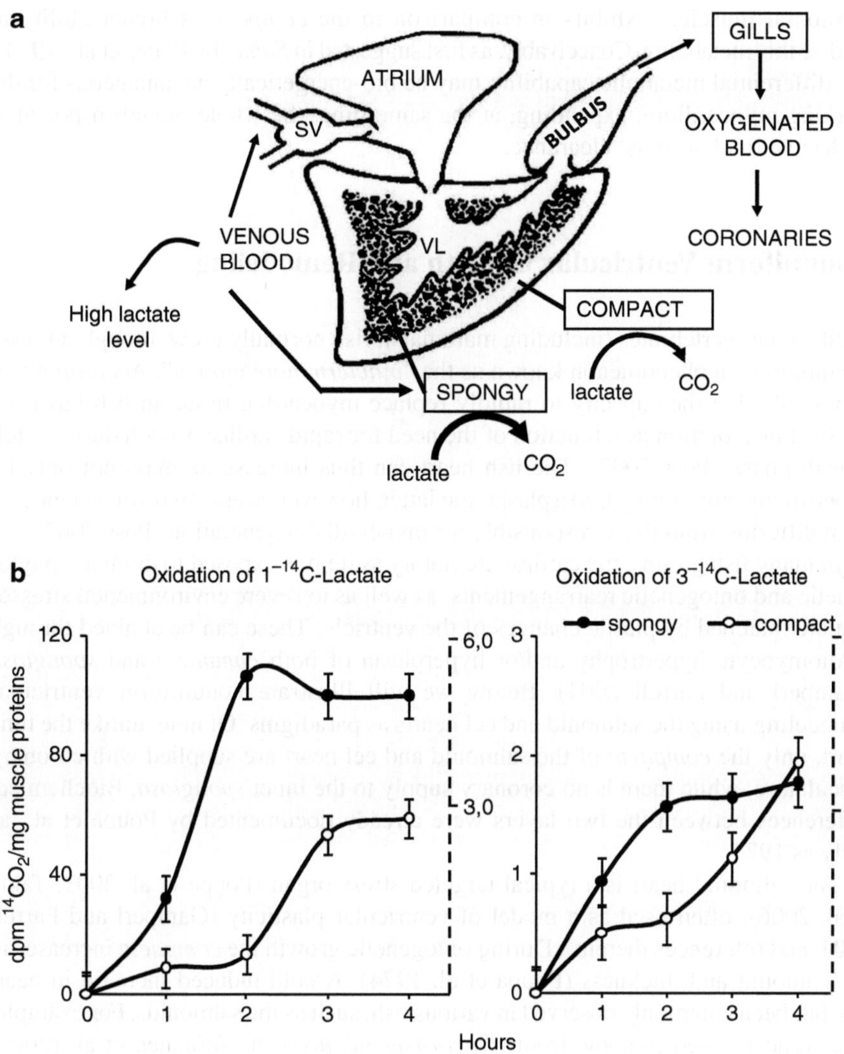

Fig. 3.2 (**a**) Structural and vascular arrangement in the mixed type of ventricular myocardium. The inner core of the myocardium in *spongy* and is supplied mainly by venous blood within the lacunary spaces that belong to the ventricular lumen. The outer part of the ventricular wall is formed by the *compact* myocardium and is supplied by coronary capillaries. *SV* sinus venosus, *VL* ventricular lumen. (**b**) The oxidation of lactate to CO_2 by the spongious and compact myocardium of tuna, expressed as dpm (disintegration per minute)/mg muscle proteins at 30°C. The *dashed* ordinate on the right of each diagram shows the fraction of total radioactivity (dpm) which is metabolized and referred to the muscle protein content of each cardiac compartment. Each point represents the mean value from six hearts (reproduced from: Gemelli et al. 1980)

myotomal muscles, exhibits in comparison to the *compacta* a higher ability to oxidize this metabolite. Conceivably, as first suggested in *Salmo* by Poupa et al. (1974), this differential metabolic capability may be bio-energetically advantageous for the working myocardium, expanding, at the same time, the whole organism potential for lactate (and protons) clearance.

Nonuniform Ventricular Growth and Remodeling

Unlike other vertebrates (including mammals), fish normally grow throughout most of adulthood, a phenomenon known as the "*indeterminate growth*". Accordingly, it is possible that the capacity to rapidly replace myocardial tissue in fish has been retained in evolution as a function of the need for rapid cardiac growth during adult growth phase (Poss 2007). The fish heart can thus increase its mass not only by hypertrophy, but also by hyperplasia; the latter, however, seems to involve a mechanism differing from those responsible for myocardial regeneration (Poss 2007).

In many fish groups, the cardio-circulatory system is exposed to dramatic phylogenetic and ontogenetic rearrangements, as well as to severe environmental stresses that are matched by plastic changes of the ventricle. These can be attained through cardiomyocyte hypertrophy and/or hyperplasia of both *compacta* and *spongiosa* (Gamperl and Farrell 2004). Below we will illustrate nonuniform ventricular remodeling using the salmonid and eel hearts as paradigms. Of note, unlike the tuna heart, only the *compacta* of the salmonid and eel heart are supplied with coronary vasculature, while there is no coronary supply to the inner *spongiosa*. Biochemical differences between the two layers were already documented by Poupa et al., as early as 1974.

The salmonid heart is a typical targeted stress organ (Poppe et al. 2003; Takle et al. 2006), often used as a model of ventricular plasticity (Gamperl and Farrell 2004, and references therein). During ontogenetic growth the *compacta* increases in both amount and thickness (Poupa et al. 1974). A cold-induced increase in heart size has been commonly observed in various fish, such as the salmonids. For example, in cold-acclimated rainbow trout (*Oncorhynchus mykiss*), Vornanen et al. (2005) documented a 33% increase of heart mass associated with enhanced expression of Ventricular Myosin Heavy Chain (VMHC), Smooth Muscle Like Cells 2 (SMLC2) and a fivefold increase in the expression of MLP (MARCKS-like protein) mRNA, indicating an upregulation of genes involved in the hypertrophic gene program. Very recently, in the same species, Johansen et al. (2011) documented cortisol-induced heterogeneous heart ventricle remodeling. By examining two genetically distinct strains of trouts that respond to stress with either a high (high responsive, Hr) or low (low responsive, Lr) cortisol production (Pottinger and Carrick 1999), they found that the divergent post-stress cortisol levels are associated to a nonuniform ventricular remodeling. In fact, the mean cardio-somatic index of adult Hr trout was 34% higher than in Lr counterparts and this enlargement was mainly due to hypertrophy of the *compacta*. Cortisol, known to stimulate protein synthesis and hypertrophy of cardiomyocytes in vivo and in vitro (Nichols et al. 1984; Lumbers et al. 2005), is a major stress hormone in teleosts (Bern and Madsen 1992; Wendelaar Bonga 1997). Therefore, it is of interest that at the transcriptional level the Hr hearts with *compacta* hypertrophy exhibit higher mRNA levels of cortisol receptors (MR, GR1, and GR2) and increased Vascular Endothelial Growth Factor (VEGF) gene expression (an indicator of enhanced angiogenesis), as well as increased RCAN1 levels. The latter indicates enhanced pro-hypertrophic Nuclear Factor of Activated T-cells (NFAT) signaling that is an essential pathway in the development of cardiac hypertrophy (Molkentin et al. 1998), including pathological hypertrophy in rodent and human hearts (Wilkins et al. 2004). Thought to have arisen following recombination about 500 million years ago, the calcium/calcineurin-regulated NFATc family produced a new group of signaling and transcription factors (the NFATc genes), only present in vertebrate genomes (Molkentin et al. 1998). Likely, this recombination allowed Ca^{2+} signals to be redirected to a new transcriptional program incorporated in the groundwork for vertebrate morphogenesis and organogenesis, including the cardiovascular system (Wu et al. 2007). The increased hypertrophy of the *compacta* with almost threefold higher regulator of calcineurin1 (RCAN1) expression in Hr hearts, consistent with increased NFAT signaling, suggests a greater adaptive hypertrophic potential of the compact layer. In addition, elevated collagen (Col 1a2) expression and deposition in Hr hearts point to enhanced fibrosis (Johansen et al. 2011).

Among teleosts, the eel (*Anguilla anguilla* L.) heart provides another striking example of growth-related plasticity of regional ventricular structure and integrated functional performance (Fig. 3.3). In fact, by comparing juvenile small-size fish with adult, large size counterparts, Cerra et al. (2004) evaluated from organ to cellular and subcellular levels the ventricular morpho-dynamic changes occurring during growth. The two groups show similar responses at increasing filling pressure (preload), but differ remarkably at increasing afterload. The latter reflects the heart's ability to adjust pressure development in response to increased peripheral resistances, such as under intense exercise. Small eel hearts decreased stroke volume (SV) at afterload greater than 3 kPa, in contrast to larger hearts, which maintained constant SV up to 6 kPa. These changes in mechanical performance are related to structural differences. In fact, compared with the small eels, the large eels show increase in the *compacta* thickness and in the diameter of the trabeculae in the *spongiosa*, together with reduction of the lacunary spaces. The increased *compacta* thickness is attained by enlargements of both the muscular and vascular compartments and reduction of the interstitium; consequently, this layer appears more compacted. Therefore, due to these growth-related morpho-dynamic changes, the cardiac ventricle of small eels, with its limited response to pressure overload and large lacunary spaces, appears better suited to produce volume work, while that of the large eels is better adapted to produce pressure work (for reference, Icardo et al. 2005). Increased myocardial hyperplasia in both *compacta* and *spongiosa* is indicated by the higher number of myocytes together with reduced cross-sectional area and myofibrillar compartment. However, the detailed morphometric and ultrastructural analyses reveal that the growth pattern of the two layers is slightly different. For example, *compacta* and *spongiosa* myocytes become smaller during growth, reducing their myofibrillar compartment, however, this reduction is threefold smaller in the *compacta* than in

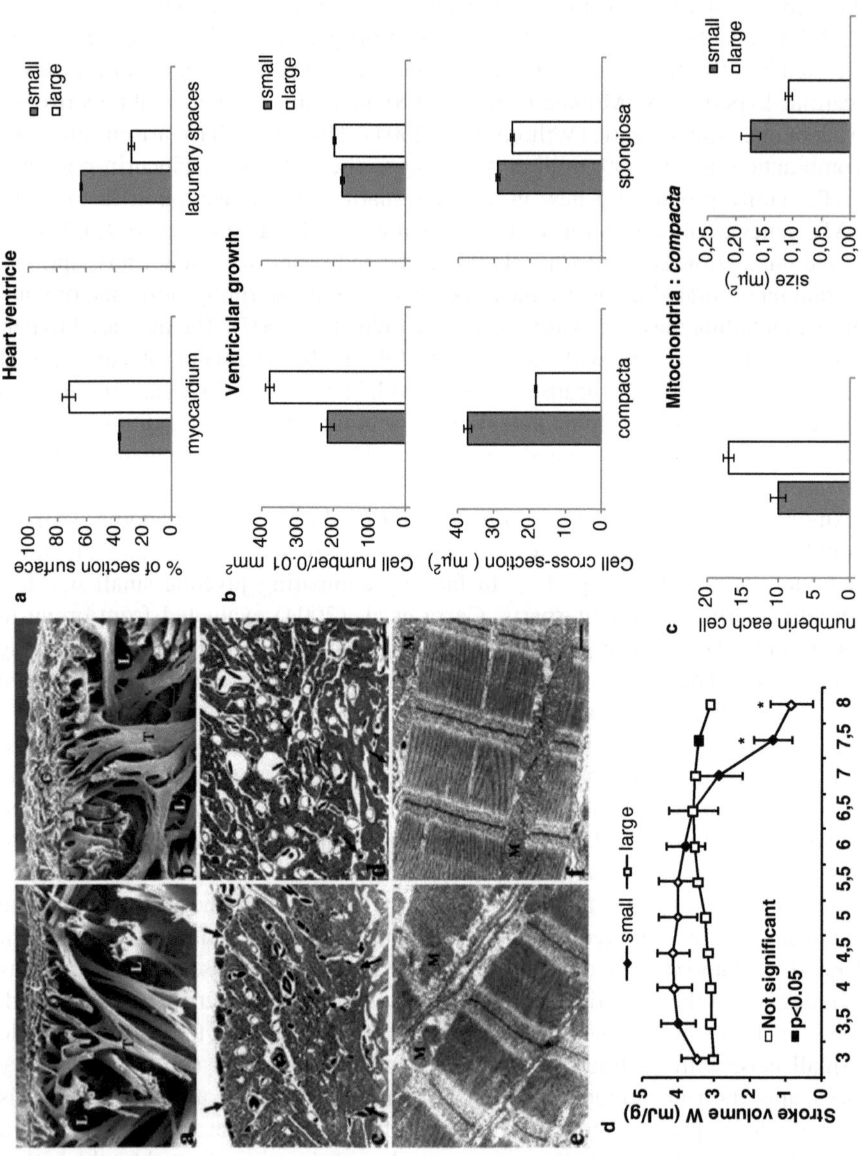
a
Heart ventricle
% of section surface
myocardium
lacunary spaces
small
large
b
Ventricular growth
Cell number/0.01 mm2
Cell cross-section (mμ2)
compacta
spongiosa
small
large
c
Mitochondria : compacta
numberin each cell
size (mμ2)
small
large
d
small
large
Stroke volume W (mJ/g)
Not significant
p<0.05
Afterload (kPa)

Fig. 3.3 (**a–f**): Morphological changes that occur in the eel ventricle during growth. (**a, b**) Scanning, (**c, d**) light and (**e, f**) transmission-electron microscope images of small (**a, c, e**) and large (**b, d, f**) eels. (**A–C**): *Heart ventricle*: in large eels, the ventricle becomes more "muscularized" by increasing both the thickness of the *compacta* and the diameter of the trabeculae in the *spongiosa*; this is accompanied by a reduction of the lacunary spaces and in increase in vascular supply which copes with the metabolic and energetic demands of the deeper myocardial cells. *Mitochondria*: the enlargement of mitochondrial compartment provides myofibrils with adequate energetic compounds. In *compacta* myocytes the increased mitochondrial density and the reduced myofibrillar compartment provide a higher workload ability in comparison with the *spongiosa*. *Cardiac growth* is characterized by an increased cellularity (expressed as cell number/units of surface area) and by a decrease in cross-sectional area of the myocytes. This suggests that the myocardial ventricular mass increases through hyperplasia. (**D**) Effects of afterload elevation Stroke Volume (SV) in isolated and perfused small (*triangles*; $N=7$) and large (*squares*; $N=7$) eel hearts. Values are mean ± SEM. Comparison within group: *open symbols*, not significant; *closed symbols*, $P<0.05$. Comparison between groups: $*P<0.05$ (reproduced from: Cerra et al. 2004)

the *spongiosa*. Moreover, unlike the *spongiosa*, the *compacta* shows an increase in the mitochondrial compartment, which appears bio-energetically important for the myofibrillar apparatus to match the accelerated contractile rhythm and a higher pumping capacity of large eels. The possible redox-dependent signaling mechanisms and transcription factors known to orchestrate the increase in mitochondrial content which is associated with workload-induced cardiac growth have been discussed in mammals by Leary et al. (2002) and in fish by Birkedal et al. (2006) and Urschel and O'Brien (2008).

In conclusion, the nonuniformity of the adult fish ventricle depends on various factors that include anatomy (e.g., three-dimensional myocyte and perimysial collagen organization of the two compartments), intramyocardial pressure resulting from *spongiosa* or *compacta* contraction pattern (isovolumic or shortening), elastic behavior and stiffness of the interstitial network, perfusion pressures (coronary and lacunary blood supply) and perfusate composition (pO_2, pH, substrates, etc.) (Cerra et al. 2004; Icardo et al. 2005 and references therein). It is not known how global synergic performance is achieved; however, the different patterns of ventricular growth and remodeling illustrated above strongly support the idea that heterogeneity also exists in mechanical cross-talk. Some aspects of this cell-molecular integration will be discussed in the last part of this article.

Cell–Cell Interactions in Cardiac Morphogenesis, Growth and Regeneration

The mature heart is composed of diverse cell types: working cardiomyocytes, pacemaker and conduction tissue cells, myo-endocrine cells, EE and vascular endothelial cells, EPI cells, valvular cells, smooth muscle cells and interstitial mesenchymal fibroblasts. They are housed in a dynamically active entity, the cardiac extracellular matrix (ECM), which comprises collagen, elastin, fibronectin, enzymes and signaling molecules, all surrounded by basal substance. ECM is able to change continually in response to local cell signaling, providing not only a passive support but also fundamental physiological influence for the resident cells (Pope et al. 2008). Their shape, division and location are controlled by highly regulated morphogenetic movements and interactions that, in turn, affect further morphogenesis, leading to the complex functional structure of the adult heart.

The lineage of each of these cell types is linked to the corresponding cell-fate decision: for example, while the progenitor cell-fate results from the behavior of its progeny, the one of a terminally differentiated cell consists in its integrated behavior and patterns of gene expression (Olson 2006). The vertebrate cardiogenic lineage is one of the first to be established. At least in early development, consensus exists that MYO cells, EE cells, vascular endothelial cells and blood cells share a common progenitor cell referred to as hemangioblast. Unfortunately, since the timing and location of cell fate decisions in later developmental stages still need to be clarified, lineage relations of the cardio-vasculature are presently murky (Fishman and Stainier 1994;

Chien et al. 2008). However, in the zebrafish embryo, the cardiogenic region can be defined before gastrulation and different blastomeres can be identified by the late blastula. Their progeny appears restricted to a single chamber suggesting that different chambers may arise from different progenitors (Stainier et al. 1993; Fishman and Stainier 1994).

The structural design of the primitive heart tube is common to all early vertebrate embryos and independent of species variations. Derived from the cardiogenic mesoderm, it is made up of two concentric cell layers, the inner EE and the outer MYO, separated by the extracellular amorph elastic matrix (*cardiac jelly*). In a later step, the formation of the outermost EPI makes the chamber three-walled. Already in this early morphogenetic scenario, cell–cell interactions play a crucial role, providing signals that direct an uncommitted cell either to respond to other differentiating signals, or to take a certain cell-fate decision, or to increase its propensity toward it.

Since the morphogenetic scenario of the beating heart is dominated by its early contractile activity, it is conceivable that crucial epigenetic chemical and mechanical events at cell and tissue levels alter, more than in other organs, gene expression, which, in turn, can modify these events through morpho-regulatory signaling.

Below we will briefly emphasize the obligatory growth-modulating cross-talk between the EE, the myocardium and the epicardium.

Endocardial–Myocardial Cross-talk

An important example of the EE-MYO cross-talk is illustrated by myocardial signals that trigger the invasion of EE cells into the cardiac jelly, directing their migration toward the MYO (Brutsaert 2003). Soon after, the EE cells, arranged in monolayer, line the cardiomyocytes so that together they form the primitive double-walled spongy heart tube, characterized by the first rhythmic myocardial contractions. EE-myocardial signaling is also a prerequisite for the subsequent trabeculation of the spongy heart. This essential and highly regulated morphogenetic process is initiated by clusters of beating differentiated myocytes that migrate toward the EE expanding into the cardiac jelly to form sheet-like protrusions, i.e. the trabeculae (see for references, Sedmera et al. 2003). Conceivably, the nascent trabeculae may result from delamination, namely a directional cardiomyocyte migration and/or oriented cell division, a process described in detail in the zebrafish ventricle by Liu et al. (2010). In wild-type larvae (5dpf) the ventricle is extensively trabeculated (Liu et al. 2010). Appropriate ventricular trabeculation depends on the normal development of the EE, which, as uncovered by zebrafish embryo studies, involves the control exerted by the *cloche* gene (Stainier et al. 1995). In fact, if the EE layer is lacking, as in *cloche* mutant zebrafish heart, the outer myocardial layer appears smaller, dysfunctional and unable to form trabeculation (Liao et al. 1997). Another important EE-confined paracrine signaling is represented by the neuregulin-1 growth factor (Nrg1), which targets the corresponding tyrosine kinase ErbB2 and ErbB4 receptors localized in nearby cardiomyocytes thereby driving trabecular initiation.

ErbB2 may play a dual role in zebrafish ventricular trabeculation, not only eliciting cardiomyocyte proliferation, but also inducing directional cardiomyocyte migration. In fact, ErbB2 mutants lack trabeculation and develop severe cardiac dysfunction associated with poorly developed conduction system (Liu et al. 2010). Similarly, the zebrafish cardiac mutant *dco* is also characterized by disorganized ventricular conduction. Its abnormal cardiomyocyte morphology and subsequent dysmorphic ventricle emphasize the epigenetic role played by the electrical forces in directing cardiac growth and orienting the various cell types (Chi et al. 2010). Another important EE-dependent epigenetic factor that influences cardiogenesis is represented by the intracardiac hemodynamic flow (Hove et al. 2003).

EE-spatially dependent induction cascades have also been identified in the atrioventricular canal and in the outflow tract and are involved in valve formation. As expected, the *cloche* mutant zebrafish heart, lacking the EE, also lacks cardiac cushions and valve formation (Stainier et al. 1995). In addition to *cloche* gene and Nrg1-ErbB2, the paracrine signaling pathways, still to be clarified, include VEGF and its receptors and angiopoietin (Fishman and Stainier 1994; Brutsaert 2003).

Epicardial–Myocardial Cross-talk

According to avian and murine models, the embryonic MYO secretes growth factors such as TGFB and EGF that induce an epithelial–mesenchymal transformation in the EPI to harvest cells that become the endothelial and smooth muscle cells of the coronary arteries and the cardiac fibroblast population (Morabito et al. 2001). Reciprocally, as indicated by studies mostly based on mouse and chick embryos, mitogen signals secreted from the EPI, such as erythropoietin and retinoic acid, specifically target the myocytes inducing their proliferation and survival (Stuckmann et al. 2003). Furthermore, EPI contributes to the establishment of compact layer vs. trabecular gene expression patterns, an important step consisting in the adherence of the EPI to the MYO. This is attained through molecules such Vascular Cell Adhesion Molecule-1 (VCAM-1), a receptor for integrins, expressed by EPI cells (Kang and Sucov 2005). Thus, according to the emerging scenario, the EPI paracrine signaling is an important player in maintaining the highest rate of MYO proliferation in the adjacent compact zone and in arranging the ventricular chamber into the trabecular and compact myocardium.

Epicardial Role in Adult Heart Ventricle Regeneration

Unlike the mammalian heart, certain adult teleost and urodele (frogs, newts and axolotls) hearts possess high regenerative capacity, even after major injury (Poss 2007). Zebrafish ventricle is typically restored within 30 days after 20% apical resection of the wall without scar tissue (fibrosis) (Poss et al. 2002). The regenerative proliferation occurs at the distal (apical) edge of the regenerate, the proliferative domain being maintained in a specific region as regeneration proceeds distally. The EPI responds in a rapid and robust manner to the injury. In fact, it expresses markers of the embryonic EPI, proliferates and covers the injured apex within 7–14 days, invading the regeneration tissue with a subpopulation of cells typically expressing the Fibroblast growth factor (Fgf) receptors *fgfr2* and *fgfr4* (Lepilina et al. 2006). Fgf signaling operates to support the regeneration of progenitor cells, which is achieved by directing the neo-vascularizing activity of developmentally activated EPI cells (Lepilina et al. 2006). Moreover, members of the Platelet-derived growth factor (PDGF) family have also been included in the set of "regeneration core molecules" (injury-activated cardiomyocyte mitogens) that are proposed to be mechanistically involved in zebrafish heart regeneration (Lien et al. 2006). Likely, molecular genetic studies using microarray and in situ hybridization screens will soon expand our knowledge on the candidate genes implicated in heart regeneration and how they orchestrate the corresponding intracardiac cell–cell interactions.

Cell–Molecular Integration: The Adult Eel Heart Paradigm

Unlike the larval heart, which is hemo-dynamically designed as a low-pressure pump, the adult fish heart is remodeled as a high-output pump adapted to perfuse a larger body size. The consequent higher organ complexity and heterogeneity (ventricular myoarchitecture and blood supply, zonal myocardial nonuniformity, etc.) call for multilevel homeostatic compensations, including cell–cell interactions among the principal cardiac cell components. To illustrate some functionally important cellular and molecular aspects of integration, we will use the adult eel ventricle paradigm, showing how the intracardiac NOS/NO system is able to link space and time in the fish heart.

Due to the multiple tissue (EE, vascular endothelium, contractile MYO, EPI, etc.) and subcellular (plasma membrane, endoplasmic reticulum, Golgi apparatus) localizations of nitric oxide synthase (NOS), and the freely diffusible property of NO as gaseous molecule, this system is well suited to modulate cardiac mechanical function both at basal and stimulated conditions. This regulation involves temporally different responses, i.e. the beat-to-beat (Starling's mechanism), the short-term (phasic control through chemical modulation) and the long-term (tonic control by altering gene expression) responses. As summarized in Fig. 3.4, the paracrine or autocrine modulations (discussed below) span multiple spatial and temporal scales, contributing to integrate the mechanical performance of the whole ventricle.

NO-Mediated Paracrine Modulation

After being synthesized by the EE or the vascular endothelium NO diffuses and affects the biology of the nearby cardiomyocytes. The EE-induced contractile modulation of myocardial performance represents a hallmark of the paracrine NO

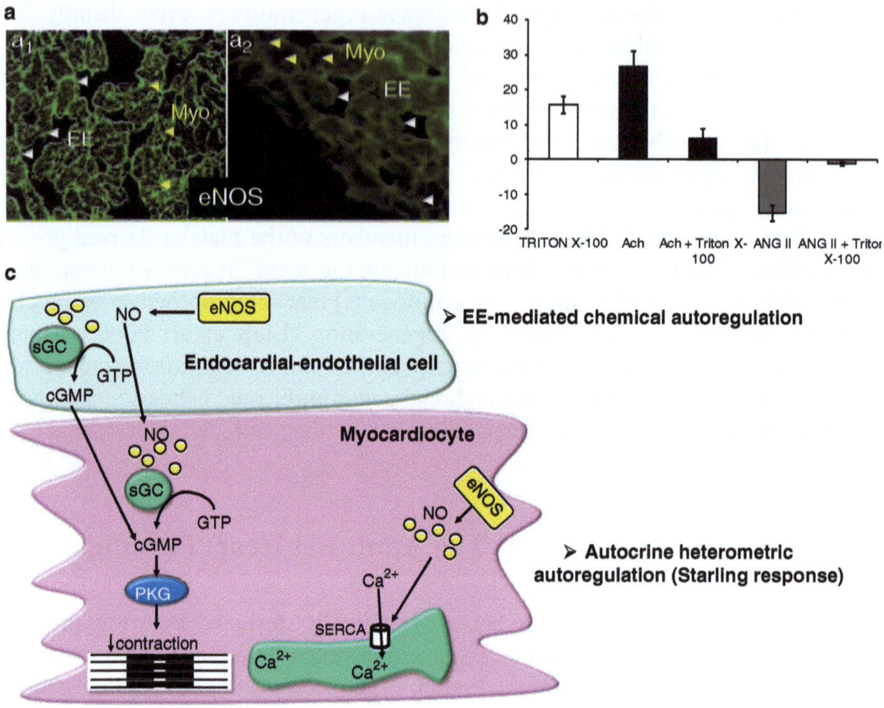

Fig. 3.4 (**a**) Representative fluorescence images of eNOS immunolabeling (polyclonal anti-eNOS antibody; 1:100) on ventricular sections (**a**) control (**b**) after Triton X-100 treatment. (**b**) Effects of Angiotensin II and Acetylcholine before and after injection with Triton X-100. (**c**) Schematic illustration showing the autocrine–paracrine NO signaling in the heart

signal and can be best evaluated in the avascular, or poorly vascularized, fish heart. Under basal conditions, the EE-generated NO exerts a mild tonic negative contractile influence. In contrast, the functional damage of the EE with consequent interruption of the EE–eNOS–cGMP axis elicits positive contractile effects (Imbrogno et al. 2003). As shown by our studies, this EE–eNOS signaling, located at the crossroad of many extrinsic and intrinsic neuro-endocrine pathways, coordinates many chemically activated cascades involved in the short-term (phasic) regulation. For example, endoluminal chemical stimuli such as Acetylcholine (Ach), Angiotensin II (ANG II), Vasostatin 1 (VS-1), Endothelin-1 (ET-1), as well as β3 adrenoceptor activation, all converge in their contractile effects via NO signaling, which requires the obligatory involvement of the EE (Imbrogno et al. 2001, 2003, 2004, 2006). Therefore, the EE not only represents a quantitatively important source of NO, but appears also necessary for coupling the intracavitary stimuli to eNOS/NO system. A striking case is provided by the mechanisms underlying the cholinergic-induced biphasic contractile effects in the eel and in the fully trabeculated and avascular frog (*Rana esculenta*) ventricle. While in the eel heart only the positive inotropism induced by nanomolar ACh concentrations requires the functional integrity of the

EE (Imbrogno et al. 2001), in the frog, the EE integrity is a prerequisite for transducing both the Ach-induced positive and negative contractile effects (Gattuso et al. 1999). The positive effects of Ach in eel, and both cholinergic responses in frog are mediated by the M1 receptor subtype. This suggests that in eel and frog this receptor subtype is located at the EE level, resembling the situation in mammals, i.e. M2 and M4 muscarinic receptors preferentially located on the cardiomyocytes, and M1, M3 and M5 principally expressed by endothelial cells (Brodde and Michel 1999).

The EE is a major target of the virus that causes ISA (Infectious Salmon Anemia) in the Atlantic salmon (Gattuso et al. 2002). It is characterized by cardiac dysfunction evidenced by depressed contractility to the 'beat-to-beat' response (Frank–Starling's Law of the Heart). This dysfunction appears correlated with the severity of the disease. The finding that the decreased myocardial performance is completely reverted by the inducible NOS (iNOS) inhibition is consistent with a pathogenetic role of higher iNOS-derived NO levels (Gattuso et al. 2002).

NO-Mediated Autocrine Modulation

Since both constitutive and inducible NOSs are expressed in cardiomyocytes, NO may also act as an autocrine molecule (Seddon et al. 2007), e.g., directly influencing ion channels and pumps. The subtle NOSs spatial confinement allows coordination of signaling networks between cardiac NO and specific intracellular effectors (Hare 2003), achieving local control of different cellular functions (Iwakiri et al. 2006). Myocardial eNOS is mainly localized in specific intracellular membrane domains (plasma membrane, particularly the caveolae, and Golgi apparatus) (Iwakiri et al. 2006). Autocrine nNOS is localized in the sarcoplasmic reticulum (SR) (Xu et al. 1999) and also at the sarcolemmal protein level (Williams et al. 2006). The subcellular NOS isoform localization orchestrates finely tuned cardiac responses. For example, the caveolar eNOS compartmentation with β-adrenergic receptors and L-type Ca^{2+} channels permits NO to inhibit the β-adrenergic-induced contractility. On the other hand, nNOS on SR allows it to induce calcium release via Ryanodine (RYRs), thereby eliciting an opposite, facilitatory inotropic effect.

NOS compartmentation allows the generation of a relatively high local concentration of NO equivalents, creating a favorable environment for protein S-nitrosylation. The latter recently emerged as a novel mechanism of autocrine NO signaling (Lima et al. 2010, and references therein). In the mammalian heart, the stretch-induced increase of contractility mediated by eNOS is attained through S-nitrosylation of thiol residues in the RYR Ca^{2+} release channels (Massion et al. 2005). In the eel heart we demonstrated that the regulation of the 'beat-to-beat' response is directly modulated by myocardial autocrine NO through a nonclassical cGMP-independent pathway involving an Akt-mediated activation of eNOS-dependent NO production that, in turn, modulates the rate of SR Ca^{2+} reuptake through Phospholamban (PLN) S-nitrosylation (Garofalo et al. 2009) (Fig. 3.5). This first fish evidence supports the idea that the NO ability of fine-tuning ventricular

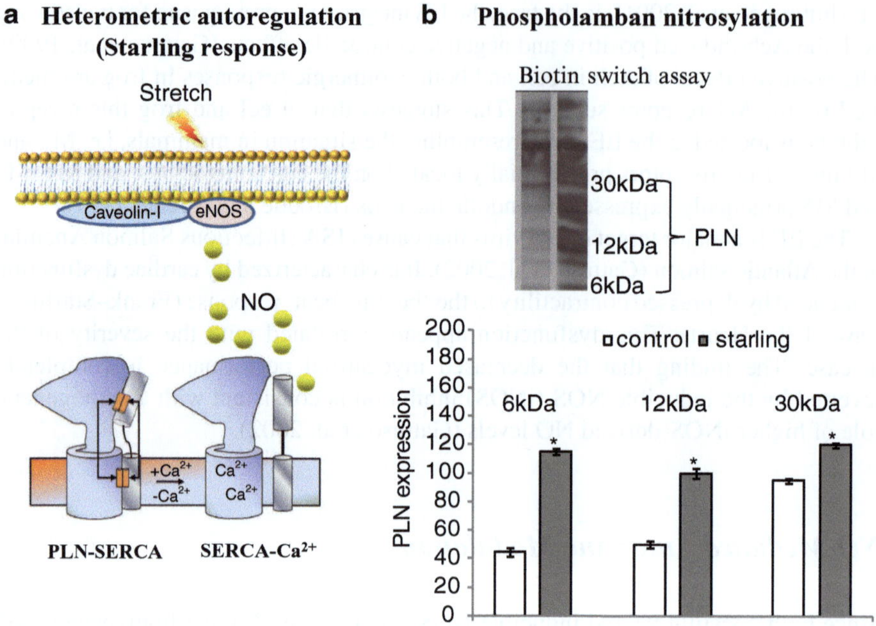

Fig. 3.5 (**a**) Schematic illustration showing the NO modulation of the rate of SR Ca^{2+} reuptake through Phospholamban (PLN) S-nitrosylation. (**b**) *Upper panel*: western blot of S-nitrosylated proteins in heart homogenates; control heart (*left lane*) and Starling-treated heart (*right lane*). *Lower panel*: amount of S-nitrosylation at the migration position corresponding to the PLN as a monomer (6 kDa), dimer (12 kDa) and pentamer (30 kDa). Data are mean + SEM of five determinations for each animal (n = 3). Statistical differences were evaluated by Student's *t* test; *p, 0.001 (Reproduced from: Garofalo et al. 2009)

performance through NOS isoforms compartmentation, as well as differences in their mode of stimulation and recruitment of distinct downstream pathways (S-nitrosylation or cGMP production) within various cell micro-domains, was a crucial and early event during the evolution of the vertebrate heart. Conceivably, this integration system paved the way toward the more complex, high-pressure pump designs of tetrapods.

Conclusions

During its developmental remodeling, the vertebrate heart grows from the primitive, three-layered low-pressure beating tube to the mature, structurally complex high-pressure pump, adjusted to perfuse the larger body size of the adult organism. The acquisition of different cardiac functions and the maintenance of the organ integrity are attained by cell division, cell differentiation and cell growth, all processes that are also fundamental for cardiomyocyte specialization, positional identity and zonal nonuniformity. Cardiac morphogenesis, growth and nonuniformity are intimately linked to the whole organ function, although the underlying genetic and epigenetic mechanisms are only partially understood, as are the mechanisms whereby heart size is coordinated to body size. However, there is mounting evidence of the central role played by cardiac cell–cell interactions. These interactions from early cardiogenesis establish dynamic regulatory loops that range from the gene to the cells and tissues, going back to regulate the expression of the same or different genes.

In an attempt to incorporate this knowledge into the morpho-dynamics and functional integration of the cardiac ventricular chamber, we sketched out some examples using as a paradigm the fish heart, the ancestor vertebrate heart. As documented in zebrafish, the cross-talk between the EE, the MYO and the EPI provide obligatory growth-modulating signals that direct with different mechanisms either the growth of the developing heart or the regeneration of the injured adult ventricle. Growth-related remodeling of the ventricle wall, as evidenced in the adult tuna, salmonid and eel hearts, highlight striking aspects of cardiac nonuniformity. A facet of spatio-temporal integration of ventricle function, as in the adult eel heart, shows that the multiple NOS localizations allow NO to interact not only with a number of intracellular targets (autocrine modulation), but also with sites of signal integration/diversification of other cell types (EE-mediated paracrine modulation).

These examples give a specific temporal dimension to cardiac function and, at the same time, a spatial framework of upgrading performance incorporated into the organ system. On the other hand, from an evolutionary perspective, the cell–cell crosstalk revealed by these fish studies indicate the deep evolutionary conservation of molecular signaling in the heart. They can be also useful in identifying aspects of novelty and diversity in the way heart ventricle nonuniformity was established and shaped during vertebrate phylogeny.

Acknowledgements We thank D. Amelio for advices and figures.

References

Bern HA, Madsen SS (1992) A selective survey of the endocrine system of the rainbow trout (*Oncorhynchus mykiss*) with emphasis on the hormonal regulation of ion balance. Aquaculture 100:237–262

Birkedal R, Shiels HA, Vendelin M (2006) Three dimensional mitochondrial arrangement in ventricular myocytes: from chaos to order. Am J Physiol Cell Physiol 291:C1148–C1158

Brodde O-E, Michel MC (1999) Adrenergic and muscarinic receptors in the human heart. Pharmacol Rev 51:651–689

Brutsaert DL (2003) Cardiac endothelial-myocardial signaling: its role in cardiac growth, contractile performance, and rhythmicity. Physiol Rev 83(1):59–115

Cerra MC, Imbrogno S, Amelio D, Garofalo F, Colvee E, Tota B, Icardo JM (2004) Cardiac morphodynamic remodelling in the growing eel (*Anguilla anguilla* L.). J Exp Biol 207:2867–2875

Chi NC, Bussen M, Brand-Arzamendi K, Ding C, Olgin JE, Shaw RM, Martin GR, Stainier DY (2010) Cardiac conduction is required to preserve cardiac chamber morphology. Proc Natl Acad Sci USA 107(33):14662–7

Chien KR, Domian IJ, Parker KK (2008) Cardiogenesis and the complex biology of regenerative cardiovascular medicine. Science 322:1494–1497

Di Maio A, Block BA (2008) Ultrastructure of the sarcoplasmic reticulum in cardiac myocytes from Pacific bluefin tuna. Cell Tissue Res 334(1):121–34

Fishman MC, Olson EN (1997) Parsing the heart: genetic modules for organ assembly. Cell 91:153–156

Fishman MC, Stainier DYR (1994) Cardiovascular development: prospects for a genetic approach. Circ Res 74:757–763

Gamperl AK, Farrell AP (2004) Cardiac plasticity in fishes: environmental influences and intraspecific differences. J Exp Biol 207:2539–2550

Garofalo F, Parisella ML, Amelio D, Tota B, Imbrogno S (2009) Phospholamban S nitrosylation modulates Starling response in fish heart. Proc Biol Sci 276(1675):4043–52

Gattuso A, Mazza R, Pellegrino D, Tota B (1999) Endocardial endothelium mediates luminal ACh-NO signaling in isolated frog heart. Am J Physiol 276(2 Pt 2):H633–H641

Gattuso A, Mazza R, Imbrogno S, Sverdrup A, Tota B, Nylund A (2002) Cardiac performance in *Salmo salar* with Infectius Salmon Anemia (ISA): putative role of nitric oxide. Dis Aquat Org 52:11–20

Gemelli L, Martino G, Tota B (1980) Oxidation of lactate in the compact and spongy myocardium of tuna fish (*Thunnus thynnus thynnus* L.). Comp Biochem Physiol B 65(2):321–326

Gerhart J, Kirschner M (1997) Cells, embryos, and evolution: toward a cellular and developmental understanding of phenotypic variation and evolutionary adaptability. Blackwell Science, Oxford

Hare JM (2003) Nitric oxide and excitation–contraction coupling. J Mol Cell Cardiol 35:719–729

Hochachka PW, Somero GN (2002) Biochemical adaptation: mechanism and process in physiological evolution. Oxford University Press, New York, p 31

Hove JR, Koster RW, Forouhar AS, Cevedo-Bolton G, Fraser SE, Gharib M (2003) Intracardiac fluid forces are an essential epigenetic factor for embryonic cardiogenesis. Nature 421:172–177

Icardo JM (2006) Conus arteriosus of the teleost heart: dismissed, but not missed. Anat Rec A 288:900–908

Icardo JM, Colvee E (2001) The atrioventricular region of the teleost heart. A distinct heart segment. Anat Rec A 294:236–242

Icardo JM, Imbrogno S, Gattuso A, Colvee E, Tota B (2005) The heart of *Sparus auratus*: a reappraisal of cardiac functional morphology in teleosts. J Exp Zool A Comp Exp Biol 303:665–675

Imbrogno S, De Iuri V, Mazza R, Tota B (2001) Nitric oxide modulates cardiac performance in the heart of *Anguilla anguilla*. J Exp Biol 204:1719–1727

Imbrogno S, Cerra MC, Tota B (2003) Angiotensin II-induced inotropism requires an endocardial endothelium-nitric oxide mechanism in the in-vitro heart of *Anguilla anguilla*. J Exp Biol 206:2675–2684

Imbrogno S, Angelone T, Corti A, Adamo C, Helle KB, Tota B (2004) Influence of vasostatins, the chromogranin A-derived peptides, on the working heart of the eel (*Anguilla anguilla*): negative inotropy and mechanism of action. Gen Comp Endocrinol 139:20–28

Imbrogno S, Angelone T, Adamo C, Pulerà E, Tota B, Cerra MC (2006) Beta3 Adrenoceptor in the eel (*Anguilla anguilla*) heart: negative inotropy and NO-cGMP-dependent mechanism. J Exp Biol 209:4966–4973

Iwakiri Y, Satoh A, Chatterjee S, Toomre DK, Chalouni CM, Fulton D, Groszmann RJ, Shah VH, Sessa WC (2006) Nitric oxide synthase generates nitric oxide locally to regulate compartmentalized protein S-nitrosylation and protein trafficking. Proc Natl Acad Sci USA 103:19777–19782

Johansen IB, Lunde IG, Røsjø H, Christensen G, Nilsson GE, Bakken M, Øverli Ø (2011) Cortisol response to stress is associated with myocardial remodeling in salmonid fishes. J Exp Biol 214(8):1313–1321

Kang JO, Sucov HM (2005) Convergent proliferative response and divergent morphogenic pathways induced by epicardial and endocardial signaling in fetal heart development. Mech Dev 122:57–65

Leary SC, Michaud D, Lyons CN, Hale TM, Bushfield TL, Adams M, Moyes CD (2002) Bioenergetic remodeling of heart during treatment of spontaneously hypertensive rats with enalapril. Am J Physiol 283:H540–548
Lepilina A, Coon AN, Kikuchi K, Holdway JE, Roberts RW, Burns CG, Poss KD (2006) A dynamic epicardial injury response supports progenitor cell activity during zebrafish heart regeneration. Cell 127:607–619
Liao W, Bisgrove BW, Sawyer H, Hug B, Bell B, Peters K, Grunwald DJ, Stainier DY (1997) The zebrafish gene cloche acts upstream of a flk-1 homologue to regulate endothelial cell differentiation. Development 124:381–389
Lien CL, Schebesta M, Makino S, Weber GJ, Keating MT (2006) Gene expression analysis of zebrafish heart regeneration. PLoS Biol 4(8):e260
Lima B, Forrester MT, Hess DT, Stamler JS (2010) S-nitrosylation in cardiovascular signaling. Circ Res 106:633–646
Liu J, Bressan M, Hassel D, Huisken J, Staudt D, Kikuchi K, Poss KD, Mikawa T, Stainier DY (2010) A dual role for ErbB2 signaling in cardiac trabeculation. Development 137(22):3867–75
Lumbers ER, Boyce AC, Joulianos G, Kumarasamy V, Barner E, Segar JL, Burrell JH (2005) Effects of cortisol on cardiac myocytes and on expression of cardiac genes in fetal sheep. Am J Physiol Regul Integr Comp Physiol 288:R567–574
Massion PB, Pelat M, Belge C, Balligand J-L (2005) Regulation of the mammalian heart function by nitric oxide. Comp Biochem Physiol 142A:144–150
Molkentin JD, Lu JR, Antos CL, Markham B, Richardson J, Robbins J, Grant SR, And Olson EN (1998) A calcineurin-dependent transcriptional pathway for cardiac hypertrophy. Cell 93:215–228
Moorman AF, Christoffels VM (2003) Cardiac chamber formation: development, genes, and evolution. Physiol Rev 83:1223–1267
Morabito CJ, Dettman RW, Kattan J, Collier JM, Bristow J (2001) Positive and negative regulation of epicardial-mesenchymal transformation during avian heart development. Dev Biol 234(1):204–15
Nichols NR, Tracy KE, Funder JW (1984) Glucocorticoid effects on newly synthesized proteins in muscle and non-muscle cells cultured from neonatal rat hearts. J Steroid Biochem 21:487–496
Olson EN (2006) Gene regulatory networks in the evolution and development of the heart. Science 5795:1922–1927
Pope AJ, Sands GB, Smaill BH, LeGrice IJ (2008) Three-dimensional transmural organization of perimysial collagen in the heart. Am J Physiol Heart Circ Physiol 295:H1243–H1252
Poppe TT, Johansen R, Gunnes G, Torud B (2003) Heart morphology in wild and farmed Atlantic salmon *Salmo salar* and rainbow trout *Oncorhynchus mykiss*. Dis Aquat Organ 57(103):108
Poss KD (2007) Getting to the heart of regeneration in zebrafish. Semin Cell Dev Biol 18(1): 36–45
Poss KD, Wilson LG, Keating MT (2002) Heart regeneration in zebrafish. Science 298:2188–2190
Pottinger TG, Carrick TR (1999) Modification of the plasma 581 cortisol response to stress in rainbow trout by selective breeding. Gen Comp Endocrinol 116:122–132
Poupa O, Lindström L (1983) Comparative and scaling aspects of heart and body weights with reference to blood supply of cardiac fibers. Comp Biochem Physiol A 76(3):413–421
Poupa O, Ostadal B (1969) Experimental cardiomegalies and cardiomegalies in free-living animals. Ann NY Acad Sci 156(31):445–468
Poupa O, Gesser H, Jonsson S, Sullivan L (1974) Coronary-supplied compact shell of ventricular myocardium in salmonids: growth and enzyme pattern. Comp Biochem Physiol A 48:85–95
Poupa O, Lindstrom L, Maresca A, Tota B (1981) Cardiac growth, myoglobin, proteins and DNA in developing tuna (*Thunnus thynnus thynnus*). Comp Biochem Physiol 70A:217–222
Seddon M, Shah AM, Casadei B (2007) Cardiomyocytes as effectors of nitric oxide signalling. Cardiovasc Res 75:315–326
Sedmera D, Reckova M, deAlmeida A, Sedmerova M, Biermann M, Volejnik J, Sarre A, Raddatz E, McCarthy RA, Gourdie RG, Thompson RP (2003) Functional and morphological evidence

for a ventricular conduction system in zebrafish and *Xenopus* hearts. Am J Physiol Heart Circ Physiol 284:H1152–H1160

Shiels HA, Di Maio A, Thompson S, Block BA (2011) Warm fish with cold hearts: thermal plasticity of excitation-contraction coupling in bluefin tuna. Proc Biol Sci 278(1702):18–27

Stainier DY, Lee RK, Fishman MC (1993) Cardiovascular development in the zebrafish. Myocardial fate map and heart tube formation. Development 119:31–40

Stainier DYR, Welnstein BM, Detrich HW, Zon LI, Fishman MC (1995) *Cloche*, an early acting zebrafish gene, is required by both the endothelial and hematopoietic lineages. Development 121:3141–3150

Stuckmann I, Evans S, Lassar AB (2003) Erythropoietin and retinoic acid, secreted from the epicardium, are required for cardiac myocyte proliferation. Dev Biol 255:334–349

Takle H, Baeverfjord G, Helland S, Kjorsvik E, Andersen O (2006) Hyperthermia induced atrial natriuretic peptide expression and deviant heart development in Atlantic salmon *Salmo salar* embryos. Gen Comp Endocrinol 147:118–125

Tota B, Cimini V, Salvatore G, Zummo G (1983) Comparative study of the arterial and lacunary systems of the ventricular myocardium of elasmobranch and teleost fishes. Am J Anat 167:15–32

Urschel M, O'Brien KM (2008) High mitochondrial densities in the hearts of Antarctic icefishes are maintained by an increase in mitochondrial size rather than mitochondrial biogenesis. J Exp Biol 211:2638–2646

Vogel S (1988) Life's devices: the physical world of animals and plants. Princeton University Press, New Jersey

Vornanen M, Hassinen M, Koskinen H, Krasnov A (2005) Steady-state effects of temperature acclimation on the transcriptome of the rainbow trout heart. Am J Physiol Regul Integr Comp Physiol 289:R1177–R1184

Wendelaar Bonga SE (1997) The stress response in fish. Physiol Rev 77:591–625

Wilkins BJ, Dai YS, Bueno OF, Parsons SA, Xu J, Plank DM, Jones F, Kimball TR, Molkentin JD (2004) Calcineurin/NFAT coupling participates in pathological, but not physiological, cardiac hypertrophy. Circ Res 94:110–118

Williams JC, Armesilla AL, Mohamed TM, Hagarty CL, McIntyre FH, Schomburg S, Zaki AO, Oceandy D, Cartwright EJ, Buch MH, Emerson M, Neyses L (2006) The sarcolemmal calcium pump, a-1 syntrophin, and neuronal nitric-oxide synthase are parts of a macromolecular protein complex. J Biol Chem 281:23341–23348

Wu H, Peisley A, Graef IA, Crabtree GR (2007) NFAT signaling and the invention of vertebrates. Trends Cell Biol 17:251–260

Xu KY, Huso DL, Dawson TM, Bredt DS, Becker LC (1999) Nitric oxide synthase in cardiac sarcoplasmic reticulum. Proc Natl Acad Sci USA 96:657–662

Chapter 4
A Perspective on the Evolution of the Coronary Circulation in Fishes and the Transition to Terrestrial Life

A.P. Farrell, N.D. Farrell, H. Jourdan, and G.K. Cox

Introduction

The rhythmic heartbeat of all vertebrates requires a more or less continuous supply of oxygen to the cardiac muscle. There are two ways of doing this: one is by removing oxygen from the venous blood passing through the lumen of the heart, and the other is by supplying blood vessels directly to the cardiac muscle from the outside, the coronary circulation.

As a dedicated blood supply for the heart, the coronary circulation is the principal oxygen supply route used by adult birds and mammals. In fact, these animals have evolved such an obligate dependence on the coronary circulation that an interruption of coronary blood flow for even a few minutes leads to permanent cardiac damage and even death. The origin of the coronary artery lies at the root of the aorta, the first arterial branch of the systemic circulation as it leaves the heart. As a result, these hearts supply themselves with oxygen before sending oxygen to any other organ, including the brain.

In contrast, many vertebrate hearts do not need a coronary circulation to function properly. Instead, they depend on the oxygen remaining in the blood after it has circulated the body. The luminal venous blood is the principal oxygen supply route for most chambered vertebrate hearts. Even so, a large number of vertebrates, especially fishes, use both oxygen supply routes, but with varying degrees of dependence. Yet, no vertebrate appears to be as dependent on its coronary circulation as birds and mammals. Amazingly, hearts in a few exceptional species, such as crucian carp and Pacific hagfish, tolerate temporary interruptions of their cardiac oxygen supply that can last for periods of hours, days and even months—but this is a topic covered elsewhere (Farrell and Stecyk 2007). This review focuses on how the

A.P. Farrell (✉) • N.D. Farrell • H. Jourdan • G.K. Cox
Department of Zoology, University of British Columbia, Vancouver, BC, Canada V6T 1Z4
e-mail: farrellt@interchange.ubc.ca

D. Sedmera and T. Wang (eds.), *Ontogeny and Phylogeny of the Vertebrate Heart*,
DOI 10.1007/978-1-4614-3387-3_4, © Springer Science+Business Media, LLC 2012

coronary circulation has evolved in its role of delivering oxygen to working cardiac muscle. As we see, the starting point in this evolution, is an avascular heart; a heart without coronary circulation.

The oxygen demand and oxygen supply routes of adult hearts clearly vary considerably among the more than 50,000 vertebrate species. Yet, all chordate hearts share a common embryonic beginning, an avascular heart, before embarking on highly varied developmental trajectories. Thus, the following brief consideration of cardiac development in birds, mammals and fishes serves two purposes. First, it sets a background framework for understanding cardiac diversity. Second, it illustrates why it is often suggested that the heart provides an excellent example of recapitulation theory, where phylogeny is recapitulated in the ontogeny of higher vertebrates (Farrell 1997; Sedmera and Watanabe 2006; Grimes and Kirby 2009). Thus, understanding developmental processes may provide insights into which selective forces have driven the evolutionary appearance of the coronary circulation among the vertebrates.

Cardiac and Coronary Development in Birds and Mammals

The vertebrate heart always begins embryonically as a heart tube. As the heart grows, myocardial thickening of the cardiac chambers initially takes the form of trabecular ridges that progress into a sponge-like sinusoidal arrangement, hereafter referred to as a spongy myocardium. These muscular trabeculae are devoid of blood vessels, and their oxygen supply comes from the blood passing through the endocardial-lined lumen of the heart chamber. This spongy architecture ensures a very large surface area, keeping individual cardiomyocytes as close as possible to their luminal oxygen supply and allowing effective gas exchange.

A challenge for the growing heart is to increase cardiac muscle mass and volume without distancing individual cardiomyocytes too far from their luminal oxygen supply. Thus, the avascular, spongy arrangement of the developing avian and mammalian heart is only a temporary embryonic state (except in certain pathologies; Angelini et al. 1999) until the coronary circulation takes over as the primary oxygen supply route. This takeover begins when the ventricle is expanding its 1–2 cell thick layer of epicardial myocardium, which up to this point has merely acted as an outer mantle around the spongy myocardium (Sedmera et al. 2000). Because the process of myocardial thickening and compaction progressively moves epicardial cardiomyocytes further from their endocardial oxygen supply route, a more proximate oxygen supply is needed: a dedicated coronary circulation from the outside surface of the compact myocardium. Coronary vessels first penetrate compact myocardium from the heart's epicardial surface, the onset of which coincides with molecular markers for hypoxia being expressed in the cardiac outflow tract (Sugishita et al. 2004). Thus, a hypoxic myocardium is likely the stimulus for angioblast invasion that forms the coronaries (Sedmera et al. 2000). Coronary vessels then coalesce into a more organized capillary network and the coronary artery grows toward the aorta. As a last major step in coronary development, the coronary artery penetrates into the root of the aorta. At this point, the coronary circulation become fully functional, as early as one-fifth of the way into gestation in humans, or as late as two-thirds into gestation in mice (Tomanek 1996).

Cardiac growth (both chamber volume and wall thickness) allows for stroke volume to increase nearly in proportion with body mass, accompanied by a smaller reduction in heart rate and a smaller increase in arterial blood pressure. The mechanism by which the heart achieves this cardiac growth has very important consequences for coronary capillary density. In mammals, cardiac growth in utero is principally through cell division (cardiac hyperplasia), but postnatal cardiac growth occurs principally through cardiomyocyte enlargement (cardiac hypertrophy), which makes cardiomyocytes wider and longer. Hypertrophy however, increases the distance between capillaries whereas hyperplasia can decrease the number of capillaries per myocyte unless new capillaries are formed. Perinatal cardiac growth in mammals sees a large increase in capillary density (Tomanek 1996). In the adult heart, the capillary-to-mitochondria diffusion distance for oxygen must be near optimal given that the coronary venous oxygen partial pressure is low and that increases in cardiac oxygen demand are closely paralleled by increased coronary blood flow (by as much as fivefold during maximum activity; Duncker and Bache 2008). Consequently, Mammalian hearts undergo a remarkable ontogenetic switch from being comprised entirely of avascular spongy myocardium to an obligate dependence on its coronary circulation because they are comprised almost entirely of highly vascularized compact myocardium.

However, the coronary circulation is rarely, if ever, evenly distributed within or among cardiac chambers. Cardiac chambers may have different degrees of compaction and therefore require different levels of perfusion. The degree of ventricular compaction is driven in a large part by its cardiac work rate (which can be estimated from the product of cardiac output and mean arterial blood pressure). Cardiac output is the product of stroke volume and heart rate. The adult mammalian right and left ventricles pump similar volumes of blood each minute, and with the same heart rate. However, each side pumps blood at different arterial pressure, and therefore the cardiac stroke work rate differs considerably. Because of the high systemic blood pressures, the left ventricle performs about four times more work than the right ventricle and, correspondingly, ventricular compaction is three times greater for the left versus the right in mammals, and five times greater in birds (Sedmera et al. 2000). As a result, the thicker free wall of the left ventricle requires more oxygen and receives a greater proportion of the total coronary blood flow than the right ventricle (Duncker and Bache 2008). In fact, differences between the left and right ventricles appear even before the development of compaction, with the left ventricle having thicker and more sheet-like trabeculae than the right ventricle (Sedmera et al. 2000). Below, we can see an analogous situation where extensive ventricular compaction evolved in athletic fishes along with a high cardiac work rate.

At an even finer level, blood flow distribution also differs within the layers of the compact myocardium of the mammalian heart. For example, the microvascular organization in the subendocardium has compensated for the greater wall tension and greater vascular compression it experiences as compared to the subepicardium

(Duncker and Bache 2008). Another variation in blood flow distribution is the Thebesian system, an extension of the coronary vascular supply to the trabeculae. The Thebesian vessels drain directly into the lumen of the heart (Thebesius 1708; Pratt 1898) rather than to the coronary sinus, where the majority of the coronary blood flow returns. Still, the Thebesian vessels are another variation of the blood supply to heart muscle. Thus, the coronary blood supply takes many forms and distributions during mammalian development.

At this juncture it is important to clarify our use of the terms spongy and compact myocardium. Sedmera et al. (2000), in describing cardiac development, suggested that a compact layer is considered present in all hearts because it starts as an avascular 1–2 cell thick mantle around the developing heart and has the potential to change markedly in structure and purpose. In the forthcoming discussion, we deal largely with the gross cardiac anatomy of adult forms. Therefore, for clarity we restrict the use of "compact myocardium" to when the muscle layer is vascularized by a functional coronary vessel. "Spongy myocardium" refers to endocardial myocytes that take on a trabecular architecture, are bathed by the luminal blood and can be avascular or vascularized. The reason for these functional definitions should become clearer below.

Cardiac and Coronary Development in Fishes

There are not many modern studies of cardiac development in fish (for example, Icardo 1990, 1996; Icardo et al. 2004, 2008, 2009; Stainier et al. 1993; Wanga et al. 2001; Santer 1985; Guerrero et al. 2004; Hu et al. 2000). Even less is known regarding the ontogenic development of the coronary circulation, despite the fact that adult fishes have evolved a fascinating array of solutions to supply their cardiac muscle with oxygen (Fig. 4.1).

For many fish species, unlike mammals, spongy myocardium remains throughout the entire adult life, without any further development of the mantle of compact myocardium (Fig. 4.1b). Other fish species undergo partial ventricular compaction and develop an associated coronary circulation (Fig. 4.1a, c, f). The degree of ventricular compaction has been quantified by expressing the mass of compact ventricle as a percentage of total ventricular mass for Atlantic salmon (Poupa et al. 1974), rainbow trout (Farrell et al. 1988) and Pacific tarpon (Farrell et al. 2007). In each case, ventricular compaction stops well short of the almost 100% ventricular compaction seen in adult birds and mammals (Fig. 4.2), even though the ventricle grows isometrically with body mass.

The upper limit for compaction varies enormously across fish species. Ventricular compaction varies from a low of 5% to a high of 70%, but more typically accounts for 15–30% of ventricular mass (Santer 1985). Thus, depending on the fish species, the vascularized shell of compact myocardium that encases spongy myocardium can be either very thin or very thick and is apparently genetically fixed within species. The percentage of compact myocardium within a species does change within

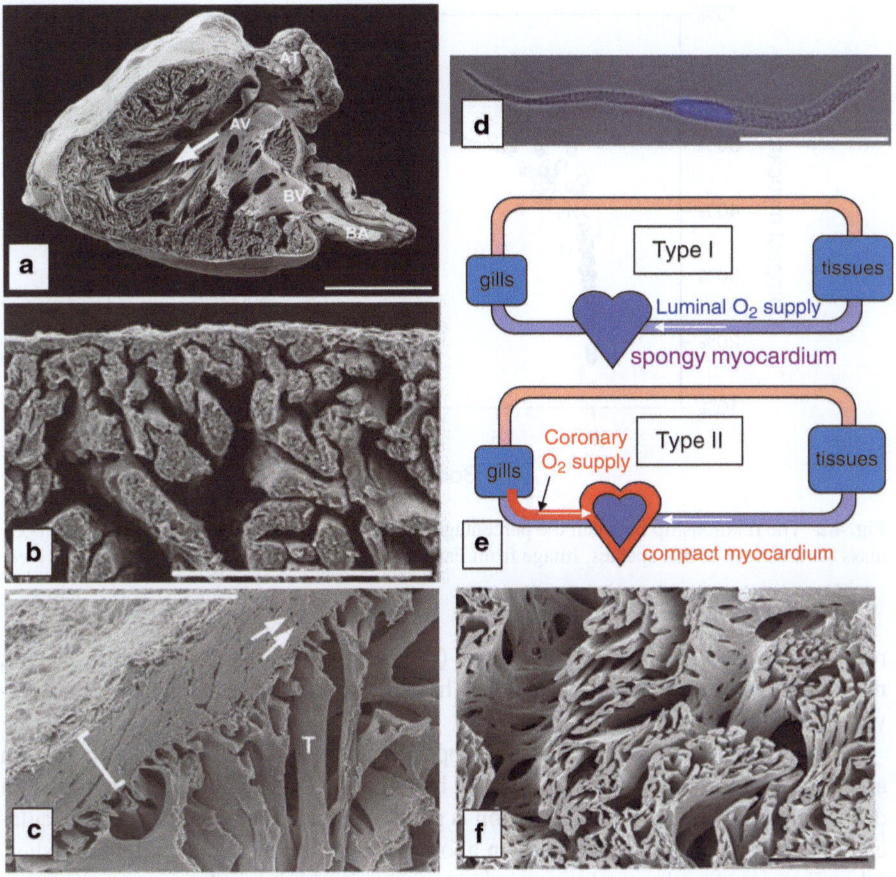

Fig. 4.1 Various muscular arrangements of fish hearts. (**a**) Scanning electron micrograph of a longitudinal sectioned heart of a juvenile sockeye salmon heart where the compact myocardium is quite thin (bar = 2 mm). (**b**) Scanning electron micrograph of a sectioned spongy Type I ventricle of tilapia (bar = 0.10 mm). (**c**) Scanning electron micrograph of the sectioned Type II ventricle of a juvenile sockeye salmon showing the junction of the spongy and compact myocardium, and the location of coronary vessels in the compact layer (*arrows*) (bar = 0.20 mm). (**d**) An isolated cardiomyocyte from a sockeye salmon illustrating its thin spindle shape (bar = 20 μm). (**e**) Cartoon representations of the circulatory arrangements for Type I and Type II ventricles, which illustrate the two oxygen supply routes used by fish heart, a luminal oxygen supply and a coronary oxygen supply. (**f**) A higher magnification scanning electron micrograph of a sectioned ventricle of a juvenile sockeye salmon showing a sheet-like arrangement of the trabeculae (bar = 0.25 mm). SEM images courtesy of Dr. S. Pieperhoff and modified from Pieperhoff et al. (2009)

limits because of phenotypic plasticity (Gamperl and Farrell 2004). All the same, ventricular compaction in lower vertebrates rarely exceeds 50% of ventricular mass, even in athletic fishes such as salmon, trout and marlin; this is a puzzle revisited later in this chapter.

Cardiac growth in fish is primarily a result of cardiac hyperplasia (an increase in number of cells) (Farrell et al. 1988), and individual cardiomyocytes retain a

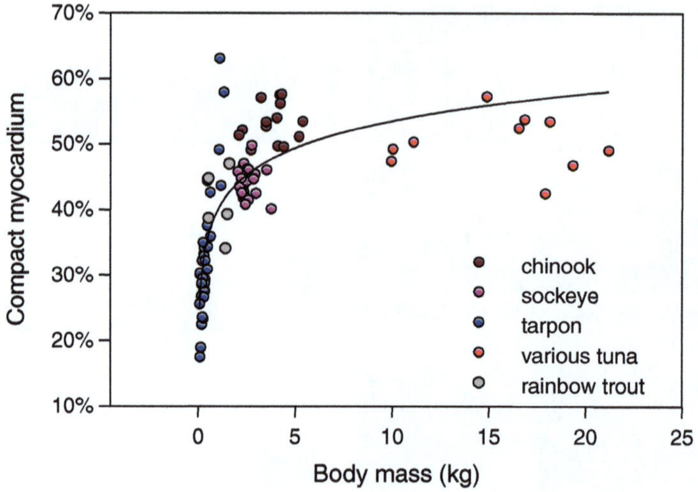

Fig. 4.2 The relationship between the percentage compact myocardium in the ventricle and body mass for a variety of fish species. Image from Farrell et al. (2009)

relatively thin (<10 μm) spindle shape (Fig. 4.1d). This situation contrasts with mammals, where the postnatal cardiac hypertrophy produces much thicker individual cardiomyocytes.

In summary, coronary arteries develop in concert with myocardial compaction and this development is likely stimulated by hypoxia. During avian and mammalian development, the extent of compaction is greatly influenced the work of a particular cardiac chamber. Very little is known about coronary artery development in fishes, other than cardiac hyperplasia persisting even in adults and compaction being highly variable. The review now considers the many coronary forms and distributions in adult fishes, the largest and most diverse group of extant vertebrates. It will become apparent that the same driving forces (hypoxia and cardiac work) also could have featured prominently in the evolution of the coronary circulation among chordates.

The Diverse Coronary Arrangements of Fishes

Information on the coronary circulation of fishes is scant relative to the nearly 30,000 species. Therefore, the generalizations made below should be treated with great caution. Detailed information regarding coronary vasculature exists for less than a dozen fish species, which means deductions concerning the coronary circulation are often based on gross ventricular anatomy. As ventricular compaction is always associated with a coronary circulation in adult fishes, as in mammals, the percent compact myocardium in the ventricle is more often than not used as a surrogate measure for coronary development, rather than any direct observation of coronary capillaries. It is for this reason that we prefer the term compact myocardium to be reserved for a vascularized, outer myocardium. Below, we briefly outline cardiac anatomy, coronary arrangements and cardiac oxygen needs in fish before providing a perspective on the evolution of the coronary circulation in fishes. What will emerge is that the arrangements for supplying oxygen to fish hearts are far more diverse, possibly more complex and certainly more mysterious than those of the adult mammalian heart.

Most adult fish have four, separate cardiac chambers contained within the pericardium; a sinus venosus, an atrium, a ventricle and an outflow tract (either a *bulbus arteriosus* or a *conus arteriosus*). Each cardiac chamber differs to some degree in its structure, oxygen demand and oxygen supply route.

The thin-walled sinus venosus has no dedicated coronary circulation in any fish species studied to date. As the simples of the four cardiac chambers it relies on its luminal venous blood for the oxygen supply to its sparse arrangement of cardiac muscle, smooth muscle and specialized cardiac peacemaker fibers.

The next chamber is the atrium. The atrium is more muscular, but because the thin wall generates a relatively low blood pressure, its oxygen demand is relatively low. The atrial myocardium is entirely spongy in most fishes and its trabeculae typically arch across short segments of the wall. Coronary vessels are reported in the atrium of only a few fish species, suggesting a predominantly luminal oxygen supply route for the atrium in the majority of fishes. However, atrial vascularization has not been systematically documented, and so future research should investigate the degree to which coronary vessels penetrate atrial trabeculae, and the role fish size plays in atrial vascularity.

Following the atrium is the single ventricle, which generates the arterial blood pressure in fish. Ventricular mass and wall thickness are up to five times greater than those of the atrium (Farrell and Jones 1992). Fish ventricles show enormous diversity in terms of shape, thickness, composition and oxygen supply among species (Santer 1985). Coronary vessels are present in the ventricle of a large variety of fish species. Tota (1989) provided the first classification of ventricular design in fishes in relation to the coronary circulation, which is now summarized (Fig. 4.3).

An entirely spongy ventricle is called Type I, and a strictly avascular, spongy ventricle is given the subclass Type Ia. The Type Ib subclass is used for a spongy ventricle with a nonpenetrating, superficial vessel. A large proportion of fish species have a Type I ventricle (one claim is 80% of species studied; Santer (1985); Santer and Greer Walker (1980), but our studies place this at closer to 50% of teleost species). Lacking cardiac capillaries, the Type I ventricle has a luminal oxygen supply. In the case of fishes, luminal venous blood has a low oxygen partial pressure, i.e., whatever oxygen the systemic circulation (Fig. 4.1e).

The three remaining ventricle classifications all possess coronary capillaries (Fig. 4.3), which supply at least part of the ventricle with blood having a much higher oxygen partial pressure than the luminal oxygen supply has left in the venous blood! (Fig. 4.1e; see below). Type II has compact and spongy myocardium, with coronary vessels restricted to the compact layer. The Type II designation is very

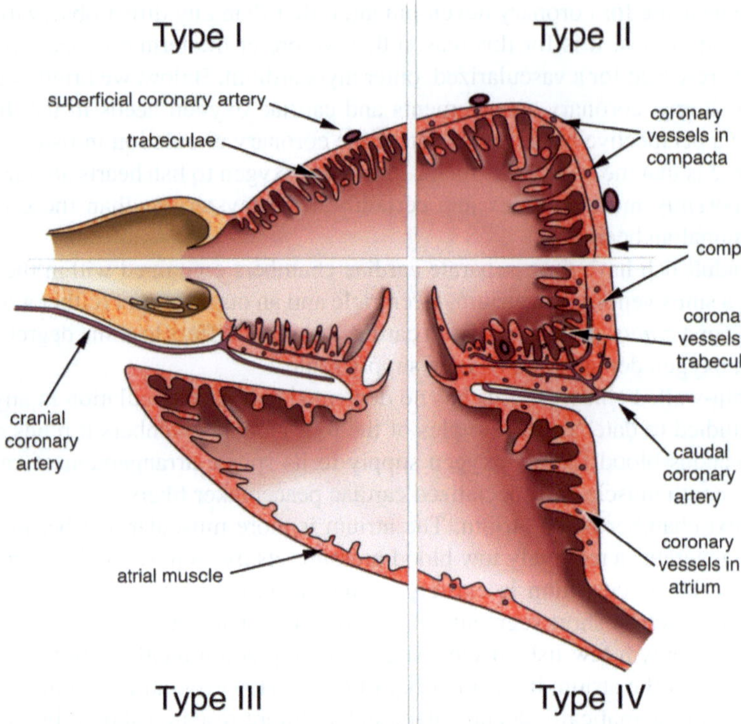

Fig. 4.3 Classification of fish ventricle based on Tota (1989). Type I has a spongy ventricle with no coronary circulation. Type II has compact and spongy myocardium, but coronary vessels are restricted to the compact layer. Types III and IV also have compact and spongy myocardium, but coronary vessels penetrate both layers, and possibly the spongy atrial myocardium. The Type IV has >30% compact myocardium. Original image adapted by Grimes and Kirby (2009) from Davie and Farrell (1991)

similar to the earliest stage of compaction in the developing mammalian heart when coronary vessels are restricted to compact myocardium. The connective tissue layer between the spongy and compact myocardium was once thought to be a barrier for coronary penetration to the trabeculae, but this connective tissue layer is not always well expressed in salmonids (Pieperhoff et al. 2009; Santer 1985). In fact, the compact and spongy myocardia may be so weakly inter-connected that a simple method to estimate the percentage of compact myocardium is to peel off the compact myocardium from the spongy myocardium, much like peeling the rind off an orange (Farrell et al. 2007).

Types III and IV are like Type II, but with coronary vessels penetrating spongy as well as the compact myocardium; vascularized trabeculae are a characteristic of the Type III and Type IV designations (Fig. 4.3). The hearts of Types III and IV are arbitrarily distinguished from one another by Type III having less than 30% heart mass composed of compact myocardium and Type IV having more than 30% total heart mass as compact myocardium. As such, most elasmobranchs fall into the

Type III classification because they vascularize the trabeculae and possess 5–30% compact myocardium. Endothermic sharks and tunas have sufficient compact myocardium to be classified as Type IV. Species such as mako shark, swordfish and tuna have multiple (2–3) layers of compact myocardium (Santer 1985; Sanchez-Quintana and Hurle 1987), but how coronary vessels are arranged relative to this layering is unknown.

To date, vascularization of the spongy myocardium in the fish ventricle remains a physiological curiosity in need of detailed study. Still needed is a characterization of the extent of the vascular connections to vessels in the compact myocardium, how trabecular blood flow is regulated and what proportion of coronary blood flow drains into the lumen rather than the coronary veins. Generally, the vascular architecture of trabeculae has been loosely described as having centrally located arteries, terminal arterioles, capillaries and no venous system (Voboril and Schieble 1970; Ostadal and Schieble 1971; Tota et al. 1983). Yet, these arrangements have never been quantified. Trabecular diameters in elasmobranchs can be quite large (ranging from 18 to 304 μm; Santer 1985), and more than one capillary has been reported in large trabeculae (Tota et al. 1983). Of course, the trabeculae do not have free ends, which raises a question as to the direction of blood flow in them. One possibility is that they drain directly into the ventricular lumen because Thebesian-like vessels have been identified in fish trabeculae (Tota 1989).

In Type II, III and IV ventricles, superficial coronary veins concentrate near the atrio-ventricular canal, where the coronary circulation drains back into. Potentially, any oxygen remaining in the coronary veins could benefit the luminal oxygen supply. However, given that systemic venous return is nearly two orders of magnitude greater than coronary venous return (Farrell 1987; Gamperl et al. 1994; Axelsson and Farrell 1993), this huge dilution is likely to reduce any benefit to a minimum.

In addition to the considerable differences in the oxygen supply routes, relative ventricular mass varies by an order of magnitude among fish. It is ~0.4% body mass in a highly active, endothermic skipjack tuna and just 0.035% body mass in a bottom-dwelling flounder (Farrell 1991). This large difference partially reflects species differences in cardiac work, which has consequences for the coronary supply. For example, the large, thick-walled tuna ventricle generates an arterial blood pressure similar to humans, although its maximum cardiac stroke volume (~1 mL/kg body mass) is not unlike that of other fishes (Farrell and Jones 1992; Brill and Bushnell 2001). Hence, routine cardiac work of a skipjack tuna (26 mW/kg body mass) is an order of magnitude greater than that of a smaller (~0.1% body mass) rainbow trout ventricle (2 mW/kg body mass). Thus, ventricular compaction is greater for tunas because cardiac oxygen needs in fishes are linearly related to cardiac power output (Farrell and Jones 1992), as in mammals (Duncker and Bache 2008). Indeed, compact myocardium accounts for 70% of the big eye tuna ventricle (Brill and Bushnell 2001), the most observed in any fish species (Santer 1985).

In fishes less athletic than tunas, the typically avascular Type I ventricle can be up to eight times smaller (typically 0.5–0.9% of body mass and even smaller for deep-sea fishes; Greer Walker et al. 1985). Antarctic icefishes, which have relative ventricular mass (~0.4% body mass) that rivals tunas, are an exception to the rule that a relatively large ventricle has a coronary circulation. They are unusual in that their thin-walled ventricle holds an enormous blood volume. The massive cardiac stroke volume (10 mL/kg body mass) (Tota et al. 1991; Farrell and Jones 1992) results in an extremely high resting cardiac output that compensates for an absence of red blood cells, a genetic adaptation that reduces blood viscosity in a frigid, near 0°C environment. Even so, cardiac work is low because of a low vascular resistance and arterial blood pressure (Axelsson 2005), which means that the large ventricle of the icefishes can rely on a luminal supply of oxygen dissolved in plasma. Thus, the Type I ventricle of icefishes may have the most precarious cardiac oxygen supply of any vertebrate, but one that supports a sluggish lifestyle in well-oxygenated water where there are relatively few predators.

The contrasting examples of tuna and icefish support the idea that workload of the heart, rather than its mass, is a primary determinant of ventricular compaction in fishes. Even so, cardiac workload is unlikely to be the sole determinant of ventricular compaction because arterial blood pressure does not vary nearly as much as compaction among teleosts, or for that matter between elasmobranchs and teleosts. Therefore, much of the large variation ventricular in compaction among fishes has a genetic basis that is not entirely related to work load.

The outflow tract is the fourth cardiac chamber within the pericardium of all fishes, with the exception of hagfishes (Farrell 2007). In all elasmobranchs, basal teleosts and dipnoans the outflow tract is called the *conus arteriosus*, while in all teleosts it is called a *bulbus arteriosus*. The *conus* and *bulbus* differ structurally in many regards. Importantly, the *conus* has a layer of compact cardiac muscle, which contracts with each heartbeat (Satchell 1991). In most elasmobranchs, conal myocardium covers the entire length of the outflow tract, but in many basal teleosts the conal myocardium occupies only a shorter, posterior portion of the outflow tract. The compact myocardium of the outflow tract, like that of the ventricle, is always associated with a coronary circulation, and the coronary artery is clearly visible on the surface of the outflow tract (see below Fig. 4.6). Coronary blood flow to the conal myocardium has never been measured. Yet, as shown below, conal myocardium represents a critical consideration in the evolution of the coronary circulation. In contrast, the smooth muscle and connective tissues of the outflow tract may be serviced by a *vasa vasorum*, a blood vessel derived from the lumen (Santer 1985).

The Evolution of the Coronary Circulation in Water-Breathing Fishes

Anatomical descriptions of the coronary circulation in water-breathing fishes have existed for well over a century (e.g., Parker 1886; Parker and Davis 1899; Grant and Regnier 1926; O'Donogue and Abbot 1928; Foxon 1950). An evolutionary perspective of the coronary circulation was first provided over 50 years ago, but was based on very few species (Halpern and May 1958). Recent information has been added for elasmobranchs (De Andres et al. 1990, 1992; Tota et al. 1983; Emery et al. 1985), deep-sea fishes (Greer Walker et al. 1985), tunas (Tota 1978; Tota et al.

1983), air-breathing fishes (Graham 1997) and fishes in general (Santer 1985; Sanchez-Quintana and Hurle 1987; Farrell and Jones 1992).

The primary challenge in erecting a phylogenetic pattern for the evolution of the coronary circulation in water-breathing fishes is the sheer number and diversity of species. For most fish species, we do not even know whether a coronary circulation is present or not. There are significant hurdles in determining the presence or absence of coronaries among fishes, which may have become apparent from the previous section. We know that a coronary circulation is always associated with compact myocardium (both for the ventricle and the *conus*). However, a vessel on the surface of the ventricle is not a fool-proof indicator of a functional coronary circulation because superficial vessels do not always penetrate into the myocardium. Conversely, the spongy appearance of a ventricle is not a reliable indicator for the absence of a coronary circulation because fish trabeculae are not always avascular. The coronary arterial plumbing adds another level of complexity to classifying coronary circulations in fishes; arteries reach the heart via anterior (cephalic or cranial) and posterior (caudal or pectoral) routes. Both bring oxygenated blood directly from efferent gill arteries and the former is clearly visible on the surface of *bulbus* or *conus* as a branch from the hypobranchial artery (see below Fig. 4.6). The caudal arrangement is less common among fishes and also is a less obvious branch from the coracoid artery. Furthermore, the two supply routes may be interconnected at the arterial level in some fish, like marlin (Davie 1990). Many mysteries remain regarding the coronary circulation in fishes.

More recently, fixed fish specimens from London's Natural History Museum were dissected by one of us (N.F.) and classified as either possessing a coronary vessel (when an artery was visible either going to or on the surface of the heart), or lacking a coronary (if nothing was visible). While this simple investigative approach allowed observations for at least one species (over 600 species in total) from the majority of the ~480 fish families, it could not distinguish among Type I subtypes. Also, there was the danger that a coronary vessel was missed as a result of poor specimen preservation. We are in the process of verifying and extending this database using more consistent fish preservation techniques. Regardless of that, the molecular phylogenetic tree shown in Fig. 4.4 is revealing. It was created from a small portion of these observations (the required molecular information was available for just 123 species from different families). By combining this molecular tree with existing literature, a phylogenetic framework is provided below (summarized in Fig. 4.5). There is good evidence for the coronary circulation first appearing with the advent of the first jawed vertebrates, the Gnathostomata, and there is good evidence that the coronary circulation also disappeared when certain new teleost lineages evolved.

The Jawless Chordate Ancestors: Hagfishes and Lampreys

If we presume that extant species are representative of ancestral lineages, we can reasonably conclude that the most primitive chambered heart of the vertebrate

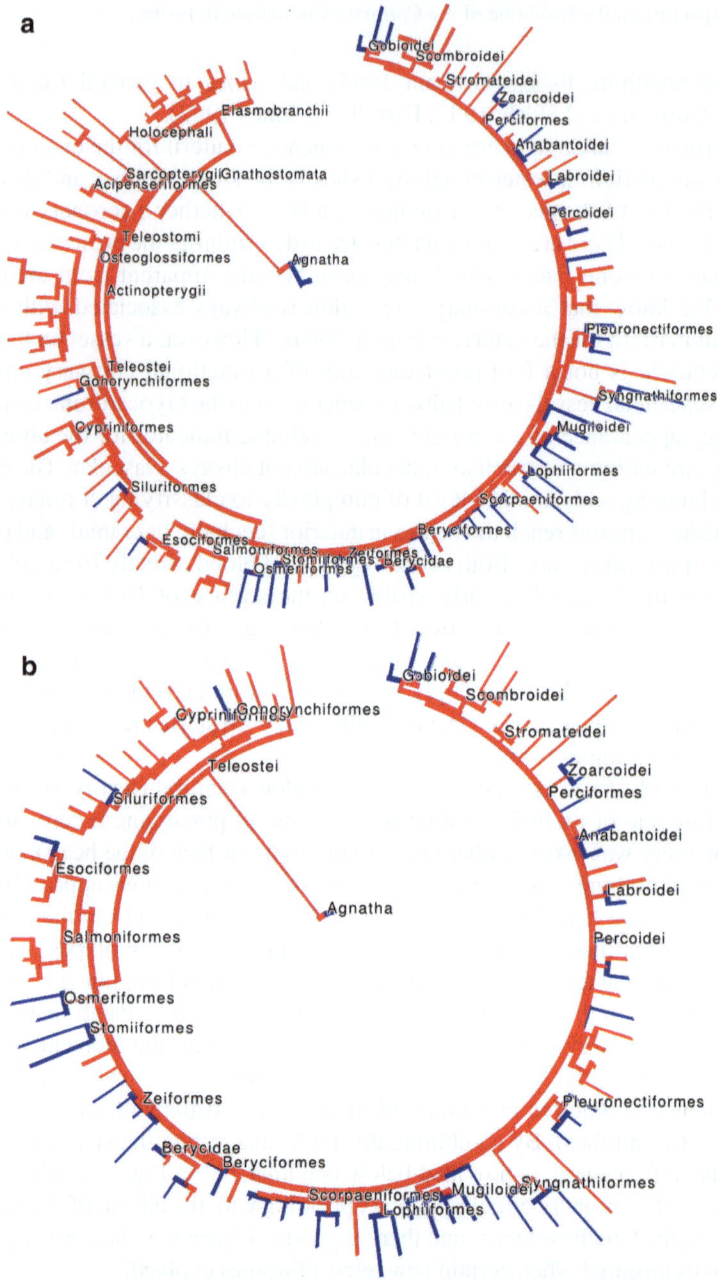

Fig. 4.4 Molecular tree for 123 fish species representing different families to illustrate the presence (*red*) or absence (*blue*) of a coronary artery, which was based on inspection of preserved fish specimens.. The phylogeny was constructed based on molecular sequences available on GenBank using concatenation of cytochrome b and 16S sequences consisting of 400-bp sequences. The concatenated sequences were aligned using ClustalX with default parameters and then imported to PHYML (http://atgc.lirmm.fr/phyml) for maximum likelihood analyses. MEGA software was also used to produce phylogenies with the Neighbor Joining, Maximum Parsimony and Minimum Evolution methods, which yielded similar results that are not shown here

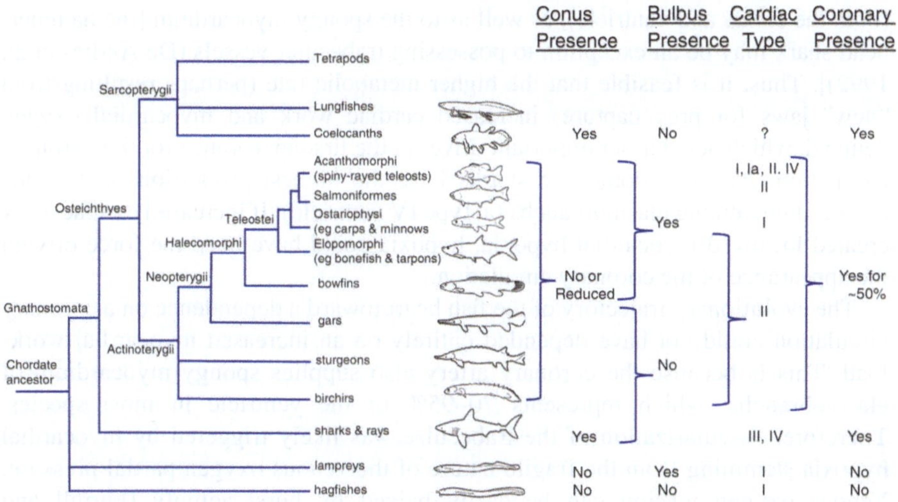

Fig. 4.5 A summary of the presence or absence of a *conus*, *bulbus* and coronary artery as well as the type of ventricle as a function of fish phylogeny

lineage was Type I and avascular. Hagfishes and lampreys are extant representatives of the most primitive chordates, the cyclostomes, which lack vertebrae and a jaw. A coronary circulation has never been observed in any agnathan species (Figs. 4.4 and 4.5). Their ventricles are composed of entirely spongy myocardium.

The agnathan heart is strikingly analogous to the embryonic vertebrate heart. However, while the agnathan heart remains as the last organ to receive oxygen from the circulatory system throughout adult life, the mammalian heart progresses during its embryonic development from being the last organ to the first organ to receive oxygen! While this is fascinating, it would be incorrect to regard the agnathan heart as having remained in a state of suspended development. The evolution of the cardiac system reflects the needs imposed by an animal's lifestyle, and the avascular heart, as represented by extant agnathans, seems to have continued to serve thousands of fish species very well for many millennia.

The Arrival of Jawed Vertebrates: Chimeras, Sharks, Rays and Skates

Today's chondrichthyan fishes are extant representatives of the gnathostomes and we can reasonably conclude that ventricular compaction and coronary vessels emerged early in the vertebrate lineage, probably in association with the appearance of a backbone and a jaw. Every elasmobranch examined to date has been classified as either Type III or Type IV, with a coronary supply to the compact myocardium (both the *conus* and ventricle), as well as to the spongy myocardium [the hammerhead shark may be an exception to possessing trabecular vessels (De Andres et al. 1992)]. Thus, it is feasible that the higher metabolic rate (perhaps resulting from "new" jaws for prey capture) increased cardiac work and myocardial oxygen demand, which acted as an important driver in the first appearance for the coronary circulation. Active endothermic sharks have the highest proportion of compact myocardium among elasmobranchs (a Type IV ventricle). If increased cardiac work created localized myocardial hypoxia, hypoxia could have been the force driving the appearance of the coronary circulation.

The evolutionary trajectory of the fish heart toward a dependence on a coronary circulation could not have depended entirely on an increased myocardial workload. This is because the coronary artery also supplies spongy myocardium of elasmobranchs, which represents 70–95% of the ventricle in most species. Therefore, vascularization of the trabeculae was likely triggered by myocardial hypoxia stemming from the fragile nature of the venous oxygen partial pressure. Venous oxygen tension can be easily halved by burst activity (Farrell and Clutterham 2003) and environmental hypoxia (Holeton and Randall 1967), which then halves the rate of oxygen diffusion into ventricular trabeculae and jeopardizes luminal oxygen supply. Clearly, an obvious remedy for this acute oxygen supply problem is to evolve a Type III ventricle with coronary vessels that supply fully oxygenated blood directly to ventricular trabeculae. This advantage is accentuated whenever myocardial oxygen demand increases during activity (Farmer 1997). What remains unclear is whether or not vascularized trabeculae represent a derived or a primitive condition for the elasmobranchs. If the cardiac ontogeny seen in higher vertebrates truly does recapitulate phylogeny in this regard, wagers might be better placed on the derived condition. Studies on trabecular vascularization in Chimeras, the most primitive extant elasmobranchs, might prove useful in resolving this difficult question.

Given that a hypoxic myocardium feasibly played a central role in driving the evolution of a coronary circulation, we must return to the most primitive chordates, the cyclostomes to ask the question: Why is it that agnathans like the hagfish have remained without a coronary circulation, especially given their propensity toward hypoxic environments and burrowing into rotting carcasses to feed? To make matters worse for hagfish, they have an extremely high blood oxygen affinity (P_{50} = 1–2 torr; Forster et al. 1992; Wells et al. 1986), and so the partial pressure gradient driving oxygen from luminal venous blood into the myocardium is exceptionally low. An answer may lie with the extremely low cardiac work rate of hagfish, which is the lowest of all chordates including elasmobranchs and less than 1/10th that of birds, mammals and tunas (Farrell and Jones 1992; Farrell 1991; Cox et al. 2010a). Such a low myocardial oxygen demand can then be supported by a relatively low oxygen flux from venous blood despite an extraordinarily low partial pressure; their extremely low cardiac work rate can be supported even by anaerobic glycolysis (Forster 1989, 1991, 1992; Cox et al. 2010b; Hansen and Sidell 1983), allowing hagfish to easily switch between normoxic and anoxic environments. Indeed, Pacific hagfish quickly recover from 30 h of anoxia, during which cardiac output and arterial blood pressure remain unchanged (Cox et al. 2010a). With the hagfish heart able to function so well during short-term anoxia, perhaps there was no selective advantage for it to develop a coronary circulation. Conversely, a benefit of cardiac trabeculae is that they are the very first tissue in line to receive the fermentable substrate (glucose) released from the liver whenever hagfish hearts become anoxic and are forced to temporarily perform using anaerobic glycolysis.

The role of conal myocardium in driving the evolution of the coronary circulation of elasmobranchs is largely unknown. It is feasible that myocardial compaction of the conus was an evolutionary predecessor of ventricular compaction as Chimeras have a fully developed, compacted conal myocardium, but only 5–7% of the ventricle is compact myocardium (Santer 1985). Developmental studies and studies examining the proportions of conal to ventricular compact myocardium might be useful in this regard. Venous drainage from the conal myocardium is equally intriguing and another area for further study because superficial coronary veins must travel across the ventricle before draining into the heart (Muñoz-Chapuli et al. 1994).

In summary, the elasmobranchs, the very first vertebrate lineage, were likely the first chordates to possess a coronary circulation. The coronary circulation took many forms during elasmobranch evolution: ventricular compaction is greatest in the more recent endothermic sharks and least in the primitive Chimeras. Yet, we know virtually nothing of coronary physiology in this group. Coronary blood flow has been measured only in an anesthetized shark (Davie and Franklin 1992) and has never been measured as a proportion of cardiac output. However, the coronary vessels have been perfused in working, isolated heart preparations (Davie and Farrell 1991a). Information on the relative distribution of coronary blood flow to the conal, compact and trabecular myocardium of elasmobranchs would greatly inform the debate on the evolution of the coronary circulation.

The Arrival of Boney Fishes: Teleost Fishes

The majority of teleost species lack a coronary circulation (Figs. 4.4 and 4.5; Santer 1985; Santer and Greer Walker 1980). Thus, by accepting that the coronary circulation first appeared in elasmobranchs, we introduce the possibility that certain species in the teleost lineage must have reverted back to an ancestral Type I heart. The molecular phylogenetic tree (Fig. 4.4) suggests that the loss of a coronary circulation could have occurred first among the cyprinid lineage, and then occurred repeatedly in several other teleost families (Fig. 4.4b).

The precise environmental, developmental or physiological factors that drove cardiac evolution in teleosts are not known. We have already suggested that hypoxia and increased myocardial work/oxygen demand played a role in the appearance and expansion of the coronary circulation during elasmobranch evolution. However, the role of environmental hypoxia and cardiac work in relation to the subsequent loss of the coronary circulation in teleosts is enigmatic. The highly successful radiation of teleosts related in part to a greater agility than elasmobranchs (Ilves and Randall

2007), possibly explaining why arterial blood pressure, cardiac power output and cardiac oxygen demand are typically somewhat higher for teleosts compared with elasmobranchs. Yet, most teleosts do not have a coronary circulation to support this higher cardiac work rate. Furthermore, some teleosts, like the flounder, have no coronary circulation yet they tolerate a hypoxic benthic environment. In contrast, hypoxia increases the importance of the coronary circulation for salmonids because hypoxia increases coronary blood flow (Axelsson and Farrell 1993). Examination of far more teleost species is needed before a definitive picture can be drawn of the factors that shaped the evolution of the coronary circulation in the teleost lineage.

Despite this uncertainty regarding the evolutionary loss of the coronary circulation among teleosts, cardiac work has remained a very important driver for the degree of ventricular compaction and for coronary blood flow. The very high proportion of ventricular compaction in tunas and anchovy attests to this (Fig. 4.2). Furthermore, when salmon swim and increase cardiac work rate over threefold, coronary blood flow increases up to threefold (Gamperl et al. 1994, 1995; Axelsson and Farrell 1993). In fact, the tuna heart may even be obligatorily dependent on its coronary circulation, unlike other fishes (Farrell 1996; Farrell et al. 1992), likely because arterial blood pressures approach those of mammals and a very high cardiac output supports a high metabolic rate (Brill 1996; Korsmeyer et al. 1996; Brill and Bushnell 2001).

Unlike elasmobranchs, teleosts with a coronary circulation, such as the salmonid and pike lineages, have characteristically lost vascularized trabeculae (the Type IV ventricles of marlin and tunas are exceptions). This important evolutionary transition to a Type II ventricle likely occurred more than once (Fig. 4.4b). Also, compact myocardium was essentially lost from the outflow tract with the development of a *bulbus arteriosus* in teleosts. As a result, the coronary artery typically becomes a single, cephalad superficial vessel running along the length ventral aorta within the pericardium of teleosts (Figs. 4.6 and 4.7). However, basal teleosts show important transitional forms in the design of the outflow tract.

Basal Teleosts: Sturgeons, Polypterids, Bowfins, Garfish and Tarpons

Observations on basal teleosts, which were abundant lineages that predated the teleosts, lend additional support to the idea that the coronary circulation was lost in the teleost lineage after first appearing in the Gnathostomata. All of today's basal teleosts possess a coronary artery (Farrell 2007). Included in this list are water-breathing sturgeons, as well as polypterids, bowfin, tarpons (Fig. 4.6) and garfish (Fig. 4.7), which also breathe air, a lifestyle that is considered further in the next section.

Conal myocardium features prominently but variably among basal fishes. Conal myocardium covers the entire outflow tract in sturgeon and polypterids, appearing as early as 4 days posthatch for sturgeon (Guerrero et al. 2004). However, it is found

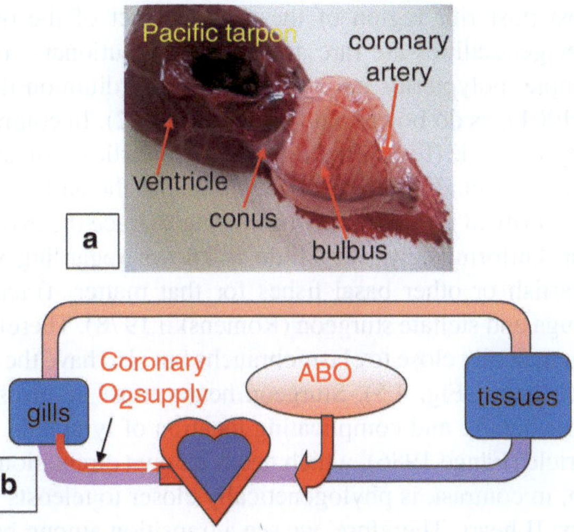

Fig. 4.6 (**a**) A photograph of the ventricle and outflow tract of the Pacific tarpon, illustrating that the *bulbus* occupies almost the entire outflow tract and the very reduced *conus* is barely visible near the ventricle, but a cephalic and divided coronary artery is clearly visible lying on the surface of the *bulbus*. Photograph courtesy of Roger Seymour. (**b**) A cartoon representation of the oxygen supply to the heart of the tarpon, illustrating the importance of the air-breathing organ (ABO) in creating a more secure luminal oxygen supply for the heart

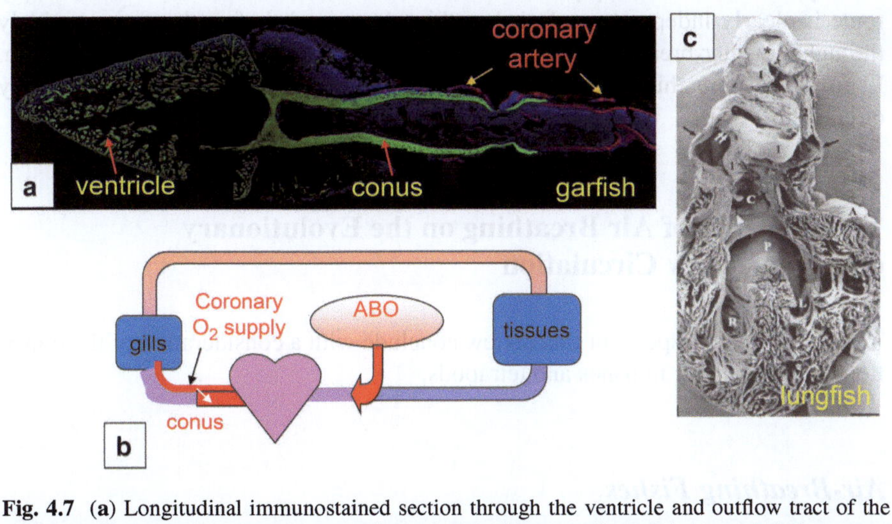

Fig. 4.7 (**a**) Longitudinal immunostained section through the ventricle and outflow tract of the garfish to illustrate the spongy myocardium, the conal myocardium and the cephalad coronary artery. Image taken using a confocal laser scanning microscope with META module and modified from Zaccone et al. (2011). (**b**) A cartoon representation of the oxygen supply to the heart of the garfish, illustrating the importance of the air-breathing organ (ABO) in creating a more secure luminal oxygen supply for the heart. (**c**) Scanning electron micrograph of a longitudinal sectioned ventricle of the African lungfish heart to illustrate its spongy nature. Image taken from Icardo et al. (2008)

only in the most posterior region of the outflow tract of the other basal groups (Fig. 4.7). Other generalities are rare among this evolutionary important group of fishes. For example, polypterids have compact myocardium on the conus and ventricle (Budgett 1901), as do bowfin and tarpon (Fig. 4.2). In contrast, garfish have a distinctly spongy ventricle (Fig. 4.7) along with a cephalic coronary artery and conal myocardium (Zaccone et al. 2011). Knowing whether the garfish ventricle is vascular or avascular is critical information when trying to trace the evolution of the coronary circulation. Unfortunately, very little is known regarding vascularization of trabeculae in garfish or other basal fishes for that matter. Trabecular vessels are reported for beluga and stellate sturgeon (Romenskii 1978). Therefore, sturgeons are not only phylogenetically close to elasmobranchs but also have the prototypical Type III elasmobranch heart (Fig. 4.5). Sturgeon hearts must be investigated further to explore the rather bizarre and complicating location of lymphoid tissue on the outside of the ventricle (Fänge 1986), which needs its own circulation to function properly. The tarpon, in contrast, is phylogenetically closer to teleosts (Fig. 4.5) and has a teleostean Type II heart. Therefore, we see a transition among basal teleosts away from the primitive Type III ventricle toward a Type II ventricle, as well as the loss of conal myocardium. Consequently, the overall evolutionary trend among fishes has been toward the heart becoming less dependent on a coronary circulation.

The anatomical direction of the coronary arterial supply also varies among the basal teleosts. Paddlefish have a cephalic arterial supply, shovelnose sturgeon and garfish have both cephalic and caudal arterial supplies, and polypterids have an entirely caudal arterial supply (Graham 1997). Thus, a dual coronary arterial supply route is clearly independent of air breathing because it is found scattered widely throughout water-breathing fish lineages from rays through basal teleosts to marlin. Even so, air breathing had a major impact on the evolution of the coronary circulation.

The Influence of Air Breathing on the Evolutionary of the Coronary Circulation

The phylogenetic aspects of this review conclude with a consideration of the transition to air breathing in fishes and tetrapods.

Air-Breathing Fishes

Air breathing evolved many times among fishes (Graham 1997). At last count, 374 species from 49 known families are air breathers, which makes the possibility of generalities seem remote. Yet, an important consequence of air breathing for almost all of the eight basic circulatory patterns found in these fishes (Graham 1997) is that systemic venous blood becomes better oxygenated because of either extracardiac or intracardiac mixing with oxygenated blood from the air-breathing organ or lung (Figs. 4.6 and 4.7). A higher venous partial pressure of oxygen would then facilitate a faster rate of oxygen diffusion to cardiac trabeculae. Thus, the air-breathing organ results in a more secure luminal oxygen supply for the hearts of almost all air-breathing fishes.

The piscine coronary circulation likely predates air breathing in fishes given that there are no known air-breathing elasmobranchs and all extant elasmobranchs have a coronary supply to the entire ventricle and conal muscle. Thus, an important question is: How did improving the luminal oxygen supply with air breathing influence the coronary circulation? Graham (1997) wisely counseled caution when trying to answer this question given the lack of detailed information on the coronary circulation in air-breathing fishes. Nevertheless, Farmer (1997, 1999) proposed that air breathing evolved to supply the fish heart with oxygen under exercise conditions, a suggestion supported by the observation of increased air breathing frequency when garfish swim faster (Shipman 1989). A more circumspect view is needed when other air-breathing fish species are considered. For example, Pacific tarpon, a prized and athletic sportfish, have a Type II ventricle with well-developed compact myocardium (Fig. 4.2). Yet, they do not breathe air at all when they swim in normoxic water. Although they increase air-breathing frequency when they are in hypoxic water, the frequency of air breathing does not increase any further when they swim in hypoxic water (Seymour et al. 2007). Clearly, the Type II ventricle is adequate for the exercising tarpon, provided it is not in hypoxic water. Another is that many air-breathing fishes have conal compact myocardium, which needs a coronary circulation to supply oxygen independent of any exercise state. Furthermore, the air-breathing organ of Anabantoids and the electric eel provides no direct benefit to the heart because the oxygenated blood from the air-breathing organ goes directly into the systemic circulation. In these species the air-breathing organ lies in series with the gills rather than in parallel with the gills as in most air-breathing fish. Thus, Farmer's suggestion (Farmer 1999) clearly has a restricted application. Graham (1997) suggests that air breathing has affected coronary circulation patterns. While air breathing has created a more secure luminal oxygen supply for the heart, it clearly has not supplanted the need for a coronary circulation. The benefit of a coronary circulation is no better seen than in the African and South American lungfishes. Their cephalic coronary circulation is derived from filament-free anterior gill arches that are preferentially perfused with oxygenated blood from the pulmonary circulation. For all other air-breathing fishes, and even in the Australian lungfish, the potential for oxygen loss at the gills during aquatic hypoxia could easily diminish the ability of air breathing to deliver fully oxygenated blood into the coronary artery.

Lungfishes and coelacanths, which are the closest extant relatives to the tetrapod lineage, resemble basal teleosts in two respects: the ventricle is remarkably spongy (Fig. 4.7c; Millot et al. 1978; Farrell 2007) and a cephalic coronary artery services a reduced conal myocardium en route to the ventricle. A curious observation is that the lungfish ventricle is remarkably spongy (Fig. 4.7c). The apparent absence of

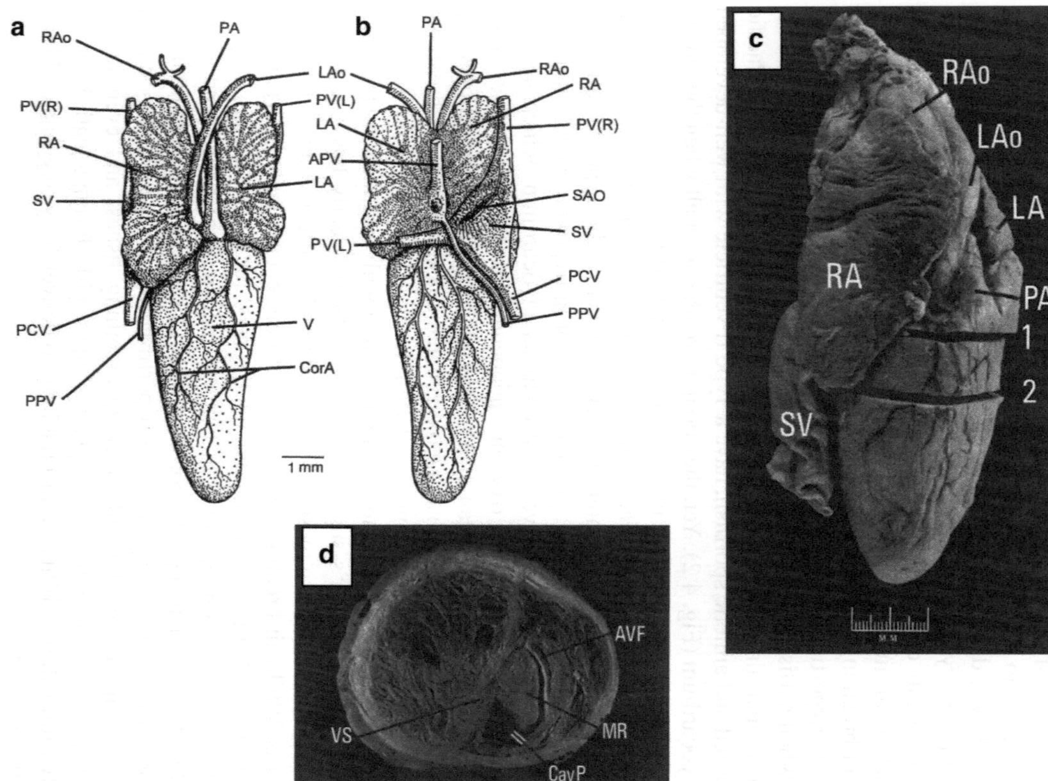

Fig. 4.8 (**a**) and (**b**) Hand-drawn illustrations of the worm snake heart and its vasculature to illustrate that the coronary artery arises at the root of the aorta. (**c**) A photograph of a heart from a python. (**d**) A view of the python ventricle sectioned at "2" to illustrate the outer layer of compact myocardium. Images taken from Farrell et al. (1998)

compact myocardium in lungfishes again raises an unresolved question as to the degree to which coronary vessels penetrate ventricular trabeculae. Also, as the right atrium and part of the ventricle receives predominantly systemic venous return, it would be interesting to learn if these parts of the heart have an enhanced coronary circulation, as seen in some reptiles (Farrell et al. 1998).

Tetrapods

Invasion of the terrestrial environment by early tetrapods provided mixed blessings compared with the aquatic environment: a greater concentration of oxygen in the respiratory medium, but a desiccating environment with potentially wider temperature fluctuations. Amphibians continue to sit on the fence, breathing air like other tetrapods yet requiring a moist epidermis like their piscine predecessors. The amphibian heart has neither compact myocardium, nor coronary vessels that penetrate the trabecular myocardium. Instead, the left atrium and a part of the ventricle benefit from oxygenated pulmonary venous return for the luminal oxygen supply. The right atrium and remainder of the ventricle can similarly benefit from cutaneous respiration. In addition, the relatively low workload and slow heart rate of the amphibian heart keep myocardial oxygen demand relatively low. While amphibians provide the first extant vertebrate example of air breathing supplanting the need for a functional coronary circulation, a nonpenetrating coronary vessel, with its origin at the root of the outflow tract, is still visible on the surface of the amphibian ventricle.

The reptilian ventricle has compact myocardium and a coronary artery derived at the root of the aorta, as in mammals (Fig. 4.8). Coronary phylogeny among reptiles was summarized in the work of Farrell et al. (1998), but the vascularization of spongy myocardium has not been studied. New information includes an association between the extensive compact myocardium of the python heart and its ability to generate high systemic blood pressures (Jensen et al. 2010). A second important discovery is that ligation of the coronary artery does not kill rattlesnakes (Hagensen et al. 2008), a finding that segues into the final section of this review.

Why Retain a Trabecular Myocardium?

This review concludes with a brief perspective on the advantages that a trabecular cardiac design might offer. A spongy heart is the evolutionary and developmental forerunner for cardiac design among chordates. Furthermore, no lower vertebrate has evolved the level of ventricular compaction seen in adult mammals, nor do they exhibit the same dependence on coronary circulation as adult mammals. In fact, all fish hearts are comprised of at least 30% and often 100% spongy myocardium. Thus, a ventricle comprised predominantly of spongy myocardium has provided some form of selective advantage for all fish species. So why is it that all fish, amphibians and reptiles rarely develop more than 50% compact myocardium in their ventricle, and why did some fishes revert back to (or retain) the ancestral spongy myocardium. Inasmuch as the forces underlying early trabeculation of the avian and mammalian myocardium are unclear (Sedmera et al. 2000), so are those that underlie the evolutionary retention of a trabecular myocardium.

A case has been made that the first chordates did not need a coronary circulation because they frequented anoxic environments and their heart's workload is low relative to its capacity for anaerobic glycolysis. A suggestion was also made that under anoxic conditions cardiac trabeculae benefit from being the very first tissue to receive glucose released from the liver. Elasmobranchs then evolved compact myocardium along with coronary vessels distributed throughout the myocardium. Exploitation of hypoxic environments by elasmobranchs could have been important in this transition because the partial pressure of oxygen in the coronary circulation still remains higher than in venous blood during hypoxia, even if both are reduced. However, this coronary arrangement was not retained in many teleosts.

Johansen (1965) advanced a biophysical benefit of spongy myocardium, suggesting that trabeculae created thousands of "miniature hearts," which could then generate higher wall tensions because of the Laplacian relationship between wall tension and radius of a sphere. However, those lower vertebrates that generate high blood pressures also have developed an extensive compact myocardium.

Instead, a spongy heart may offer the advantage of a very low end-systolic blood volume because trabeculae that arc across wall segments can bring opposing walls close together with their contractions. Indeed, the cardiac ejection fraction for rainbow trout is 90–95% (Franklin and Davie 1992) versus 55–70% for a human. A low anatomical dead space resulting from a high ejection fraction then allows for almost complete refilling with fresh venous blood, which keeps venous oxygen partial pressure as high as possible with each heartbeat. Perhaps, for a given wall tension, a sponge can squeeze itself more empty than a cardiac chamber.

Despite these advantages, cardiac growth in fish presents a difficulty for trabeculae because they must get thicker and branching is finite. Thicker trabeculae lead to a slower rate of oxygen diffusion to the myocytes contained in them. There are several possible solutions to this problem. The evolutionary solution used by all elasmobranchs is to vascularize trabeculae. Teleosts have adopted another solution, creating trabecular sheets rather than uniform tubes of muscle, such as those observed in salmonids (Fig. 4.1b) (Pieperhoff et al. 2009). A muscular sheet offers the benefit of shorter diffusion distances while simultaneously increasing muscle cross-section area for tension development (sheet-like trabeculae are even seen in the developing left ventricle of chickens, increasing the surface area to volume ratio; Sedmera et al. 2000). Although, no one has studied the potential benefits of an eccentric trabecular cross-section for lower vertebrates, Goo et al. (2009) demonstrated that the eccentricity of rat trabeculae carneae provided a considerable benefit for oxygenation. They also showed that while small trabeculae were avascular, a capillary served about two myocytes in wider trabeculae, a ratio that is about half of that in the free wall of the ventricle. Sheet-like trabeculae can have the added advantage of possibly directing blood flow within a single ventricular chamber, perhaps helping to keep oxygenated and deoxygenated streams of blood separated, much like the cava of reptilian hearts, but this idea needs experimental verification. In terms of the minimum venous oxygen partial pressure for adequate diffusion, Davie and Farrell (1991b) suggested a range of 6–10 torr for elasmobranchs and teleosts, based on a very limited database. Farrell and Clutterham (2003) showed a minimum venous oxygen partial pressure when rainbow trout were swimming maximally and proposed this minimum was important for maintaining luminal oxygen delivery. Obviously, cardiac workload and blood oxygen affinity have important consequences for minimum venous oxygen partial pressure across species. Venous oxygen partial pressure could increase and cardiac oxygen demand could decrease to offset an increase in trabecular thickness. While, neither possibility has been studied, a decrease in the oxygen affinity of blood with warming is favorable in association with myocardial oxygen demand also increasing.

In addition to the potential benefits of retaining trabeculae, switching completely to compact myocardium carries potential disadvantages. While mammalian hearts are obligatorily dependent on the coronary circulation, other vertebrates are not nearly as sensitive to an interruption in the coronary oxygen supply. For example, coronary artery ablation or ligation does not kill rattlesnakes (Hagensen et al. 2008) or salmonids (Daxboeck 1982; Farrell and Steffensen 1987; Gamperl et al. 1994). In fact, the coronary-ligated salmonids can swim, albeit with a lower blood pressure, even in hypoxic water (Farrell and Steffensen 1987). Thus, a coronary circulation in the lower vertebrate heart appears to only supplement the cardiac oxygen supply. Perhaps this is why adult salmon can still migrate upstream to their spawning grounds in high velocity rivers despite developing coronary arteriosclerosis (Van Citters and Watson 1967; Saunders et al. 1992).

Summary

New information has been integrated with existing literature to provide a framework for the evolution of the coronary circulation in both water- and air-breathing fishes. Throughout we acknowledge that data are limited and so conclusions needed to be treated with caution. Also, the reader is recommended to explore more detailed information on specific aspects of morphology and innervation of the fish heart in Santer's (1985) superb review, on air-breathing fishes in Jeff Graham's (1997) monograph, on primitive fishes (McKenzie et al. 2007), on oxygen gradients in fish hearts (Davie and Farrell 1991), on cardiac classification in fish (Tota 1983) and on the development and the phylogenetic differences in the design of the outflow tract of fishes (Santer 1985; Bushnell et al. 1992; Icardo et al. 1999; Grimes and Kirby 2009).

A number of important points emerge, provided we accept that extant species and developmental patterns are a reasonable guide for speculation on the evolutionary emergence of the coronary circulation. (a) An avascular, spongy ventricle was the most primitive form of chambered heart. (b) The first vertebrates used a fully vascularized ventricle with two types of cardiac muscle, compact and spongy.

They vascularized the compact myocardium on the outflow tract as well. Thus, in terms of the association between coronary development and myocardial compaction, early phylogeny is recapitulated in the ontogeny of birds and mammals. However, any semblance of recapitulation becomes lost among teleosts, with a reversion to an avascular Type I heart and the loss of trabecular vessels in Type II hearts. (c) Classifying fish ventricles as I to IV remains very useful, but evolution clearly did not follow this numerical order because, based on extant species, Type III predates Type II. (d) A Type IV ventricle is a modern refinement, satisfying an increased myocardial oxygen demand associated in part with regional endothermy and supporting the idea that the degree of ventricular compaction is linked to workload of the heart. (e) Why an entirely or predominantly trabecular heart dominates the lower vertebrate world is still a mystery, and a great curiosity in this regard is whether or not the spongy hearts of basal teleosts and lungfishes are vascularized.

Acknowledgments We are deeply indebted to the courtesy shown to us by the staff of the Natural History Museum of London, UK, who provided open access to their fish collection and especially Oliver Crimmen, Patrick Campbell, James Maclain, and Ralf Britz. This work was by funded by the Natural Sciences and Engineering Council of Canada.

References

Angelini A, Melancini P, Barbero F, Thiene G (1999) Evolutionary persistense of spongy myocardium in humans. Circulation 99:2475

Axelsson M (2005) The circulatory system and its control. In: Farrell AP, Steffensen JF (eds) Physiology of polar fishes. Fish physiology, vol 22. Elsevier, San Diego

Axelsson M, Farrell AP (1993) Coronary blood flow *in vivo* in the coho salmon (*Oncorhynchus kisutch*). Am J Physiol 264:R963–971

Brill RW (1996) Selective advantages conferred by the high performance physiology of tunas, bilfishes, and dolphin fish. Comp Biochem Physiol 113A:3–15

Brill RW, Bushnell PG (2001) Tuna metabolism and energetics. In: Block BA, Stevens ED (eds) Tuna: physiology, ecology and evolution. Fish physiology, vol 19. Academic, San Diego, pp 79–120

Budgett JS (1901) On some points in the anatomy of *Polypterus*. Trans Zool Soc Lond 15:323–341

Bushnell P, Jones DR, Farrell AP (1992) The arterial system. In: Hoar WS, Randall DJ, Farrell AP (eds) Fish physiology, vol 12A. Academic, New York, pp 89–139

Cox GK, Sandblom E, Farrell AP (2010a) Cardiac responses to anoxia in the Pacific hagfish, *Eptatretus stoutii*. J Exp Biol 213:3692–3698

Cox GK, Sandblom E, Richards JG, Farrell AP (2010b) Anoxic survival of the Pacific hagfish (*Eptatretus stoutii*). J Comp Physiol B 181:361–371

Davie PS (1990) Pacific Marlins. Anatomy and Physiology. Massey University Printery, Palmerston North, New Zealand

Davie PS, Farrell AP (1991a) Cardiac performance of an isolated heart preparation from the dogfish (*Squalus acanthias*): the effects of hypoxia and coronary artery perfusion. Can J Zool 69:1822–1828

Davie PS, Farrell AP (1991b) The coronary and luminal circulations of the myocardium of fishes. Can J Zool 69:1993–2001

Davie PS, Franklin CE (1992) Myocardial oxygen consumption and mechanical efficiency of a perfused dogfish heart preparation. J Comp Physiol B 162:256–262

Daxboeck C (1982) Effect of coronary artery ablation on exercise performance in *Salmo gairdneri*. Can J Zool 60:375–381

De Andres AV, Muñoz-Chapuli R, Sans-Coma V, Garcia-Garrido L (1990) Anatomical studies of the coronary system in elasmobranchs: I. Coronary arteries in lamnoid sharks. Am J Anat 187:303–310

De Andres AV, Muñoz-Chapuli R, Sans-Coma V, Garcia-Garrido L (1992) Anatomical studies of the coronary system in elasmobranchs: II. Coronary arteries in hexanchoid, squaloid, and carcharhinoid sharks. Anat Rec 233:429–439

Duncker DJ, Bache RJ (2008) Regulation of coronary blood flow during exercise. Physiol Rev 88:1009–1086

Emery SH, Mangano C, Randazzo V (1985) Ventricle morphology in pelagic elasmobranch fishes. Comp Biochem Physiol A 82:635–643

Fänge R (1986) Lymphoid organs in sturgeons (Acipenseridae). Vet Immunol Immunop 12:153–161

Farmer C (1997) Did lungs and the intracardiac shunt evolve to oxygenate the heart in vertebrates? Paleobiology 23:358–372

Farmer CG (1999) Evolution of the vertebrate cardio-pulmonary system. Annu Rev Physiol 61:573–592

Farrell AP (1987) Coronary flow in a perfused rainbow trout heart. J Exp Biol 129:107–123

Farrell AP (1991) From hagfish to tuna: a perspective on cardiac function in fish. Physiol Zool 64:1137–1164

Farrell AP (1996) Features heightening cardiovascular performance in fishes, with special reference to tunas. Comp Biochem Physiol A 113:61–67

Farrell AP (1997) Evolution of cardiovascular systems: insights into ontogeny. In: Burggren WW, Keller B (eds) Development of the cardiovascular system: molecules to organisms. Cambridge University Press, Cambridge, pp 101–113

Farrell AP (2007) Cardiovascular systems in primitive fishes. In: McKenzie DJ, Farrell AP, Brauner CJ (eds) Primitive fishes. Fish physiology, vol 26. Academic, San Diego, pp 53–120

Farrell AP, Clutterham SM (2003) On-line venous oxygen tensions in rainbow trout during graded exercise at two acclimation temperatures. J Exp Biol 206:487–496

Farrell AP, Jones DR (1992) The heart. In: Hoar WS, Randall DJ, Farrell AP (eds) Fish physiology, vol 12A. Academic, San Diego, pp 1–88

Farrell AP, Stecyk JA (2007) The heart as a working model to explore themes and strategies for anoxic survival in ectothermic vertebrates. Comp Biochem Physiol A 147:300–312

Farrell AP, Steffensen JF (1987) Coronary ligation reduces maximum sustained swimming speed in Chinook salmon, *Oncorhynchus tshawytscha*. Comp Biochem Physiol A 87(1):35–37

Farrell AP, Hammons AM, Graham MS, Tibbits GF (1988) Cardiac growth in rainbow trout, *Salmo gairdneri*. Can J Zool 66:2368–2373

Farrell AP, Davie PS, Franklin CE, Johansen JA, Brill RW (1992) Cardiac physiology in tunas I: in vitro perfused heart preparations from yellowfin and skipjack tunas. Can J Zool 70:1200–1210

Farrell AP, Gamperl AK, Francis ETB (1998) Comparative aspects of heart morphology. In: Gans C, Gaunt AS (eds) Biology of the Reptilia, vol 19. Society for the Study of Amphibians and Reptiles, Ithaca, pp 375–424

Farrell AP, Simonot DL, Seymour RS, Clark TD (2007) A novel technique for estimating the compact myocardium in fishes reveals surprising results for an athletic air-breathing fish, the Pacific tarpon. J Fish Biol 71:389–398

Farrell AP, Eliason EJ, Sandblom E, Clark TD (2009) Fish cardiorespiratory physiology in an era of climate change. Can J Zool 87:835–851

Forster ME (1989) Performance of the heart of the hagfish, *Eptatretus cirrhatus*. Fish Physiol Biochem 6:327–331

Forster ME (1991) Myocardial oxygen consumption and lactate release by the hypoxic hagfish heart. J Exp Biol 156:583–590

Forster ME, Davison W, Axelsson M, Farrell AP (1992) Cardiovascular responses to hypoxia in the hagfish, *Eptatretus cirrhatus*. Respir Physiol 88:373–386

Foxon GE (1950) A description of the coronary arteries in dipnoan fishes and some remarks on their importance from the evolutionary standpoint. J Anat 84:121–131

Franklin CE, Davie PS (1992) Dimensional analysis of the ventricle of an in situ perfused trout heart using echocardiography. J Exp Biol 166:47–60

Gamperl AK, Farrell AP (2004) Cardiac plasticity in fishes: environmental influences and intraspecific differences. J Exp Biol 207:2539–2550

Gamperl A, Pinder A, Grant R, Boutilier R (1994) Influence of hypoxia and adrenaline administration on coronary blood flow and cardiac performance in seawater rainbow trout (*Oncorhynchus mykiss*). J Exp Biol 193:209–232

Gamperl AK, Axelsson M, Farrell AP (1995) Effects of swimming and environmental hypoxia on coronary blood flow in rainbow trout. Am J Physiol 269:R1258–1266

Goo S, Joshi P, Sands G, Gerneke D, Taberner A, Dollie Q, LeGrice I, Loiselle D (2009) Trabeculae carneae as models of the ventricular walls: implications for the delivery of oxygen. J Gen Physiol 134:339–350

Graham JB (1997) Air-breathing fishes. Evolution, diversity and adaptation. Academic, San Diego

Grant RT, Regnier M (1926) The comparative anatomy of the cardiac coronary vessels. Heart J Stud Circ 8:285–317

Greer Walker M, Santer RM, Benjamin M, Norman D (1985) Heart structure of some deep-sea fish (Teleostei: Macrouridae). J Zool Lond A 205:75–89

Grimes AC, Kirby ML (2009) The outflow tract of the heart in fishes: anatomy, genes and evolution. J Fish Biol 74:983–1036

Guerrero A, Icardo JM, Duran AC, Gallego A, Domezain A, Colvee E, Sans-Coma V (2004) Differentiation of the cardiac outflow tract components in alevins of the sturgeon *Acipenser naccarii* (Osteichthyes, Acipenseriformes): implications for heart evolution. J Morphol 260:172–183

Hagensen MK, Abe AS, Falk E, Wang T (2008) Physiological importance of the coronary arterial blood supply to the rattlesnake heart. J Exp Biol 211:3588–93

Halpern MH, May MM (1958) Phylogenetic study of the extracardiac arteries to the heart. Am J Anat 102:469–481

Hansen CA, Sidell BD (1983) Atlantic hagfish cardiac muscle: metabolic basis of tolerance to anoxia. Am J Physiol 244:R356–R362

Holeton GF, Randall DJ (1967) The effect of hypoxia upon the partial pressure of gases in the blood and water afferent and efferent to the gills of rainbow trout. J Exp Biol 46:317–327

Hu N, Sedmera D, Yost HJ, Clark EB (2000) Structure and function of the developing zebrafish heart. Anat Rec 260:148–157

Icardo JM (1990) Development of the outflow tract. A study in hearts with situs solitus and situs inversus. Ann NY Acad Sci 588:26–40

Icardo JM (1996) Developmental biology of the vertebrate heart. J Exp Zool 275:144–161

Icardo JM, Colvee E, Cerra MC, Tota B (1999) Bulbus arteriosus of the Antarctic teleosts. I. The white-blooded Chionodraco hamatus. Anat Rec 254:396–407

Icardo JM, Colvee E, Cerra MC, Tota B (1999) Bulbus arteriosus of the Antarctic teleosts. II. The red-blooded Trematomus bernacchii. Anat Rec 256:116–126

Icardo JM, Guerrero A, Duran AC, Domezain A, Colvee E, Sans-Coma V (2004) The development of the sturgeon heart. Anat Embryol 208:439–449

Icardo JM, Amelio D, Garofalo F, Colvee E, Cerra MC, Wong WP, Tota B, Ip YK (2008) The structural characteristics of the heart ventricle of the African lungfish *Protopterus dolloi*: freshwater and aestivation. J Anat 213:106–119

Icardo JM, Guerrero A, Duran AC, Colvee E, Domezain A, Sans-Coma V (2009) The development of the epicardium in the sturgeon *Acipenser naccarii*. Anat Rec 292:1593–1601

Ilves KL, Randall DJ (2007) Why have primitive fishes survived? In: McKenzie DJ, Farrell AP, Brauner C (eds) Primitive fishes. Fish physiology, vol 26. Academic, San Diego, pp 515–536

Jensen B, Nielsen JM, Axelsson M, Pedersen M, Lofman C, Wang T (2010) How the python heart separates pulmonary and systemic blood pressures and blood flows. J Exp Biol 213: 1611–1617

Johansen K (1965) Cardiovascular dynamics in fishes, amphibians, and reptiles. Ann NY Acad Sci 127:414–442

Korsmeyer KE, Dewar H, Lai NC, Graham JB (1996) The aerobic capacity of tunas: adaptation for multiple metabolic demands. Comp Biochem Physiol A 113:17–24

McKenzie DJ, Farrell AP, Brauner CJ (eds) (2007) Primitive fishes. Fish physiology, vol 26. Academic, San Diego

Millot J, Anthony J, Robineau D (1978) Anatomie de *Latimeria chalumnae*. Tome 3, Appareils digestifs, respiratoire, urogenetal, circulatoire, glandes endocrine, teguments, ecailles, conclusions generales. Paris, Editions de Centre National de Rech. Scient., 198 pp

Muñoz-Chapuli R, De Andres AV, Dingerkus G (1994) Coronary artery anatomy and elasmobranch phylogeny. Acta Zool Stockholm 75:249–254

O'Donogue CH, Abbot E (1928) The blood vascular sustem of the spiny dogfish *Squlaus acanthias* and *Squalus sucklii.* Trans R Soc Edinb 55:823–890

Ostadal B, Schieble Th (1971) Terminal blood bed in heart of fish. Z Anat Entwicklungs 134:101–110

Parker TJ (1886) On the blood vessels of *Mustelus anataricus*. Phil Trans R Soc Lond Biol 177:685–732

Parker GH, Davis FK (1899) The blood vessels of the heart of *Carachias*, *Raja* and *Amia*. Proc Boston Soc Nat Hist 29:163–178

Pieperhoff S, Bennett W, Farrell AP (2009) The intercellular organization of the two muscular systems in the adult salmonid heart, the compact and the spongy myocardium. J Anat 215:536–547

Poupa O, Gesser H, Jonsson S, Sullivan L (1974) Coronary supplied compact shell of ventricular myocardium in salmonids: growth and enzyme pattern. Comp Biochem Physiol A 48:85–95

Pratt FH (1898) The nutrition of the heart through the vessels of Thebesius and coronary veins. Am J Physiol 1:86–103

Romenskii O (1978) Blood supply of the compact and spongy myocardium of fish, amphibia and reptiles. Arkh Anat Gistol Embriol 75:91–95

Sanchez-Quintana D, Hurle JM (1987) Ventricular myocardial architecture in marine fishes. Anat Rec 217:263–273

Santer RM (1985) Morphology and innervation of the fish heart. Adv Anat Embryol Cell Biol 89:1–99

Santer RM, Greer Walker M (1980) Morphological studies on the ventricle of teleost and elasmobranch hearts. J Zool 190:259–272

Satchell GH (1991) Physiology and form of fish circulation. Cambridge University Press, Cambridge

Saunders RL, Farrell AP, Knox DE (1992) Progression of coronary arterial lesions in Atlantic salmon (*Salmo salar*) as a function of growth rate. Can J Fish Aquatic Sci 49:878–884

Sedmera D, Watanabe M (2006) Growing the coronary tree: the quail saga. Anat Rec A 288:933–935

Sedmera D, Pexieder T, Vuillemin M, Thompson RP, Anderson RH (2000) Developmental patterning of the myocardium. Anat Rec 258:319–337

Seymour RS, Farrell AP, Christian K, Clark TD, Bennett MB, Wells RM, Baldwin J (2007) Continuous measurement of oxygen tensions in the air-breathing organ of Pacific tarpon (*Megalops cyprinoides*) in relation to aquatic hypoxia and exercise. J Comp Physiol B 177:579–587

Shipman B (1989) Patterns of ventilation and acid-base recovery following exhausting activity in the air-breathing fish Lepisosteus oculatus. M.Sc. Thesis. University of Texas, Arlington

Stainier DY, Lee RK, Fishman MC (1993) Cardiovascular development in the zebrafish. I. Myocardial fate map and heart tube formation. Development 119:31–40

Sugishita Y, Watanabe M, Fisher SA (2004) Role of myocardial hypoxia in remodeling of the embryonic avian cardiac outflow tract. Dev Biol 267:294–308

Thebesius AC (1708) Dissertatio Medica de Cirulo Sanguinis in Corde. Elsevier, Lugduni Batacorum

Tomanek RJ (1996) Formation of the coronary vasculature: a brief review. Cardiovasc Res 31:E46–E51
Tota B (1978) Functional cardiac morphology and biochemistry in Atlantic bluefin tuna. In: Sharp GD, Dizon AE (eds) The physiological ecology of tunas. Academic, New York, pp 89–112
Tota B (1983) Vascular and metabolic zonation in the ventricular myocardium of mammals and fishes. Comp Biochem Physiol A 76:423–437
Tota B (1989) Myoarchitecture and vascularization of the elasmobranch heart ventricle. J Exp Zool:122–135
Tota B, Cimini V, Salvatore G, Zummo G (1983) Comparative study of the arterial and lacunary systems of the ventricular myocardium of elasmobranch and teleost fishes. Am J Anat 167:15–32
Tota B, Acierno R, Agnisola C (1991) Mechanical Performance of the isolated and perfused heart of the haemoglobinless Antarctic icefish *Chionodraco hamatus* (Lonnberg): effects of loading conditions and temperature. Philos Trans Roy Soc B 332:191–198
Van Citters RL, Watson NW (1967) Coronary disease in spawning steelhead trout, *Salmo gairdneri*. Science 159:105–107
Voboril Z, Schieble T (1970) Blood supply of turtle heart. Z Anat Entwicklungs 130:95–98
Wanga J, Eckberg WR, Anderson WA (2001) Ultrastructural differentiation of cardiomyocytes of the zebrafish during the 8-26-somite stages. J Submicrosc Cytol Pathol 33:275–287
Wells RMG, Forster ME, Davişon W, Taylor HH, Davie PS, Satchell GH (1986) Blood oxygen transport in the free swimming hagfish, *Eptatretus cirrhatus*. J Exp Biol 123:43–53
Zaccone D, Grimes AC, Sfacteria A, Jaroszewska M, Caristina G, Manganaro M, Farrell AP, Zaccone G, Dabrowski K, Marino F (2011) Complex innervation patterns of the *conus arteriosus* in the heart of the longnose gar, *Lepisosteus osseus*. Acta Histochem 113:578–584

Chapter 5
The Sarcoplasmic Reticulum in the Vertebrate Heart

Gina L.J. Galli and Holly A. Shiels

Abbreviations

$[Ca^{2+}]_i$	Intracellular Ca^{2+}
CASQ	Calsequestrin
CICR	Ca^{2+}-induced Ca^{2+}-release
DHPR	L-type Ca^{2+} channel dihydropyridine receptor
E–C coupling	Excitation–contraction coupling
ER	Endoplasmic reticulum
*f*SR	Free SR
*j*SR	Junctional SR
NCX	Na^+/Ca^{2+} exchanger
RyR	Ryanodine receptor
SERCA	SR Ca^{2+} ATPase
SR	Sarcoplasmic reticulum

The Role of the SR in E–C Coupling

The SR is a specialised form of endoplasmic reticulum (ER) which forms a system of intracellular membranes that acts as a reservoir for Ca^{2+} which can be rapidly mobilised for contraction and re-sequestered for subsequent relaxation. Briefly, E–C coupling begins with an action potential which depolarises the sarcolemmal

G.L.J. Galli (✉)
Faculty of Medical and Health Sciences, Core Technology Facility,
The University of Manchester, 46 Grafton Street, Manchester M139NT, UK
e-mail: gina.galli@manchester.ac.uk

H.A. Shiels
Faculty of Life Sciences, Core Technology Facility, The University of Manchester,
46 Grafton Street, Manchester M139NT, UK

D. Sedmera and T. Wang (eds.), *Ontogeny and Phylogeny of the Vertebrate Heart*,
DOI 10.1007/978-1-4614-3387-3_5, © Springer Science+Business Media, LLC 2012

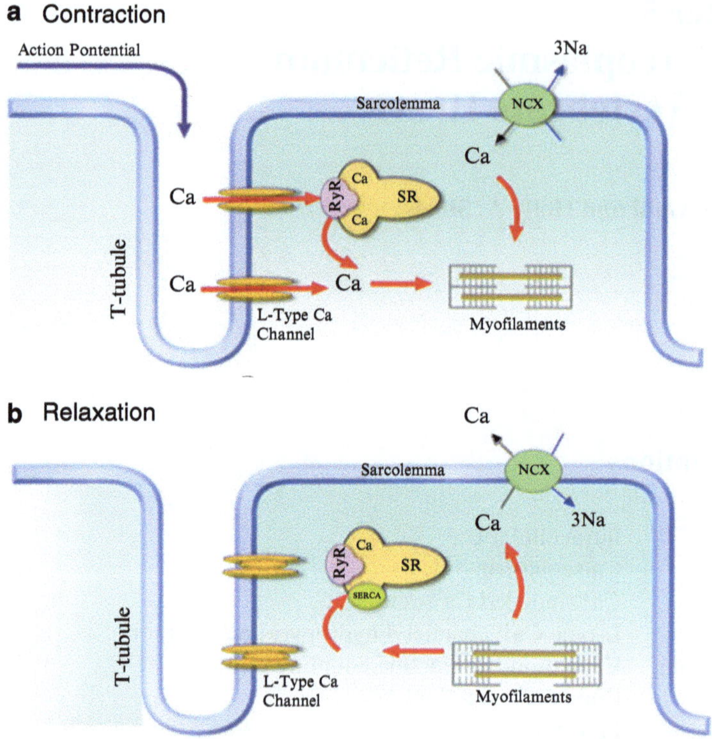

Fig. 5.1 Schematic diagram of Ca^{2+} transport during contraction (**a**) and relaxation (**b**) of a mammalian ventricular myocyte. *NCX* Na^+/Ca^{2+} exchanger, *SR* sarcoplasmic reticulum, *RyR* ryanodine receptor, *SERCA* SR Ca^{2+} ATPase. See text for details. Figure re-drawn by author from Bers (2002)

membrane and opens L-type Ca^{2+} channels (or dihydropyridine receptors, DHPRs) causing an influx of Ca^{2+} from the extracellular space (Fig. 5.1a) (Bers 2001). The rise in intracellular Ca^{2+} ($[Ca^{2+}]_i$) activates SR Ca^{2+} release channels (the ryanodine receptors, RyRs), whose opening further raises the Ca^{2+} concentration to a level which allows Ca^{2+} binding to the myofilaments, and thus contraction. This process is called Ca^{2+}-induced Ca^{2+}-release (CICR) (Bers 2001). The Na^+/Ca^{2+} exchanger (NCX) can also act as a third source of Ca^{2+}, exchanging Ca^{2+} for Na^+. Ca^{2+} is then removed from the cardiomyocyte by the SR Ca^{2+} ATPase (SERCA), and the NCX (Fig. 5.1b). Once $[Ca^{2+}]_i$ has declined to diastolic levels, the cardiomyocyte will fully relax [for a review of E–C coupling see Bers (2002)].

In this generalised scheme of E–C coupling, the SR functions as an important source and sink of activator Ca^{2+} used for contraction and relaxation. The RyR responds to small local changes in Ca^{2+} and amplifies the signal from the L-type Ca^{2+} channel (and possibly the NCX) to create a smoothly graded Ca^{2+} release. Thus, the sensitivity of the RyR to Ca^{2+} determines the gain and stability of E–C coupling (Vornanen et al. 2002). However, the density, ultrastructure and function of the SR differ greatly between vertebrate species.

Ultrastructural Studies of the SR

Each region of the SR is specialised to perform specific functions during E–C coupling. In mammalian ventricular muscle, two structural and functional types of SR exist (1) longitudinal "free" SR (*f*SR) and (2) junctional SR (*j*SR) (Rossi et al. 2008). *f*SR forms a loose tubular interconnected network surrounding the myofibrils, and is not closely associated with any membrane system (Michalak and Opas 2009). It is enriched in SERCA and is mainly responsible for the rapid removal of Ca^{2+} during relaxation. *j*SR forms peripheral couplings with either the sarcolemmal membrane or the T-tubular network (invaginations of the sarcolemma) (Michalak and Opas 2009). In these peripheral couplings where the *j*SR profile aligns with the sarcolemma or T-tubule, rows of proteins known as "feet" are visible between the two membranes. These are the RyRs, which release Ca^{2+}, and are often associated with calsequestrin (CASQ), a major Ca^{2+} buffering protein (Rossi and Dirksen 2006). The "feet" are in the immediate vicinity of the sarcolemmal L-type Ca^{2+} channels, thus providing close apposition between Ca^{2+} release components of both membrane systems. Ventricular mammalian myocytes contain both *f*SR and *j*SR that couple to the surface sarcolemmal membrane and the extensive T-tubular network (Michalak and Opas 2009). Mammalian atrial cardiomyocytes are more variable (Dibb et al. 2009), however, most small mammals, and all ectothermic cardiomyocytes (atrial and ventricular), lack T-tubules and thus *j*SR only couples with sites in the surface sarcolemma [see Di Maio and Block (2008)]. In frog myocardium, both atrial and ventricular SR form peripheral sarcolemmal couplings which co-localise with clusters of L-type Ca^{2+} channels (Tijskens et al. 2003). Interestingly, in the frog atria, these peripheral membrane couplings show co-localization between L-type Ca^{2+} channels and RyRs, but in the ventricle, there is no evidence of RyRs at the coupling (Tijskens et al. 2003). This is consistent with the lack of internal Ca^{2+} release in the frog ventricle, but raises questions regarding the significance of L-type Ca^{2+} channels at peripheral couplings in the absence of RyRs (Tijskens et al. 2003).

The volume of *j*SR and *f*SR varies greatly among cell type and species, but some generalisations can be made. In all cardiomyocytes studied, *f*SR volume is 10–30 times greater than *j*SR volume, and atrial muscle contains 2–3 times the total SR volume of ventricular muscle (Table 5.1). It was previously thought that ectothermic vertebrates have significantly lower SR volumes than endotherms, but more recent studies have shown that ectothermic values can range from completely absent, to poorly developed, to well developed. Quantitative measurements are only available for one species of reptile, and two species of fish, and are highly variable, ranging from 0.7 to 6.1% of total cell volume (Table 5.1). Fish hearts have two distinct myocardial layers which vary in relation to SR volume; the spongy layer of the fish ventricle has a greater SR volume than the compact (bluefin tuna, Table 5.1). The temperature of ectotherms must also be taken into account, as cold acclimation increases SR density (perch and tuna ventricle, Table 5.1). All of the differences outlined above may reflect functional differences in the overall importance of the SR in Ca^{2+} handling during E–C coupling (see below).

Table 5.1 Sarcoplasmic reticulum density in a range of vertebrates, expressed as a percentage of cell volume/sarcolemmal membrane

	As a percentage of total cell volume (%)		
Species	Junctional SR	Free SR	Total SR
Finch V	0.21	0.62	0.83
Guinea pig A	0.46	9.47	9.93
Guinea pig V	0.56	7.37	7.93
Mouse A	–	–	12.3
Mouse V	–	–	6.9
Rat V	0.3	3.2	3.5
Frog V	0.03	0.35	0.38
Frog A	0.03	0.53	0.56
Lizard V	0.09	0.6	0.69
Lizard A	0.11	1.12	1.22
Perch V 20°C acclimated	–	–	4.50
Perch V 5°C acclimated	–	–	6.05
Bluefin tuna V compact	–	–	2.2
Bluefin tuna V spongy	–	–	2.5
Bluefin tuna A	–	–	3.2
	Surface area of SR/cell volume	As a percentage of sarcolemma	
Species	Junct SL tot ($\mu m^2/\mu m^3$)	Ext SL (%)	TT (%)
Rabbit V	0.068	4.6	21
Rat V	0.093	7.7	48
Mouse V	0.090	6.5	40

V ventricle, *A* atria, *SR* sarcoplasmic reticulum, *SL* sarcolemmal membrane, *TT* T-tubules, *Ext* external, *Junct* junctional. Data compiled from Bers (2001), Di Maio and Block (2008), Bowler and Tirri (1990)

In Vitro Studies on SR Function

Early attempts to elucidate the role of the SR in vertebrates focussed on in vitro multicellular cardiac preparations and the effect of inhibiting SR function on contractile force and relaxation. These studies have been reviewed previously [see Vornanen et al. (2002), Tibbits et al. (2002)] and will only be discussed briefly here. Figure 5.2 shows the effect of ryanodine (a specific inhibitor of RyR) on the steady-state contractile tension in several vertebrate cardiac preparations. By inhibiting the SR with ryanodine, it is possible to quantify the percentage of Ca^{2+} contributing to E–C coupling that originates from the SR vs. sarcolemmal transport (L-type Ca^{2+} channel and NCX). It should be noted that under these conditions, sarcolemmal Ca^{2+} entry may overcompensate, and thus, SR contribution to tension may be underestimated in some species. Furthermore, pacing frequency and temperature strongly influence the effect of ryanodine on contractile force, which complicates the interpretation of some of the species comparisons in Fig. 5.2. Nevertheless, many of the results correlate with the ultrastructural differences outlined in Table 5.1.

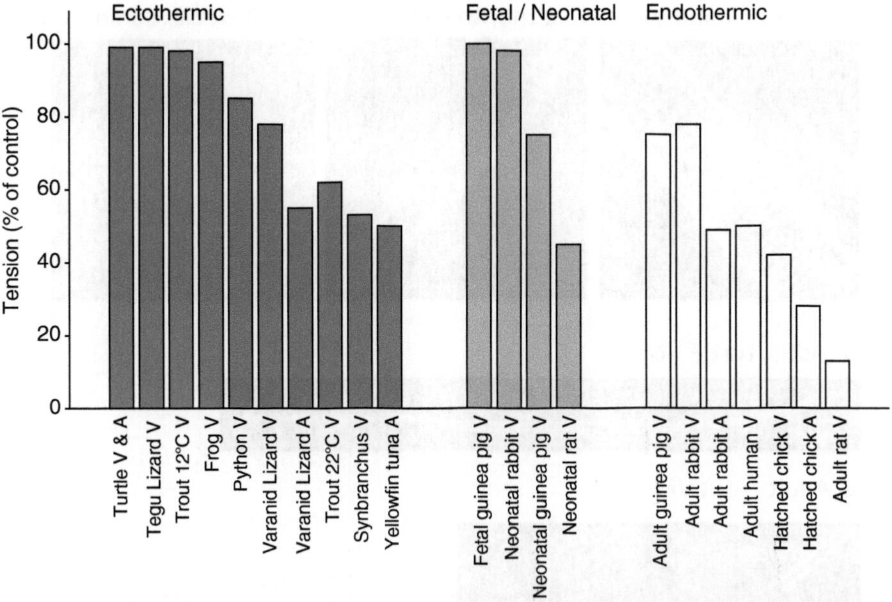

Fig. 5.2 The effects of ryanodine pre-treatment on steady-state twitch contraction in various cardiac muscle preparations. *Dark grey bars* represent ectothermic animals, *light grey bars* are foetal or newborns and *white bars* are endothermic animals. *V* ventricular muscle, *A* atrial muscle. All preparations were paced at 0.5 Hz, except for human, hatched chick and guinea pig, which were paced at 1.0 Hz. Rat (Bers 1985, 2001), rabbit (Haddock et al. 1999; Bers and Bridge 1989), frog (Bers 1985) and all reptiles (Galli et al. 2006b) were tested at 30°C. Guinea pig (Agata et al. 1994), synbranchus (Rocha et al. 2007), human (Ravens et al. 1997) and chick (Tanaka et al. 1995) were tested at 35–36°C. Trout were tested at either 12 or 22°C (Shiels and Farrell 1997), and yellowfin tuna at 25°C (Shiels et al. 1999)

With regard to endotherms, variation is apparent between different species (rat vs. rabbit), at different stages of development (neonatal or foetal vs. adult) and regionally within the heart (ventricle vs. atria) (Bers 2001). Some species-specific effects of ryanodine correlate with ultrastructural differences (rat vs. rabbit), and scale allometrically with heart rate (Table 5.1; Fig. 5.3). Smaller animals with high heart rates, and tissues that can achieve high frequencies of contraction (atrial muscle), tend to be more sensitive to ryanodine and have a better developed SR network [for a review see (Bers 2001)].

The mammalian neonatal (Haddock et al. 1999) and foetal heart (Agata et al. 1994) is far less sensitive to ryanodine and relies strongly, in some cases exclusively, on sarcolemmal Ca^{2+} cycling, while the adult mammal obtains 60–90% of systolic Ca^{2+} from the SR (Fig. 5.2) (Haddock et al. 1999; Maylie 1982). Consistent with functionality, ultrastructural studies have shown the SR system is gradually developing from the prenatal period to early life (Cordeiro et al. 2007). Furthermore, neonatal and foetal cardiomyocytes are structurally different to their adult counterparts, with a smaller diameter (5–10 μM), larger surface area and no T-tubular network (Fig. 5.3a–c). This morphology is consistent with a more important role for transsarcolemmal Ca^{2+} flux, as the small diameter will enhance the efficacy of

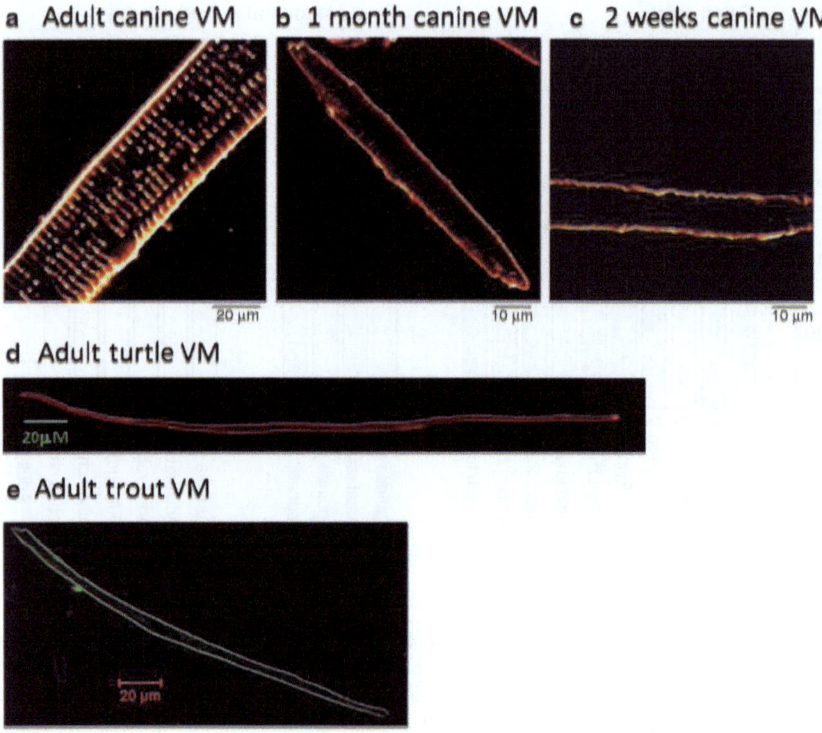

Fig. 5.3 Ventricular myocytes isolated from (**a–c**) dogs at various stages of development; (**d**) adult turtle (*Trachemys scripta*); (**e**) adult rainbow trout (*Oncorhynchus mykiss*). Cells were stained with the membrane dye di-8-ANEPPS and visualised with confocal microscopy. The T-tubular structure can only be seen in adult cells as a transverse membrane network (**a**). In contrast, ventricular cells from 2-week-old dogs, adult turtles and trout lack T-tubules and have a much thinner cell diameter (**c–e**). 4-week-old dog cells show T-tubules in a formative stage of development (**b**). Figure compiled from Cordeiro et al. (2007), Galli et al. (2006a), Shiels and White (2005)

sarcolemmal Ca^{2+} transport by reducing the diffusion distance that Ca^{2+} has to travel to the myofilaments (Vornanen et al. 2002). Thus, for neonates, sarcolemmal Ca^{2+} transport can support contraction and relaxation entirely, but as they develop into adulthood, cardiomyocyte width more than doubles, making a T-tubular/SR network necessary. Indeed, increases in SR dependence correlate with T-tubule development in the adult mammalian ventricle (Tibbits et al. 2002).

Similar to the neonate, in vitro studies of the hearts of most amphibians, fish and reptiles indicate contraction relies more heavily on transsarcolemmal Ca^{2+} influx than SR Ca^{2+} release during E–C coupling (Fig. 5.2). Interestingly, ectothermic cardiomyocytes are structurally similar to neonates, being long/thin, spindle shaped and lacking T-tubules (Fig. 5.3d, e) (Vornanen 1997; Galli et al. 2006a, 2009; Shiels et al. 2004), suggesting cell morphology may impact SR dependence. However, adult atrial mammalian myocytes do not follow the same pattern. Most atrial myocytes are long and thin and do not possess T-tubules, but in all species studied to date, atrial SR contributes a greater proportion to E–C coupling than ventricular myocytes (Fig. 5.2). Clearly, the structure of a cardiomyocyte is not necessarily a good proxy for SR dependence and contractile performance.

There are certain ectothermic species and experimental conditions, where SR dependence is greater than expected. Highly active ectotherms rely more strongly on SR Ca^{2+} stores than sedentary species. Tuna, which are considered teleost athletes, have a well-developed SR network (Di Maio and Block 2008) (Table 5.1) and SR inhibition significantly decreases contractile force in bigeye tuna ventricle [by 30% (Galli et al. 2008)], yellowfin tuna atria [by 50% (Shiels et al. 1999)] and skipjack tuna atrium [by 30% (Keen et al. 1992)] (Fig. 5.2). In the reptilian ventricle, ryanodine decreased contractile force in the python (15%) and varanid lizard (22%) (Galli et al. 2006b), which are unique amongst reptiles for having a functionally divided ventricle with high systemic blood pressures (Webb et al. 1971; Wang et al. 2003) (Fig. 5.2). In atrial muscle, contractile force was also decreased in the varanid (45%) and tegu lizard (10%), both of which have high heart rates compared with other reptiles (Galli et al. 2006b) (Fig. 5.2). By comparison, contractile force in the more sedentary turtle is insensitive to SR inhibition (Galli et al. 2006a, b), suggesting their cardiomyocytes rely exclusively on sarcolemmal Ca^{2+} transport.

A link between phylogeny and SR dependence has also been suggested, as members of the neotropical freshwater fish species from the order Characiformes, superorder Ostariophysi, appear to utilise SR Ca^{2+} stores to a large extent during E–C coupling, independent of activity level or temperature (Rivaroli et al. 2006; Anelli et al. 2004). The reasons underlying this phylogenetic SR dependence are still unclear.

Cellular Studies of SR Function

Multicellular studies can be limited in their ability to reveal mechanisms of SR function and spatiotemporal properties of the SR-induced Ca^{2+} transient. Cellular fluorescent ion imaging and electrophysiological studies have made significant advances in this regard. These techniques have only recently (in the last 15 years) been applied to ectothermic vertebrates, but the limited physiological data available do reveal striking functional differences between ectothermic and mammalian cardiomyocytes in the way Ca^{2+} is released and re-sequestered into the SR.

The Role of the SR in the Cellular Ca^{2+} Transient

The cellular Ca^{2+} transient (sometimes denoted as $\Delta[Ca^{2+}]_i$) is defined as the transient increase and decrease in cytosolic Ca^{2+} from all sarcolemmal and internal sources of Ca^{2+} during a contraction. The Ca^{2+} transient has been measured for decades in mammalian hearts to provide direct information on the role of the SR in E–C coupling [see Tibbits et al. (1991)]. Ca^{2+} influx from L-type Ca^{2+} channels in the mammalian cardiomyocyte triggers the release of 35–59% of the intracellular SR Ca^{2+} store, contributing to 70–90% of the Ca^{2+} transient (Fabiato 1983). Similar measurements in ectothermic cardiomyocytes have only very recently been imaged (see Fig. 5.4). In turtle ventricular myocytes, the Ca^{2+} transient is insensitive to

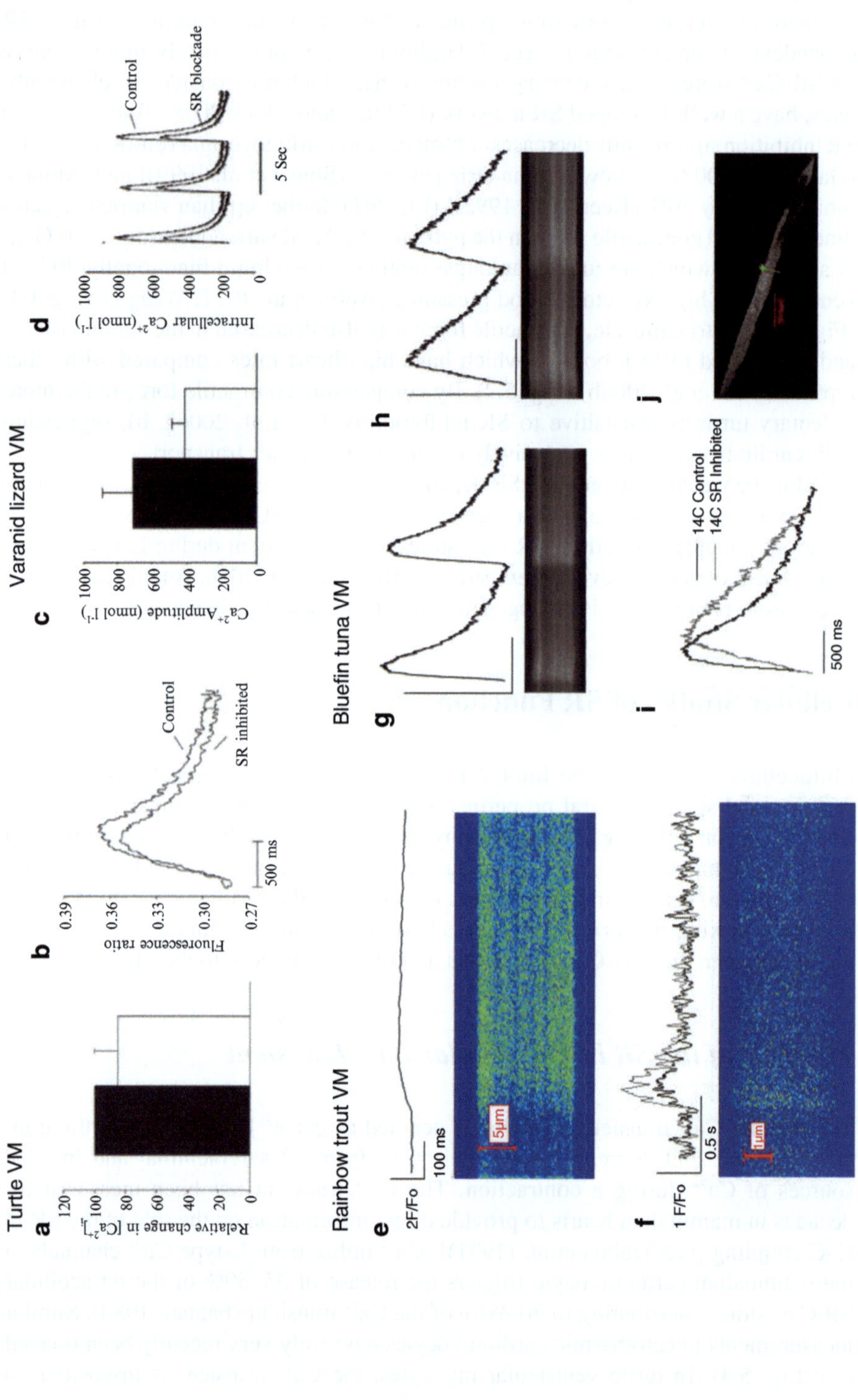
a
Turtle VM
Relative change in [Ca2+]i
120
100
80
60
40
20
0
b
Fluorescence ratio
0.39
0.36
0.33
0.30
0.27
500 ms
Control
SR inhibited
c
Varanid lizard VM
Ca2+Amplitude (nmol l-1)
1000
800
600
400
200
0
d
Intracellular Ca2+(nmol l-1)
5 Sec
Control
SR blockade
e
Rainbow trout VM
2F/Fo
100 ms
5μm
f
1 F/Fo
0.5 s
1μm
g
Bluefin tuna VM
h
i
500 ms
14C Control
14C SR Inhibited
j

Fig. 5.4 Effect of sarcoplasmic reticulum (SR) inhibition with ryanodine (10 μmol) and thapsigargin (1 μM) on the intracellular Ca^{2+} transient ($\Delta[Ca^{2+}]_i$) in isolated ventricular myocytes from the turtle (*Trachemys scripta*, **a, b**), the varanid lizard (*Varanus exanthematicus*, **c, d**), the rainbow trout (*Oncorhynchus mykiss*, **e, f**) and the bluefin tuna (*Thunnus orientalis*, **g–j**). In all cases, cardiomyocytes were field stimulated at 0.2 Hz at room temperature, except tuna which was measured at 14°C. (**a**) Relative reduction in turtle $\Delta[Ca^{2+}]_i$ due to SR inhibition, values are mean ± SEM, $n=6$. (**b**) Original traces of turtle $\Delta[Ca^{2+}]_i$ measured in myocytes loaded with the Ca^{2+} sensitive dye Fura-2 AM. (**c**) Amplitude of the varanid lizard $\Delta[Ca^{2+}]_i$ in the presence and absence of SR blockade, values are mean ± SEM, $n=12$. (**d**) Original traces of varanid lizard $\Delta[Ca^{2+}]i$ also measured using Fura2-AM. The *y*-axis scale is the same in both figures. (**e, f**) The transient time course of the Ca^{2+} transient (*upper panel*) and the corresponding raw line scan image (*lower panel*) under control conditions (**e**) and after SR inhibition (**f**) for a rainbow trout ventricular myocyte (scale is indicated on the linescan). Notice the smaller increase in Ca^{2+} in the linescan image and the slower rate of rise in the cell interior shown by the *black line* in the corresponding time course compared with the cell periphery (*grey line*). (**g, h**) The transient time course of the Ca^{2+} transient (*upper panel*) and the corresponding raw line scan image (*lower panel*) under control conditions (**g**) and after SR inhibition (**h**) for a bluefin tuna ventricular myocyte loaded with Fluo-4 AM shown in (**j**). The effects of SR inhibition on Ca^{2+} cycling kinetics at 14°C are emphasised when the amplitudes are normalised (I). The *green line* in (**i**) indicates position of the scan. Time course scale is 100 nM $[Ca^{2+}]$ by 1 s. Line scans are 2,500 lines × 512 pixels. Figure compiled from Galli et al. (2006a, 2009), Shiels et al. (2011) and Shiels and White (2003)

ryanodine and thapsigargin (a specific inhibitor of SERCA), confirming the limited role of the SR in turtle E–C coupling (Galli et al. 2006a) (Fig. 5.4a, b). In contrast, up to 40% of the more active varanid lizard ventricular myocyte Ca^{2+} transient is dependent on SR Ca^{2+} release (Galli et al. 2009) (Fig. 5.4c, d). Furthermore, the rate of rise of the varanid lizard Ca^{2+} transient, and the rate of decay (tau, τ), is significantly slowed with SR inhibition (Galli et al. 2009).

In both rainbow trout and bluefin tuna, confocal microscope imaging across the width of the cardiomyocyte has provided important information on the temporal and spatial coupling of Ca^{2+} signals (Fig. 5.4e–j) (Shiels et al. 2011; Shiels and White 2005). Similar to the varanid lizard, SR inhibition reduces the amplitude of the fish Ca^{2+} transient, and slows the rate of rise and decay. This can be clearly seen in Fig. 5.4g, h by comparing the shape of the Ca^{2+} waveform of the bluefin tuna ventricular Ca^{2+} transient (Shiels et al. 2011). The slower rate of rise and fall after SR inhibition is emphasised in Fig. 5.4i, where the two transients have been normalised to highlight the effect of SR inhibition on kinetics. These studies are discussed in detail below. Clearly, further studies in more species are necessary to fully appreciate the role of the SR in the generation of the ectothermic Ca^{2+} transient, but in those species that utilise the SR, the effects are similar to those described for mammals: transients are smaller with slower time courses when SR Ca^{2+} uptake and release are inhibited.

SERCA Activity and SR Ca^{2+} Content

In line with ultrastructural and multicellular data, regional, developmental and species-specific differences exist in mammalian SERCA expression and activity. SERCA protein/mRNA expression and SERCA activity are higher in adult vs. neonatal (Harrer et al. 1997; Arai et al. 1992; Fisher et al. 1992) (Fig. 5.5a), larger vs. smaller animals (Su et al. 2003; Hove-Madsen and Bers 1993a; Minajeva et al. 1997; Walden et al. 2009), and atrial vs. ventricular cardiomyocytes (Walden et al. 2009). These differences have consequences on SR Ca^{2+} content which is determined by SERCA activity, SR density and SR luminal Ca^{2+} buffering. Indeed, steady-state SR Ca^{2+} content is greater in smaller animals and in tissues that are capable of generating high frequencies of contraction (Su et al. 2003; Walden et al. 2009).

SERCA activity in rat and trout is almost equivalent at room temperature, but at their respective physiological temperatures, the rat heart (37°C) can pump three times more Ca^{2+} than the trout heart (18°C, Fig. 5.5a) (Aho and Vornanen 1998). However, the caffeine releasable Ca^{2+} stores (representing the *maximal* Ca^{2+} content of the SR) in atrial and ventricular trout myocytes are almost an order of magnitude larger than in adult mammalian cardiomyocytes (Hove-Madsen et al. 1998; Shiels et al. 2001). Furthermore, *steady-state* SR Ca^{2+} load in trout [750 $\mu mol\ l^{-1}$ (Hove-Madsen et al. 1998)] is much higher than mammalian cardiomyocytes (~100 $\mu mol\ l^{-1}$) (Hove-Madsen and Bers 1993a, b; Negretti et al. 1995). Thus, there is a striking dichotomy between the large SR Ca^{2+} stores in trout and the relatively weak effect of ryanodine on contractile force. This dichotomy is not restricted to the trout heart,

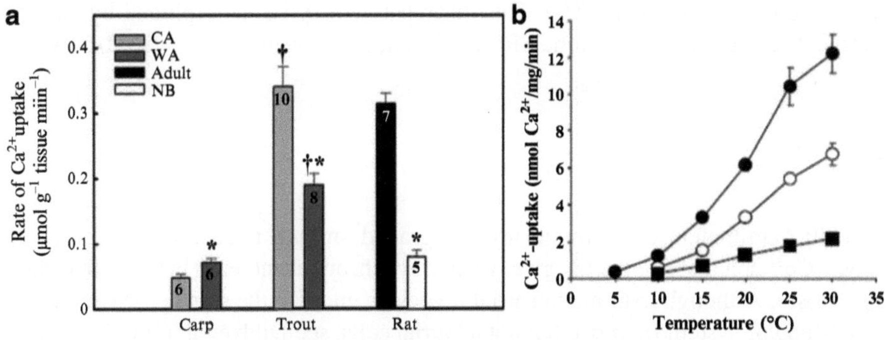

Fig. 5.5 (**a**) Sarcoplasmic reticulum (SR) Ca^{2+} uptake rate of crude cardiac homogenates from crucian carp (either cold acclimated (CA) to 4°C, or warm acclimated (WA) to 24°C); rainbow trout (either CA to 4°C or WA to 17°C) and rat [either adult or neonate (new born, NB)]. The results are mean+SEM of 5–10 preparations, as indicated. An *asterisk* indicates a statistically significant difference ($P<0.05$) between age groups or acclimation groups. A dagger indicates a statistically significant difference ($P<0.05$) between trout and carp. Figure taken from Aho and Vornanen (1998). (**b**) The acute temperature dependence of SR Ca^{2+} uptake in isolated SR membrane preparations from different tuna species. *Black circles*, Pacific bluefin (*Thunnus orientalis*); *white circles*, albacore (*Thunnus alalunga*); *black squares*, yellowfin (*Thunnus albacares*). The figure shows both species differences and temperature effects. Values are mean±SEM of 3–4 experiments [Figure taken from Landeira Fernandez et al. (2004)]

as SR Ca^{2+} content in the ventricle of the burbot, crucian carp, Pacific mackerel, Pacific bluefin tuna, yellowfin tuna and bonito is also considerably higher than in mammalian cardiomyocytes (Haverinen and Vornanen 2009; Galli et al. 2011). Interestingly, large SR Ca^{2+} loads and insensitivity to ryanodine is also a characteristic of mammalian neonatal cardiomyocytes (Bers 2001), suggesting a link between SR Ca^{2+} handling and cardiomyocyte morphology. From these studies, it appears the difference between fish or neonates and mammalian SR resides in the ability to release Ca^{2+}, rather than sequestering it (see below).

SERCA activity in ectothermic species is also species and tissue-type dependent. In line with the mammalian paradigm of activity-based SR dependence, the Pacific bluefin tuna shows an enhanced expression of SERCA, and higher SERCA activities across a range of temperatures, compared with other less active species of the family scombridae (Landeira Fernandez et al. 2004). Similarly, SERCA activity is lower in carp than the active trout regardless of thermal history (Fig. 5.5a). Atrial cardiomyocytes from a range of fishes have greater SR loads than ventricular myocytes (Haverinen and Vornanen 2009; Galli et al. 2011). However, the maximal capacity of the SR to sequester Ca^{2+}, and the steady-state SR Ca^{2+} load, does not necessarily correlate with SERCA activity or ryanodine sensitivity in all ectotherms. Within the scombrid fishes, mackerel are capable of storing equally large SR Ca^{2+} loads to the more active Pacific bluefin tuna (Galli et al. 2011), and an inverse relationship between ryanodine sensitivity and SR Ca^{2+} load was found in the rainbow trout, burbot and crucian carp

(Haverinen and Vornanen 2009). This discrepancy may be underpinned by variable luminal Ca^{2+} buffering within the SR of different ectothermic species (see below).

SR Ca^{2+} Release

In adult mammalian cardiomyocytes, the graded smooth release of Ca^{2+} from the L-type Ca^{2+} channel and the SR give rise to a synchronous increase in Ca^{2+} throughout the cytosol. Although mammalian atrial myocytes undergo the same process, the lack of a T-tubular system [in most, but not all atrial cells, see Dibb et al. (2009)] creates a systolic Ca^{2+} transient that rises initially at the cell periphery and then propagates as a wave of CICR into the cell interior (Mackenzie et al. 2001; Huser et al. 1996). Neonatal mammalian cardiomyocytes, which also lack T-tubules and have limited SR recruitment, have a spatiotemporal Ca^{2+} transient showing Ca^{2+} diffusing from the periphery to the centre of the cell with little intracellular propagation (Bers 2001). Presumably, the reduced width of atrial and neonatal cardiomyocytes facilitates the intracellular propagation of the Ca^{2+} transient in the absence of T-tubules and/or SR.

Similar information for ectothermic vertebrates is scare. Rainbow trout ventricle displays temporal and spatial properties of Ca^{2+} gradients midway between those of mammalian atrial and neonatal cardiomyocytes (Shiels and White 2005). Voltage-dependent Ca^{2+} entry at the trout cell periphery causes SR Ca^{2+} release from peripheral couplings, which increases the Ca^{2+} transient amplitude (Shiels and White 2005). Peripheral Ca^{2+} then diffuses to the cell centre where it initiates the subsequent release of more centrally located SR Ca^{2+} stores. However, SR Ca^{2+} release in the trout heart is considerably less than a mammalian heart. In fact, even with a maximally loaded SR, only 5% of the total trout SR Ca^{2+} store has been estimated to be released (Hove-Madsen et al. 1998). But despite this, inhibiting the SR has implications for the temporal and spatial properties of the cellular Ca^{2+} transient as indicated in Fig. 5.4e, f.

In bluefin tuna cardiomyocytes, which are heavily dependent upon the SR (Shiels et al. 2011), there is a homogeneous rise in the Ca^{2+} wavefront across the width of the cardiomyocyte. This indicates voltage-dependent Ca^{2+} entry triggers Ca^{2+} release from peripheral *j*SR that subsequently induces Ca^{2+} release in more central regions of the cell through propagated CICR (Fig. 5.4g). Because cardiomyocyte morphology is similar between rainbow trout and tuna cardiomyocytes, it suggests more robust coupling between SR Ca^{2+} release units in tuna than in trout (Shiels et al. 2011) (compare Fig. 5.4e, g).

Ca^{2+} Sparks and the RyRs

The difference between mammalian and ectothermic SR Ca^{2+} release can be further realised by scanning cardiomyocytes for spontaneous SR Ca^{2+} release events, or

"Ca^{2+} sparks". Ca^{2+} sparks have been reported with variable frequency from a number of species and in a number of cell types, including cardiac, skeletal and smooth muscle. Shiels and White (2003) show Ca^{2+} sparks are extremely rare in ventricular myocytes from the rainbow trout, even when conditions which highly favour spark activity were applied (high extracellular Ca^{2+} and the addition of Bay K). In contrast, a more recent study has detected Ca^{2+} sparks in trout and zebrafish cardiomyocytes which have dimensions, durations and frequencies comparable to values reported in mammalian and human cardiomyocytes (Llach et al., 2011). However, the study only observed sparks in 40% of the trout myocytes tested, and they exhibited a smaller fractional release of SR Ca^{2+} compared to mammalian cardiomyocytes. Thus, SR Ca^{2+} content can be very high in trout myocytes (~2,000 μM total Ca^{2+}) without spontaneous release.

Several possibilities exist which may explain the limited SR Ca^{2+} release relative to the high storage capacity of fish cardiomyocytes. Firstly, the number of fish RyRs could be limiting Ca^{2+} release from the SR. Indeed, the density of RyR binding sites in carp, rainbow trout and burbot is 19, 28 and 66%, respectively, of that in the rat heart (Thomas et al. 1996; Chugun et al. 1999; Vornanen 2005), which may reduce the number of functionally active "clusters" of RyR (Vornanen et al. 2002). Radioligand binding studies indicate the ratio of RyRs to L-Type Ca^{2+} channel receptors in trout ventricle is approximately 1 compared with approximately 4 in the rat ventricle, which indicates smaller clusters (Tiitu and Vornanen 2003). Furthermore, immunological studies with carp heart show small RyR clusters that are distributed in a punctuate manner throughout the cardiomyocyte, which is in contrast to the larger and more regularly organised RyR clusters observed in rat cardiomyocytes in the same study (Chugun et al. 2003).

Another factor affecting Ca^{2+} sparks and CICR is the subcellular location of the RyRs. It is well documented that in adult mammalian cardiomyocytes, co-localization of RYRs and L-type Ca^{2+} channels in the dyad is the structural prerequisite for their functional coupling (Bers 2002). Additionally, mammalian RYRs align with the sarcomeric *z*-lines (Shannon et al. 2003; Bootman et al. 2006). In neonatal cardiomyocytes, co-localization of RYRs with L-type Ca^{2+} channels is less prominent (Sedarat et al. 2004). Instead, RYRs are mainly co-localised with NCX where reverse-mode NCX is a dominant source of sarcolemmal Ca^{2+} influx (Dan et al. 2007; Huang et al. 2008). Neonatal RyRs are also associated with both the *m*-line and the *z*-line (Dan et al. 2007), and platforms of SR at the level of the *m*-line have Ca^{2+}-sparking clusters of RyRs (Lukyanenko et al. 2007). Recently it has been shown that trout ventricular RyRs were located peripherally in transverse bands aligning with the *m*-line, and centrally near the mitochondria (Birkedal et al. 2009). This pattern is in agreement with early electron microscopy studies that did not reveal a particular association of SR with sarcomeric bands in trout heart (Santer 1974). In Pacific bluefin tuna cardiomyocytes, RyRs align with the sarcomeric *z*-lines (Di Maio and Block 2008) similar to that observed in adult mammalian atrial and ventricular myocytes. Thus structural and subcellular organisation of RyRs may influence RyR function and spark frequency.

RyR Ca^{2+} Sensitivity and Ca^{2+} Buffering

Another possibility is that ectotherm RyRs are less sensitive to Ca^{2+} compared to those in the mammalian heart; i.e. the SR gain is reduced. Indeed, fish that are insensitive to SR inhibition appear to have similarly insensitive RyRs. Crucian carp, rainbow trout and the common carp require 5.9, 4.4 and 6 times more Ca^{2+}, respectively, for half-maximal activation compared with RyRs of rat (Chugun et al. 1999; Vornanen 2005). Only the RyRs of the burbot heart, which have greater RyR density (Vornanen 2005; Tiitu and Vornanen 2003) and are more SR dependent than other teleosts (Tiitu and Vornanen 2002), are equally sensitive to Ca^{2+} as the rat heart. Thus, the relatively low sensitivity of fish RyR to trigger Ca^{2+} may be a contributing factor to low SR Ca^{2+} release.

Aside from a cytosolic sensitivity to Ca^{2+} (CICR), the RyR is also sensitive to luminal Ca^{2+} (Sitsapesan and Williams 1994, 1995; Gyorke and Gyorke 1998; Laver 2007). Thus, intraluminal buffering of Ca^{2+} is a major determinant of free SR Ca^{2+}, maximal SR Ca^{2+} load and RyR open probability (Bers 2001). CASQ is the predominant Ca^{2+} buffering protein in the SR of cardiomyocytes, and it modulates SR Ca^{2+} release via two regulatory proteins: triadin and junction (Gyorke and Terentyev 2008). Although the mechanism remains poorly understood, it is thought that in the absence of Ca^{2+}, CASQ inhibits RyR via triadin and junctin, whereas binding of Ca^{2+} to CASQ relieves the inhibitory effect (Terentyev et al. 2003). In this way, SR Ca^{2+} content is maintained at a constant level, as increased SR Ca^{2+} content stimulates Ca^{2+} release, and decreased content inhibits release.

The Ca^{2+} buffering capacity of the SR may also help to explain the low SR Ca^{2+} release from the teleost and neonatal cardiomyocyte. Indeed, CASQ is upregulated in adult vs. foetal hearts (Arai et al. 1992; Mahony and Jones 1986), and triadin and junctin expressions are downregulated in atrial vs. ventricular mammalian cardiomyocytes (Luss et al. 1999). Only one study has attempted to investigate CASQ in an ectothermic vertebrate heart but the antibodies for fish (rainbow trout) and mammalian cardiac CASQ differed, and comparative CASQ expression was not faithfully determined (Korajoki and Vornanen 2009). More studies are necessary to determine the role of SR Ca^{2+} buffering in ectothermic cardiomyocytes, as this is likely to be a significant reason why SR Ca^{2+} release is attenuated.

RyR Isoform

A final explanation for the difference between RyR activity among ectotherms and endotherms may reside in the isoform of the RyR. At present, there are no single channel recordings or sequencing data of trout heart RyR that would enable determination of receptor isoform or subtype. Indeed, the use of molecular biology to determine fish RyR subtype has failed (Shiels and White 2005; Birkedal et al. 2009). However, immunohistochemistry has successfully been used to label "generic"

RyRs from three species of fish [trout (Birkedal et al. 2009), bluefin tuna (Di Maio and Block 2008) and carp (Chugun et al. 2003)].

The Effect of Temperature on the Role of the SR

Ectotherms experience acute and chronic temperature changes as part of their daily life which directly affects the temperature of their heart. Mammals, on the other hand, only experience temperature change under extreme conditions, like hypothermia or hibernation. Nevertheless, studies of the effect of cold temperatures on the mammalian heart have provided important information about the role of the SR in E–C coupling. The effects of temperature on whole heart performance [see Gamperl et al. (2011)] can be attributed, in part, to its effects on SR Ca^{2+} cycling.

Acute Temperature Change

Acute temperature changes, in the order of minutes to hours, occur during diurnal fluctuations in air temperature for terrestrial ectotherms, or when moving through thermoclines in aquatic species. In all mammals investigated, acute cold temperature increases the open probability of the cardiac RyR (Sitsapesan et al. 1991) causing Ca^{2+} to be released from the SR as temperatures approach 0°C. This property has been exploited experimentally with rapid cooling contractures which can be used to assess SR Ca^{2+} content (Bers 1989). Certainly, the RyRs from animals whose hearts function over a range of temperatures, including acutely cold temperatures, must have different thermal sensitivities. Indeed, altered sensitivity of ryanodine binding has been described for RyRs of two species of hibernating ground squirrels (Milner et al. 1991), and early studies with rainbow trout showed no change in SR Ca^{2+} leak at cold temperatures (Hove-Madsen et al. 2001) Later studies showed that the SR actively sequesters and releases Ca^{2+} during acute cold bouts in trout (Shiels et al. 2002a). Similar studies have not been conducted for reptiles but early work showed that similar to fish, amphibians do not exhibit rapid cooling contractures (Bers 1989). These differences in acute thermal sensitivity of the RyR in ectotherms and mammals calls into question whether they both use the same isoform of the RyR. As discussed above, this is still unclear.

Changes in SR function must be coupled to other temperature-induced changes in E–C coupling. These include the L-type Ca^{2+} channel (Herve et al. 1992; Shiels et al. 2000, 2003), the RyR and SERCA (Di Maio and Block 2008; Landeira Fernandez et al. 2004; Shiels et al. 2002a; Castilho et al. 2007), and the NCX (Xue et al. 1999; Llach et al. 2001) and Q_{10} effects on all the ion channels and pumps that underlie the cardiac action potential. Acute warming and acute cooling accelerate and slow the kinetics of the Ca^{2+} transient, respectively, indicating temperature limits cellular Ca^{2+} cycling in all vertebrates studied to date. This is partially because of the direct effect of temperature on Ca^{2+} influx via L-type Ca^{2+} channels which are faster and greater in amplitude at warm temperatures [Rabbit (Puglisi et al. 1999), trout (Shiels et al. 2000)] which triggers greater release from the SR. On the other hand, acute cooling tends to increase the amplitude of the Ca^{2+} transient and the force of contraction (Aho and Vornanen 1999) because the slower kinetics allow more time for Ca^{2+} influx across the membrane through L-type Ca channels (Shiels et al. 2000) which can load Ca^{2+} into the SR (Shiels et al. 2002b).

The activity of SERCA is temperature-dependent in all vertebrates studied (Aho and Vornanen 1998; Landeira Fernandez et al. 2004) (see Fig. 5.5). In tuna species, a reduction in temperature from 25 to 10°C significantly slows SERCA Ca^{2+} pumping (Fig. 5.5b) (Landeira Fernandez et al. 2004). It is important to note that this temperature range is physiologically relevant, as tuna experience these temperatures as part of their daily foraging routine in the oceans. Thus, temperature can severely affect the ability of the SR to contribute to the decay of the Ca^{2+} transient in vivo in ectotherms.

Rapid temperature change can also affect SR Ca^{2+} content and release by changing the rate at which the heart beats. Warm temperatures increase heart rate through direct actions on the pacemaker rate in fish, causing a decrease in the amplitude of the intracellular Ca^{2+} transient. This reduces the force of contraction and is termed "the negative force-frequency relationship". In mammalian hearts the negative force–frequency relationship is directly related to the SR and whether it loads or leaks Ca^{2+} at rest (Endoh 2004). It has been difficult to relate SR function with the force–frequency relationship in fish, as the outcome seems more dependent on species, stimulation rate and experimental conditions than SR function (see Shiels et al. (2002c) for a review).

Chronic Temperature Changes

In an attempt to compensate for the direct effects of temperature on heart function, some animals adjust their physiological and biochemical phenotypes during long-term exposure to a new temperature. This thermal acclimation is especially important in the hearts of ectotherms to maintain contractility and to avoid disturbances in electrical excitability. Contractile force and contractile kinetics in isolated muscle preparations from rainbow trout (Aho and Vornanen 1999), perch (*Perca flavescens*) (Bailey et al. 1999) and tuna (Shiels et al. 2011) are accelerated with chronic cold suggesting a "remodelling programme" which increases both inotropy and chronotropy. This is a common feature of chronic cold exposure in the hearts of metabolically active fish in cold waters (Haverinen and Vornanen 2008).

The SR is known to play a large role in the remodelling associated with cold acclimation in fish and cold tolerance in hibernating mammals (Milner et al. 1991; Belke et al. 1991). The proliferation of cardiac SR as a consequence of cold acclimation has been reported for the perch heart (Bowler and Tirri 1990) and bluefin tuna (Shiels et al. 2011), and may explain the increase in the ryanodine sensitivity of contraction after cold acclimation in the rainbow trout heart (Aho and Vornanen 1999; Keen et al. 1994; Shiels and Farrell 1997). In rainbow trout, cold acclimation increases SERCA activity in isolated SR vesicles (Aho and Vornanen 1998) and cold tolerant species (bluefin tuna) and stenothermic cold species (burbot) show greater SERCA activity that extend to lower temperatures compared with other tunas (yellowfin and albacore) (Landeira Fernandez et al. 2004; Castilho et al. 2007; Shiels et al. 2006). In contrast, studies have found no effect of temperature acclimation on the expression or subcellular localization of the RyR in rainbow trout (Birkedal et al. 2009) or the total number of RyRs in crucian carp (Tiitu and Vornanen 2003) Similarly, no change in the expression of CASQ with cold acclimation in the trout heart has been found (Korajoki and Vornanen 2009). Thus, perhaps within the SR, SERCA is the malleable unit of cellular Ca^{2+} flux offering a mechanism for thermal plasticity in fish hearts. Interestingly, ventricular myocytes from Siberian hamsters (*Phodopus sungorus*) showing daily bouts of torpor exhibit an increased amplitude of the Ca^{2+} transient and an acceleration of the rise and fall kinetics when compared with non-torpid littermates (Dibb et al. 2005). This may suggest animals that experience rapid changes in temperature on a daily cycle (e.g. diving fish or torpid mammals/birds) compensate for the effects of temperature via Ca^{2+} flux mechanisms (Shiels et al. 2011). This idea could be tested in future studies.

General Conclusions

The role of the SR in vertebrates is dependent on a number of factors. In mammals, a relatively clear picture has unfolded, with SR dependence being greater in smaller animals with higher heart rates, in adult vs. neonatal or foetal stages, and in tissues that are capable of achieving high frequencies of contraction (e.g. atria). However, a major question does remain, which has not yet been addressed in the mammalian literature: if the evolution of larger, stronger and faster cardiomyocytes with smaller surface area to volume ratios required a T-tubular network and greater SR dependence, then how are atrial myocytes, which are morphologically similar to ectothermic cardiomyocytes but extremely SR dependent, able to produce such high frequencies of contraction and contractile force? The bird cardiomyocyte may provide important information in this regard. Birds are endothermic, with hearts rates and cardiac output pressures as high, if not higher than mammals, but avian cardiomyocytes are also thinner and lack T-tubules. The structure of a cardiomyocyte may not necessarily be a good indicator of SR dependence and contractile performance but comparative cardiology has much to offer in uncovering the correlation between myocyte morphology and SR function.

Ectothermic cardiac SR dependence is not as clean-cut as the mammalian paradigm. While most studies suggest athletic ectotherms rely more strongly on SR Ca^{2+} cycling, phylogenetic investigations suggest other factors may be important, and environmental conditions are also major determinants. Additionally, the dichotomy of the large Ca^{2+} storage capacity of ectothermic cardiomyocytes and the relatively small role of the SR in E–C coupling on a beat-to-beat basis raises many questions. It is possible that the SR may be important during periods of environmental stress associated with increases in intracellular Ca^{2+}, such as acidosis, hypoxia and anoxia. Ca^{2+} may be sequestered in the SR during these periods to protect the heart from Ca^{2+} overload, which is known to promote cellular necrosis and apoptosis in mammalian hearts. Early studies have found this hypothesis, while attractive, difficult to demonstrate under physiologically relevant conditions. Certainly, this is an avenue of research that requires greater effort and could prove especially interesting and clinically relevant to oxygen-related diseases of the human heart.

References

Agata N, Tanaka H, Shigenobu K (1994) Inotropic effects of ryanodine and nicardipine on fetal, neonatal and adult guinea-pig myocardium. Eur J Pharm 260:47–55

Aho E, Vornanen M (1998) Ca^{2+}-ATPase activity and Ca^{2+} uptake by sarcoplasmic reticulum in fish heart: effects of thermal acclimation. J Exp Biol 201:525–532

Aho E, Vornanen M (1999) Contractile properties of atrial and ventricular myocardium of the heart of rainbow trout (*Oncorhynchus mykiss*): effects of thermal acclimation. J Exp Biol 202:2663–2677

Anelli LC Jr, Ole CD, Costa MJ, Rantin FT, Kalinin AL (2004) Effects of temperature and calcium availability on ventricular myocardium from the neotropical teleost *Piaractus mesopotamicus* (Holmberg 1887—Teleostei, Serrasalmidae). J Therm Biol 29:103–113

Arai M, Otsu K, MacLennan DH, Periasamy M (1992) Regulation of sarcoplasmic reticulum gene expression during cardiac and skeletal muscle development. Am J Physiol 262:C614–20

Bailey JR, Val AL, Almeida-Val V, Driedzic WR (1999) Anoxic cardiac performance in Amazonian and north-temperate-zone teleosts. Can J Zool 77:683–689

Belke DD, Milner RE, Wang LC (1991) Seasonal variations in the rate and capacity of cardiac SR calcium accumulation in a hibernating species. Cryobiology 28:354–63

Bers DM (1985) Ca influx and sarcoplasmic reticulum Ca release in cardiac muscle activation during postrest recovery. Am J Physiol 248:H366–H381

Bers DM (1989) SR Ca^{2+} loading in cardiac muscle preparations based on rapid-cooling contractures. Am J Physiol 256:C109–C120

Bers DM (2001) Excitation-contraction coupling and contractile force. Kluwer Academic, Dordecht

Bers DM (2002) Cardiac excitation-contraction coupling. Nature 415:198–205

Bers DM, Bridge JH (1989) Relaxation of rabbit ventricular muscle by Na-Ca exchange and sarcoplasmic reticulum calcium pump. Ryanodine and voltage sensitivity. Circ Res 65:334–342

Birkedal R, Christopher J, Thistlethwaite A, Shiels H (2009) Temperature acclimation has no effect on ryanodine receptor expression or subcellular localization in rainbow trout heart. J Comp Physiol B 179:961–969

Bootman MD, Higazi DR, Coombes S, Roderick HL (2006) Calcium signalling during excitation-contraction coupling in mammalian atrial myocytes. J Cell Sci 119:3915–3925

Bowler K, Tirri R (1990) Temperature dependence of the heart isolated from the cold or warm acclimated perch (*Perca fluviatilis*). Comp Biochem Physiol A 96:177–180

Castilho PC, Landeira-Fernandez AM, Morrissette J, Block BA (2007) Elevated Ca^{2+} ATPase (SERCA2) activity in tuna hearts: comparative aspects of temperature dependence. Comp Biochem Physiol 148:124–132

Chugun A, Oyamada T, Temma K, Hara Y, Kondo H (1999) Intracellular Ca^{2+} storage sites in the carp heart: comparison with the rat heart. Comp Biochem Physiol 123:61–67

Chugun A, Taniguchi K, Murayama T, Uchide T, Hara Y, Temma K, Ogawa Y, Akera T (2003) Subcellular distribution of ryanodine receptors in the cardiac muscle of carp (*Cyprinus carpio*). Am J Physiol 285:R601–R609

Cordeiro JM, Malone JE, Di Diego JM, Scornik FS, Aistrup GL, Antzelevitch C, Wasserstrom JA (2007) Cellular and subcellular alternans in the canine left ventricle. Am J Physiol 293: H3506–16

Dan P, Lin E, Huang J, Biln P, Tibbits GF (2007) Three-dimensional distribution of cardiac Na^{+}-Ca^{2+} exchanger and ryanodine receptor during development. Biophys J 93:2504–2518

Di Maio A, Block BA (2008) Ultrastructure of the sarcoplasmic reticulum in cardiac myocytes from Pacific bluefin tuna. Cell Tissue Res 334:121–134

Dibb KM, Hagarty CL, Loudon ASI, Trafford AW (2005) Photoperiod-dependent modulation of cardiac excitation contraction coupling in the Siberian hamster. Am J Physiol 288: R607–R614

Dibb KM, Clarke JD, Horn MA, Richards MA, Graham HK, Eisner DA, Trafford AW (2009) Characterization of an extensive transverse tubular network in sheep atrial myocytes and its depletion in heart failure. Circ Heart Fail 2:482–9

Endoh M (2004) Force-frequency relationship in intact mammalian ventricular myocardium: physiological and pathophysiological relevance. Eur J Pharm 500:73–86

Fabiato A (1983) Calcium-induced release of calcium from the cardiac sarcoplasmic reticulum. Am J Physiol 245:1–14

Fisher DJ, Tate CA, Phillips S (1992) Developmental regulation of the sarcoplasmic reticulum calcium pump in the rabbit heart. Pediatr Res 31:474–9

Galli GLJ, Taylor EW, Shiels HA (2006a) Calcium flux in turtle ventricular myocytes. Am J physiol 291:R1781–R1789

Galli GLJ, Gesser H, Taylor EW, Shiels HA, Wang T (2006b) The role of the sarcoplasmic reticulum in the generation of high heart rates and blood pressures in reptiles. J Exp Biol 209: 1956–1963

Galli GLJ, Shiels HA, Brill R (2008) Temperature sensitivity of cardiac function in pelagic fishes with different vertical mobilities: yellowfin tuna (*Thunnus albacares*), bigeye tuna (*T. obesus*), mahimahi (*Coryphaena hippurus*) and swordfish (*Xiphias gladius*). Phys Biochem Zool 82: 280–90

Galli GLJ, Warren DE, Shiels HA (2009) Ca^{2+} cycling in cardiomyocytes from a high-performance reptile, the varanid lizard (*Varanus exanthematicus*). Am J physiol 297:R1636–44

Galli GL, Lipnick MS, Shiels HA, Block BA (2011) Temperature effects on Ca2+ cycling in scombrid cardiomyocytes: a phylogenetic comparison. J Exp Biol 214:1068–76

Gamperl AK, Swafford BL, Rodnick KJ (2011) Elevated temperature, per se, does not limit the ability of rainbow trout to increase stroke volume. J Therm Biol 36:7–14

Gyorke I, Gyorke S (1998) Regulation of the cardiac ryanodine receptor channel by luminal Ca^{2+} involves luminal Ca^{2+} sensing sites. Biophys J 75:2801–10

Gyorke S, Terentyev D (2008) Modulation of ryanodine receptor by luminal calcium and accessory proteins in health and cardiac disease. Cardiovasc Res 77:245–55

Haddock PS, Coetzee WA, Cho E, Porter L, Katoh H, Bers DM, Jafri MS, Artman M (1999) Subcellular $[Ca^{2+}]_i$ gradients during excitation-contraction coupling in newborn rabbit ventricular myocytes. Circ Res 85:415–427

Harrer JM, Haghighi K, Kim HW, Ferguson DG, Kranias EG (1997) Coordinate regulation of SR Ca^{2+}-ATPase and phospholamban expression in developing murine heart. Am J Physiol 272:H57–66

Haverinen J, Vornanen M (2008) Responses of action potential and K^{+} currents to chronic thermal stress in fish hearts. Phylogeny or thermal preferences? Phys Biochem Zool 82:468–82

Haverinen J, Vornanen M (2009) Comparison of sarcoplasmic reticulum calcium content in atrial and ventricular myocytes of three fish species. Am J Physiol 297:R1180–7

Herve JC, Yamaoka K, Twist VW, Powell T, Ellory JC, Wang LC (1992) Temperature dependence of electrophysiological properties of guinea pig and ground squirrel myocytes. Am J Physiol 263:177–84

Hove-Madsen L, Bers DM (1993a) Passive Ca buffering and SR Ca uptake in permeabilized rabbit ventricular myocytes. Am J Physiol 264:C677–C686

Hove-Madsen L, Bers DM (1993b) Sarcoplasmic reticulum Ca^{2+} uptake and thapsigargin sensitivity in permeabilized rabbit and rat ventricular myocytes. Circ Res 73:820–828

Hove-Madsen L, Llach A, Tort L (1998) Quantification of Ca^{2+} uptake in the sarcoplasmic reticulum of trout ventricular myocytes. Am J Physiol 44:R2070–R2080

Hove-Madsen L, Llach A, Tort L (2001) The function of the sarcoplasmic reticulum is not inhibited by low temperatures in trout atrial myocytes. Am J Physiol 281:R1902–R1906

Huang J, Hove-Madsen L, Tibbits GF (2008) Ontogeny of Ca^{2+}-induced Ca^{2+} release in rabbit ventricular myocytes. Am J Physiol 294:C516–C525

Huser J, Lipsius SL, Blatter LA (1996) Calcium gradients during excitation-contraction coupling in cat atrial myocytes. J Physiol 494:641–651

Keen JE, Farrell AP, Tibbits GF, Brill RW (1992) Cardiac physiology in Tunas. 2. Effect of ryanodine, calcium, and adrenaline on force frequency relationships in atrial strips from skipjack tuna, Katsuwonus Pelamis. Can J Zool 70:1211–1217

Keen JE, Vianzon DM, Farrell AP, Tibbits GF (1994) Effect of temperature and temperature-acclimation on the ryanodine sensitivity of the trout myocardium. J Comp Physiol 164: 438–443

Korajoki H, Vornanen M (2009) Expression of calsequestrin in atrial and ventricular muscle of thermally acclimated rainbow trout. J Exp Biol 212:3403–14

Landeira Fernandez A, Morrisette JM, Blank JM, Block BA (2004) Temperature dependence of Ca^{2+}-ATPase (SERCA2) in the ventricles of tuna and mackerel. Am J Physiol 286: R398–R404

Laver DR (2007) Ca^{2+} stores regulate ryanodine receptor Ca^{2+} release channels via luminal and cytosolic Ca^{2+} sites. Clin Exp Pharm Physiol 34:889–96

Llach A, Tibbits GF, Sedarat F, Tort L, Hove-Madsen L (2001) Low temperature reduces Na^{+}-Ca^{2+} exchange rate but not SR Ca^{2+} release in trout atrial myocytes. Biophysical J 80:585A

Llach A, Molina CE, Alvarez-Lacalle E, Tort L, Benítez R and Hove-Madsen L (2011) Detection, properties and frequencies of local calcium release form the sarcoplasmic reticulum in teleost cardiomyocytes. PLOS one 6: e23708

Lukyanenko V, Ziman A, Lukyanenko A, Salnikov V, Lederer WJ (2007) Functional groups of ryanodine receptors in rat ventricular cells. J Physiol 583:251–269

Luss I, Boknik P, Jones LR, Kirchhefer U, Knapp J, Linck B, Luss H, Meissner A, Muller FU, Schmitz W, Vahlensieck U, Neumann J (1999) Expression of cardiac calcium regulatory proteins in atrium v ventricle in different species. J Mol Cell Cardiol 31:1299–1314

Mackenzie L, Bootman MD, Berridge MJ, Lipp P (2001) Predetermined recruitment of calcium release sites underlies excitation-contraction coupling in rat atrial myocytes. J Physiol 530:417–429

Mahony L, Jones LR (1986) Developmental changes in cardiac sarcoplasmic reticulum in sheep. J Biol Chem 261:15257–65

Maylie JG (1982) Excitation-contraction coupling in neonatal and adult myocardium of cat. Am J physiol 242:H834–43

Michalak M, Opas M (2009) Endoplasmic and sarcoplasmic reticulum in the heart. Trends Cell Biol 19:253–9

Milner RE, Michalak M, Wang LC (1991) Altered properties of calsequestrin and the ryanodine receptor in the cardiac sarcoplasmic reticulum of hibernating mammals. Biochim Biophys Acta 1063:120–128

Minajeva A, Kaasik A, Paju K, Seppet E, Lompre AM, Veksler V, Ventura-Clapier R (1997) Sarcoplasmic reticulum function in determining atrioventricular contractile differences in rat heart. Am J Physiol 273:2498–507

Negretti N, Varro A, Eisner DA (1995) Estimate of net calcium fluxes and sarcoplasmic reticulum calcium content during systole in rat ventricular myocytes. J Physiol 486:581–591

Puglisi JL, Yuan WL, Bassani JWM, Bers DM (1999) Ca^{2+} influx through Ca^{2+} channels in rabbit ventricular myocytes during action potential clamp: Influence of temperature. Circ Res 85: E7–E16

Ravens U, Gath J, Hussaini MA, Himmel H (1997) Mechanical restitution in atrial muscle from human and rat hearts: effects of agents that modify sarcoplasmic reticulum function. Pharmacol Toxicol 81:97–104

Rivaroli L, Rantin FT, Kalinin AL (2006) Cardiac function of two ecologically distinct Neotropical freshwater fish: Curimbata, *Prochilodus lineatus* (Teleostei, Prochilodontidae), and trahira, *Hoplias malabaricus* (Teleostei, Erythrinidae). Comp Biochem Physiol A 145:322–7

Rocha ML, Rantin FT, Kalinin AL (2007) Importance of the sarcoplasmic reticulum and adrenergic stimulation on the cardiac contractility of the neotropical teleost *Synbranchus marmoratus* under different thermal conditions. J Comp Physiol B 177:713–21

Rossi AE, Dirksen RT (2006) Sarcoplasmic reticulum: the dynamic calcium governor of muscle. Muscle Nerve 33:715–31

Rossi D, Barone V, Giacomello E, Cusimano V, Sorrentino V (2008) The sarcoplasmic reticulum: an organized patchwork of specialized domains. Traffic 9:1044–9

Santer RM (1974) The organization of the sarcoplasmic reticulum in teleost ventricular myocardial cells. Cell Tissue Res 151:395–402

Sedarat F, Lin E, Moore EDW, Tibbits GF (2004) Deconvolution of confocal images of dihydropyridine and ryanodine receptors in developing cardiomyocytes. J Appl Physiol 97: 1098–1103

Shannon TR, Guo T, Bers DM (2003) Ca^{2+} scraps: local depletions of free [Ca^{2+}] in cardiac sarcoplasmic reticulum during contractions leave substantial Ca^{2+} reserve. Circ Res 93:40–5

Shiels HA, Farrell AP (1997) The effect of temperature and adrenaline on the relative importance of the sarcoplasmic reticulum in contributing Ca^{2+} to force development in isolated ventricular trabeculae from rainbow trout. J Exp Biol 200:1607–1621

Shiels HA, White E (2003) Confocal imaging of Ca 2+ transients and Ca2+ sparks in cardiac myocytes from rainbow trout. J Physiol 551P:PC8

Shiels HA, White E (2005) Temporal and spatial properties of cellular Ca^{2+} flux in trout ventricular myocytes. Am J Physiol 288:R1756–R1766

Shiels HA, Freund EV, Farrell AP, Block BA (1999) The sarcoplasmic reticulum plays a major role in isometric contraction in atrial muscle of yellowfin tuna. J Exp Biol 202:881–90

Shiels HA, Vornanen M, Farrell AP (2000) Temperature-dependence of L-type Ca2+ channel current in atrial myocytes from rainbow trout. J Exp Biol 203:2771–2780

Shiels HA, Vornanen M, Farrell AP (2001) Effect of temperature on Ca^{2+} cycling in rainbow trout myocytes. Biophysical Journal Annual Meeting Abstracts, 646a

Shiels HA, Vornanen M, Farrell AP (2002a) Effects of temperature on intracellular [Ca^{2+}] in trout atrial myocytes. J Exp Biol 205:3641–3650

Shiels HA, Vornanen M, Farrell AP (2002b) Temperature dependence of cardiac sarcoplasmic reticulum function in rainbow trout myocytes. J Exp Biol 205:3631–3639

Shiels HA, Vornanen M, Farrell AP (2002c) The force-frequency relationship in fish hearts–a review. Comp Biochem Phsyiol 132:811–826

Shiels HA, Vornanen M, Farrell AP (2003) Acute temperature change modulates the response of I Ca to adrenergic stimulation in fish cardiomyocytes. Physiol Biochem Zool 76:816–24

Shiels HA, Blank JM, Farrell AP, Block BA (2004) Electrophysiological properties of the L-type Ca^{2+} current in cardiomyocytes from bluefin tuna and Pacific mackerel. Am J Physiol 286:R659–R668

Shiels HA, Paajanen V, Vornanen M (2006) Sarcolemmal ion currents and sarcoplasmic reticulum Ca^{2+} content in ventricular myocytes from the cold stenothermic fish, the burbot (*Lota lota*). J Exp Biol 209:3091–3100

Shiels HA, Di Maio A, Thompson S, Block BA (2011) Warm fish with cold hearts: thermal plasticity of excitation-contraction coupling in bluefin tuna. Proc Biol Sci 278:18–27

Sitsapesan R, Williams AJ (1994) Regulation of the gating of the sheep cardiac sarcoplasmic reticulum Ca^{2+}-release channel by luminal Ca^{2+}. J Membr Biol 137:215–226

Sitsapesan R, Williams AJ (1995) The gating of the sheep skeletal sarcoplasmic reticulum Ca^{2+}-release channel is regulated by luminal Ca^{2+}. J Membr Biol 146:133–144

Sitsapesan R, Montgomery RAP, Macleod KT, Williams AJ (1991) Sheep cardiac sarcoplasmic-reticulum calcium-release channels: modification of conductance and gating by temperature. J Physiol 434:469–488

Su Z, Li F, Spitzer KW, Yao A, Ritter M, Barry WH (2003) Comparison of sarcoplasmic reticulum Ca^{2+}-ATPase function in human, dog, rabbit, and mouse ventricular myocytes. J Mol Cell Cardiol 35:761–7

Tanaka H, Takagi N, Shigenobu K (1995) Difference in excitation-contraction mechanisms between atrial and ventricular myocardia of hatched chicks. Gen Pharmacol Vasc Syst 26:45–49

Terentyev D, Viatchenko-Karpinski S, Gyorke I, Volpe P, Williams SC, Gyorke S (2003) Calsequestrin determines the functional size and stability of cardiac intracellular calcium stores: mechanism for hereditary arrhythmia. Proc Natl Acad Sci USA 100:11759–64

Thomas MJ, Hamman BN, Tibbits GF (1996) Dihydropyridine and ryanodine binding in ventricles from rat; trout; dogfish and hagfish. J Exp Biol 199:1999–2009

Tibbits GF, Hovemadsen L, Bers DM (1991) Calcium-transport and the regulation of cardiac contractility in teleosts: a comparison with higher vertebrates. Can J Zool 69:2014–2019

Tibbits GF, Xu L, Sedarat F (2002) Ontogeny of excitation-contraction coupling in the mammalian heart. Comp Biochem Physiol A 132:691–698

Tiitu V, Vornanen M (2002) Regulation of cardiac contractility in a stenothermal fish, the burbot (*Lota lota*). J Exp Biol 205:1597–1606

Tiitu V, Vornanen M (2003) Ryanodine and dihydropyridine receptor binding in ventricular cardiac muscle of fish with different temperature preferences. J Comp Physiol 173:285–291

Tijskens P, Meissner G, Franzini-Armstrong C (2003) Location of ryanodine and dihydropyridine receptors in frog myocardium. Biophys J 84:1079–1092

Vornanen M (1997) Sarcolemmal Ca^{2+} influx through L-type Ca^{2+} channels in ventricular myocytes of a teleost fish. Am J physiol 41:R1432–R1440

Vornanen M (2005) Temperature and Ca^{2+} dependence of [3H]ryanodine binding in the burbot (*Lota lota L.*) heart. Am J Physiol 290:R345–R351

Vornanen M, Shiels HA, Farrell AP (2002) Plasticity of excitation-contraction coupling in fish cardiac myocytes. Comp Biochem Physiol A 132:827–846

Walden AP, Dibb KM, Trafford AW (2009) Differences in intracellular calcium homeostasis between atrial and ventricular myocytes. J Mol Cell Cardiol 46:463

Wang T, Altimiras J, Klein W, Axelsson M (2003) Ventricular haemodynamics in Python molurus: separation of pulmonary and systemic pressures. J Exp Biol 206:4241–4245

Webb G, Heatwole H, De Bevay DBJ (1971) Comparative cardiac anatomy of the reptilia. I. The chambers and septa of the varanid ventricle. J Morphol 134:335–50

Xue XH, Hryshko LV, Nicoll DA, Philipson KD, Tibbits GF (1999) Cloning, expression, and characterization of the trout cardiac Na^+/Ca^{2+} exchanger. Am J Physiol 277:C693–C700

Chapter 6
Evolution of the Regulatory Control of the Vertebrate Heart: The Role of the Contractile Proteins

Todd E. Gillis

Introduction

Evolutionary pressures have driven the vertebrate heart to transform from the simple, single-chambered structure found in the tunicates to the four-chambered organ with multiple valves and septa found in mammals and birds. Between these two extremes are intermediate forms found in other vertebrate groups. This includes the teleost heart where there is a single ventricle and a single atrium and the amphibian heart with a single ventricle but two atria. Along with these morphological changes, there is also significant variation in the functional capacity of hearts from different vertebrate groups. For example, the heart rate of a Brazilian hummingbird, *Amazilia lacteal*, is 1,056 beats per minute (bpm) (Bishop and Butler 1995), that of a mouse heart is greater than 600 bpm (Burggren et al. 1997; Farmer 1999; Suzuki et al. 2002), and that of a trout is 93 bpm at 15°C (Priede 1974). These differences are due to significant variation in pacemaker firing rate and in the rate of Ca^{2+} flux through the myocytes. There is also variation in the ability of the contractile element to sense, and respond to, changes in intercellular Ca^{2+}.

The evolution of endothermy in vertebrates represented a significant challenge to contractile function. This is because an increase in temperature dramatically increases the Ca^{2+} sensitivity of the heart (Harrison and Bers 1989). Such an increase in Ca^{2+} sensitivity caused by an increase in physiological temperature would reduce the rate of contraction and possibly cause the heart to remain in contracture. It was therefore necessary for the Ca^{2+} sensitivity of the heart to decrease with the evolution of endothermy in the vertebrates.

In addition to modifications that allow for high rates of contraction at high temperatures there have also been changes to the components of the cardiac myocytes

T.E. Gillis, Ph.D. (✉)
Department of Integrative Biology, University of Guelph,
Guelph, ON, Canada N1G 2W1
e-mail: tgillis@uoguelph.ca

D. Sedmera and T. Wang (eds.), *Ontogeny and Phylogeny of the Vertebrate Heart*,
DOI 10.1007/978-1-4614-3387-3_6,

that allow for greater regulatory control. Every myocyte within the vertebrate heart contracts with each heartbeat. This differs from skeletal muscle where the number of cells activated increases with the force required. As the oxygen requirements of an animal are not constant, but influenced by factors such as activity level, stress level, and reproductive status, it is important that animals can accommodate cardiac output to the fluctuating need. The mechanisms that enable such control include the Frank–Starling response where force generation of the heart increases as sarcomere length increases and the stimulatory effect of β-adrenergic stimulation (β-AS) on cardiac function. (α-adrenergic stimulation [α-AS] is also active within the vertebrate heart but it does not play as a significant role in regulating cardiac output.) As a result of these regulatory control mechanisms, there is a significant dynamic range in the functional capacity of the vertebrate heart. While these mechanisms are present in skeletal muscle, they have a larger influence on the function of cardiac muscle.

In mammals, the activation of β-receptors leads to an increase in heart rate, the strength of contraction, the rate of contraction, as well as the rate of relaxation between beats (Bers 2001; Kobayashi and Solaro 2005; Stelzer et al. 2007; Weisberg and Winegrad 1996). These changes result in a dramatic increase in cardiac output as more blood is being pumped at a faster rate. The stimulus of contractile function following β-AS is due, in part, to modifications in the function of specific Ca^{2+} handling and contractile proteins within the vertebrate heart caused by phosphorylation by protein kinase A (PKA). This kinase targets specific residues within the responsible proteins. One reason why the heart is more sensitive to β-AS than other muscle types is that there are targets for PKA in the cardiac isoforms of a number of regulatory proteins including cardiac (c) troponin I (cTnI) and cardiac myosin-binding protein C (cMyBP-C) (Shaffer and Gillis 2010). The isoforms of these proteins that are found in slow skeletal (ss) and fast skeletal (fs) muscle lack these target residues. The functional properties of the proteins and therefore of the muscle types are not as impacted as cardiac muscle following β-AS (Shaffer and Gillis 2010).

The stimulation of α-adrenergic receptors in the mammalian heart leads to a reduction in contractile function and force generation (Braz et al. 2004). These functional changes are due to the activation of protein kinase C (PKC) that in turn influences the function of a number of regulatory proteins either directly through phosphorylation or indirectly by initiating a signal cascade resulting in a change in phosphorylation state (Braz et al. 2004).

Interestingly, phylogenetic analyses of a number of different contractile proteins suggest that the ancestral form of the contractile proteins was "fast skeletal like" and that the cardiac and slow skeletal isoforms are the product of two subsequent genome duplications (Ota and Saitou 1999; Shaffer and Gillis 2010). Previous work by Cleto et al. (2003) identified a single gene for TnI expressed in the heart, the body wall, and the embryonic larval tail of the tunicate *Ciona intestinalis*. As tissue-specific TnI isoforms have been identified in teleost fast skeletal, slow skeletal, and cardiac muscle, this indicates that the gene duplication events occurred in a common ancestor to all vertebrates after the urochordates diverged from the vertebrate lineage (500 million years ago), but before the teleosts branched off (440 million years ago) (Shaffer and Gillis 2010). This means that the amino acid sequence of the TnI isoform that evolved and became specialized to function within the vertebrate heart most likely

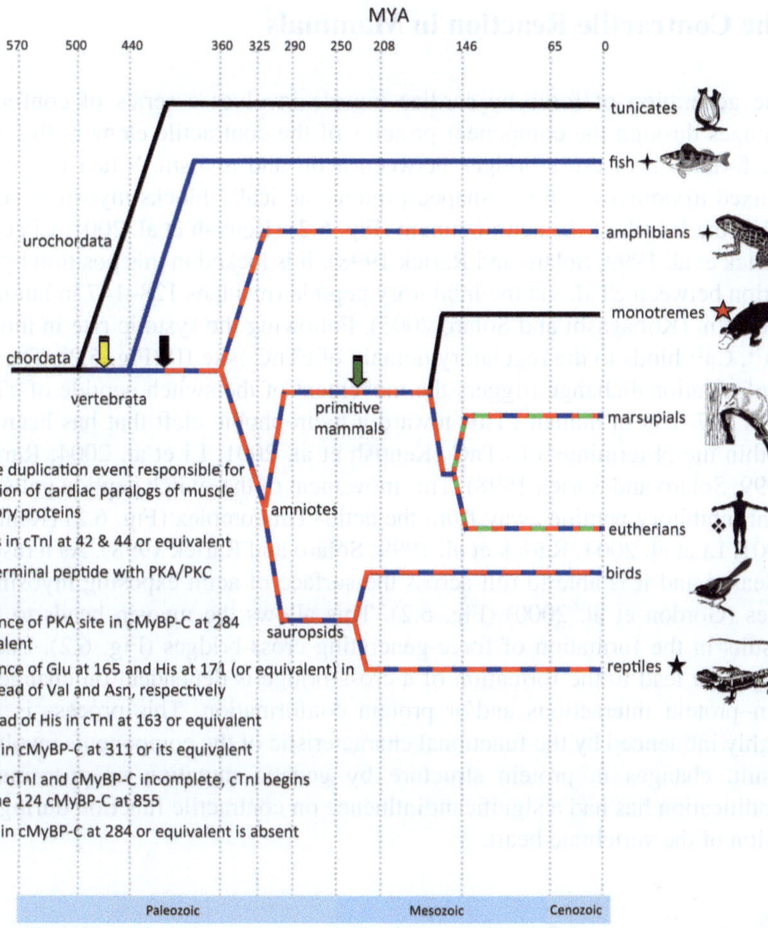

Fig. 6.1 A vertebrate phylogeny onto which critical changes in the sequence of cTnI and cMyBP-C have been mapped. Figure modified from Shaffer and Gillis (2010)

was similar to fast skeletal (fs) TnI of extant vertebrates. It was then subsequent changes to this protein that produced the tissue-specific isoform, with unique functional properties, found in the vertebrate heart today (Fig. 6.1).

The goal of this chapter is to examine the relationship between the contractile element and the functional changes that have occurred during the evolution of the vertebrate heart. This includes examining the mechanistic basis for the influence of β-AS on contractile function as well as the changes that have occurred to enable cardiac function at relatively high temperatures in endothermic animals. To accomplish this, phylogenetic data are integrated with the results of functional studies examining the Ca^{2+} activation of cardiac tissue as well as cardiac proteins from different species. It is hoped that this work will provide new insight into the role that the contractile element plays in regulating cardiac function and how this has evolved in vertebrates.

The Contractile Reaction in Mammals

The generation of force by cardiac muscle involves a series of conformational changes through the component proteins of the contractile element that results in the formation of cross-bridges between actin and myosin. When the myocyte is relaxed tropomyosin, a rod-shaped protein, sterically blocks myosin-binding sites along the length of the actin filament (Fig. 6.2) (Kentish et al. 2001; Li et al. 2004; Rarick et al. 1999; Solaro and Rarick 1998). It is locked in this position by an interaction between cTnI, via the inhibitory peptide (residues 128–147 in human cTnI), and actin (Kobayashi and Solaro 2005). Following the systolic rise in intracellular Ca^{2+}, Ca^{2+} binds to the regulatory domain of cTnC (site II) (Fig. 6.2). The resultant conformational change triggers the movement of the switch peptide of cTnI (residues 147–163 in human cTnI) toward a hydrophobic cleft that has been exposed within the N-terminus of cTnC (Kentish et al. 2001; Li et al. 2004; Rarick et al. 1999; Solaro and Rarick 1998). This movement of the switch peptide pulls the adjacent inhibitory peptide away from the actin–Tm complex (Fig. 6.2) (Kentish et al. 2001; Li et al. 2004; Rarick et al. 1999; Solaro and Rarick 1998). As a result Tm is released and it is able to roll across the surface of actin exposing myosin-binding sites (Gordon et al. 2000) (Fig. 6.2). This allows the myosin heads to bind and results in the formation of force-generating cross-bridges (Fig. 6.2). Each of the steps that lead to the formation of a cross-bridge is dependent on changes in protein–protein interactions and/or protein conformation. This process is therefore highly influenced by the functional characteristic of the components involved. As a result, changes in protein structure by genetic mutation or post-translational modification has had a significant influence on contractile function during the evolution of the vertebrate heart.

The Influence of β-Adrenergic Stimulation on Cardiac Function

In vertebrates, β-receptors (β_1 and β_2 subtypes) are activated by norepinephrine which is released from sympathetic nerve ganglia. These receptors are coupled to G proteins that activate adenyl cyclase within the cell. Once activated, adenyl cyclase catalyzes the conversion of ATP to cyclic AMP, which then activates the catalytic subunit of PKA. It is the catalytic subunit of PKA that phosphorylates specific proteins within the cell by targeting individual residues (serine, tyrosine, threonine). For each of the different changes in contractile function that occur with β-AS, there is/are separate protein(s) involved (Table 6.1). By targeting the specific proteins responsible for each of these different functions, PKA acts to coordinate the response of the cell to β-AS.

The increase in heart rate caused by β-AS is due to the stimulation of β-receptors on the surface membrane of the cells that compose the sinoatrial (SA) node. The SA node is the primary pacemaker that generates action potentials at a rate that sets heartbeat frequency. Following β-receptor activation, there is an increase in the rate of action potential firing. This is caused primarily by the phosphorylation of the

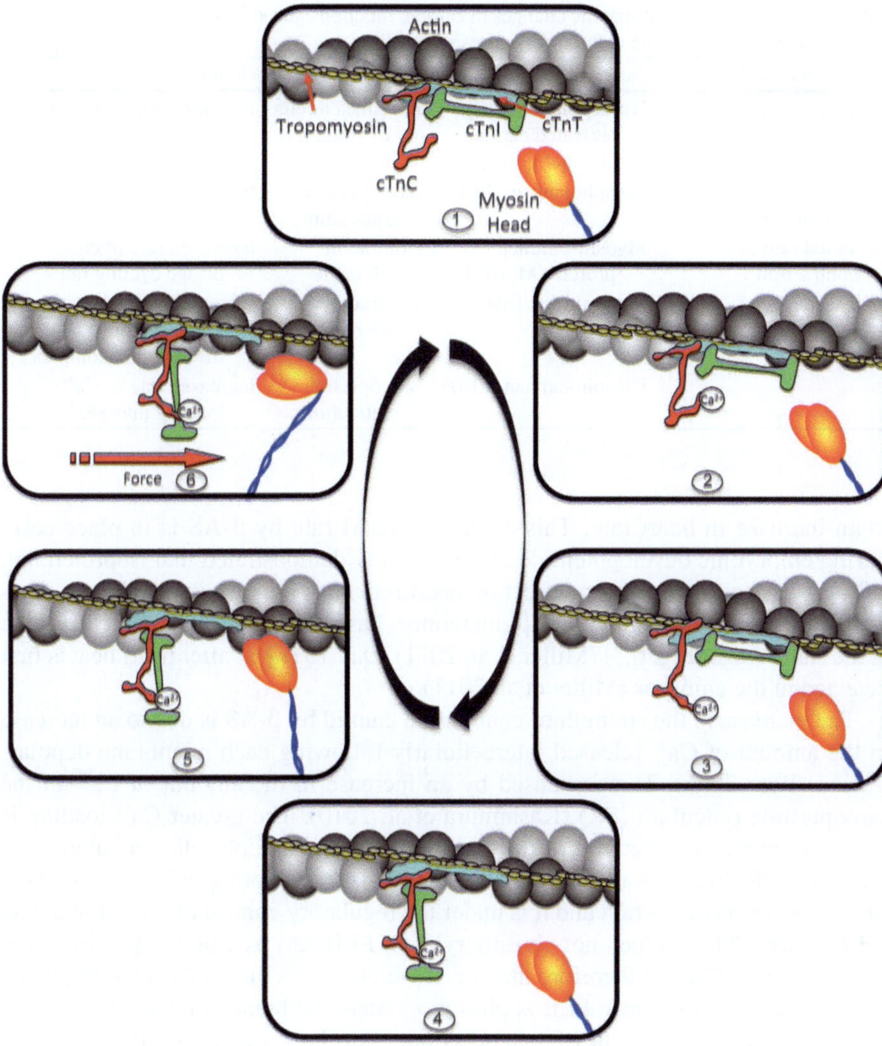

Fig. 6.2 The Ca^{2+} activation of the cardiac contractile element. (1) In the relaxed state, tropomyosin blocks the myosin-binding sites on actin. It is locked in this position by troponin I (cTnI) interacting with actin. (2) Ca^{2+} enters the cell and binds to the regulatory domain of troponin C (cTnC). (3) The conformation of cTnC changes exposing a hydrophobic patch. (4) The hydrophobic patch pulls cTnI away from the actin, releasing tropomyosin to move across the thin filament and expose myosin-binding sites. (5) The myosin head binds to actin. (6) The myosin head flexes and force is generated

hyperpolarization-activated cation channel I_f (DiFrancesco 1993). The phosphorylation of this channel increases the slope of diastolic depolarization thereby reducing the time required to reach the threshold for action potential generation (DiFrancesco 1993; Irisawa et al. 1993). As a result, the time between each action potential is reduced. This results in an increased frequency of membrane depolarization leading

Table 6.1 Summary of the specific changes in cardiac function caused by β-adrenergic stimulation and the specific proteins and cellular responses responsible

Change in heart function	Target protein(s)	Location	Cellular response
Increased heart rate	Hyperpolarization-activated cation channel I_f	Pacemaker cells of sinoatrial node	Increased frequency of membrane depolarization
Increased strength of contraction	Phospholamban (PLB)	Sarcoplasmic reticulum	Increased SR Ca^{2+} loading
Increased rate of contraction	Myosin-binding protein (MyBP-C)	Contractile element	Increased rate of cross-bridge cycling rates
Increased rate of relaxation	Troponin I (cTnI)	Contractile element	Increased rate of Ca^{2+} disassociation (k_{off}) from the troponin complex
	Phospholamban (PLB)	Sarcoplasmic reticulum	Increased rate of Ca^{2+} loading into SR

to an increase in heart rate. This control of heart rate by β-AS is in place early during embryonic development. We have recently demonstrated that isoproterenol, a β-agonist, has a stimulatory effect on heart rate in trout embryos at 15 days post-fertilization, while propranolol, a β-antagonist, has an inhibitory affect on heart rate at the same stage (Fig. 6.3) (Miller et al. 2011). Day 15 is just after heart beat is first detected in the embryos (Miller et al. 2011).

The increase in the strength of contraction caused by β-AS is due to an increase in the amount of Ca^{2+} released intercellularly following each membrane depolarization (Bers 2001). This is caused by an increase in the amount of Ca^{2+} in the sarcoplasmic reticulum (SR) (Kashimura et al. 2010). The greater Ca^{2+} loading is due to an increase in activity of the Ca^{2+} pump, sarco/endoplasmic reticulum Ca^{2+} ATPase (SERCA 2a) (Kashimura et al. 2010). This protein pumps Ca^{2+} back into the SR following a contraction and it is under the regulatory control of phospholamban (PLB) (Bers 2001). When not phosphorylated, PLB acts as a break that limits the activity of SERCA and therefore the rate at which Ca^{2+} is pumped back into the SR (McIvor et al. 1988). Once PLB is phosphorylated, the break is released and more Ca^{2+} is pumped back into the SR at a faster rate. This increased load is then available for release into the cytosol following the next membrane depolarization (Kashimura et al. 2010). The end result is an increase in the number of troponin complexes activated leading to more cross-bridges formed with each contraction. This translates into greater force production.

The increase in the rate of force generation with β-AS is caused by phosphorylation of cMyBP-C (Stelzer et al. 2007; Weisberg and Winegrad 1996). cMyBP-C is located in the A-band of the sarcomeres, where it binds to actin (Shaffer et al. 2009) and myosin (Gruen et al. 1999) via the M-domain (Fig. 6.4) (Barefield and Sadayappan 2010). This interaction is thought to limit cross-bridge formation (Ababou et al. 2008; Shaffer et al. 2009, p. 1406). The phosphorylation of the M-domain causes the release of myosin (Gruen et al. 1999) and actin (Shaffer et al. 2009) (Fig. 6.4). This change is thought to result in a change in the orientation of the

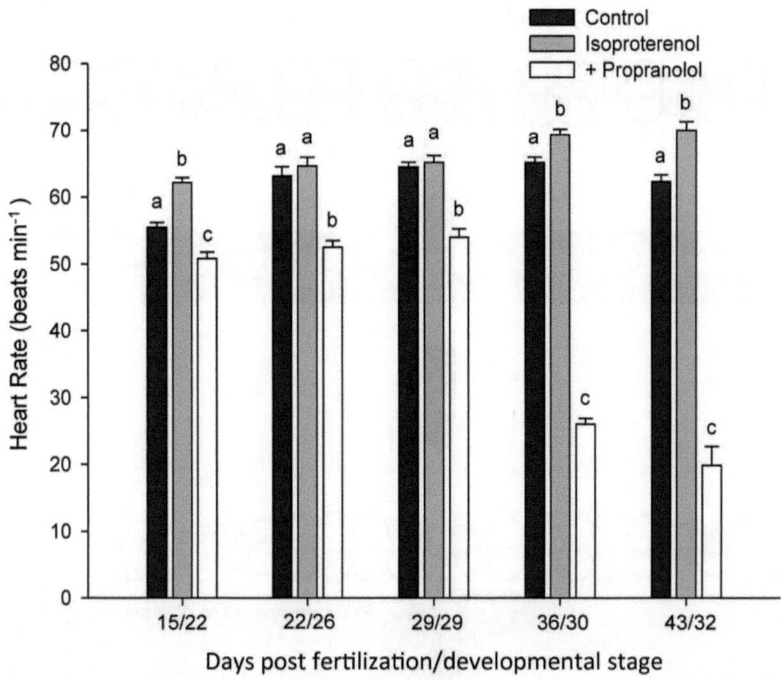

Fig. 6.3 The trout heart is responsive to adrenergic stimulation during early embryonic development. Heart rate measurements of rainbow trout embryos and larvae were measured in control conditions or following exposure to 0.1 mmol l^{-1} isoproterenol or combined 0.1 mmol l^{-1} isoproterenol + 1 mmol l^{-1} propranolol at different stages of development (Series I). Data are presented as means ± SEM ($n = 6$). *Different letters* indicate statistically significant differences between treatments at a given stage ($p < 0.05$). Figure modified from Miller et al. (2011)

myosin head and enables an increase in the rate of cross-bridge cycling (Stelzer et al. 2007; Weisberg and Winegrad 1996; Barefield and Sadayappan 2010). This leads to faster force generation within a single contraction.

The increase in the rate of muscle relaxation following a cardiac contraction with β-AS is due to the phosphorylation of cTnI (Fentzke et al. 1999; Kobayashi and Solaro 2005) and PLB (McIvor et al. 1988). On the N-terminus of cTnI cloned from mammalian hearts, there is a domain that contains two target residues of PKA/PKC (Fig. 6.5) (Noland et al. 1996). This domain, called the N-terminal extension, is only found in cTnI isoforms. It is not present in the TnI isoforms found in either slow skeletal muscle (ssTnI) or fast skeletal muscle (fsTnI). When the two PKA/PKC sites in the N-terminal extension are phosphorylated, the Ca^{2+} affinity of the troponin complex is reduced (Noland et al. 1996). This is due to changes in the interaction between cTnI and cTnC influencing the ability of cTnC to bind Ca^{2+} (Fig. 6.5) (Kobayashi and Solaro 2005). This reduction in Ca^{2+} affinity is caused by an increased rate of Ca^{2+} dissociation (k_{off}) from the regulatory domain of cTnC (Dong et al. 2007) and leads to a reduction in the Ca^{2+} sensitivity of force generation

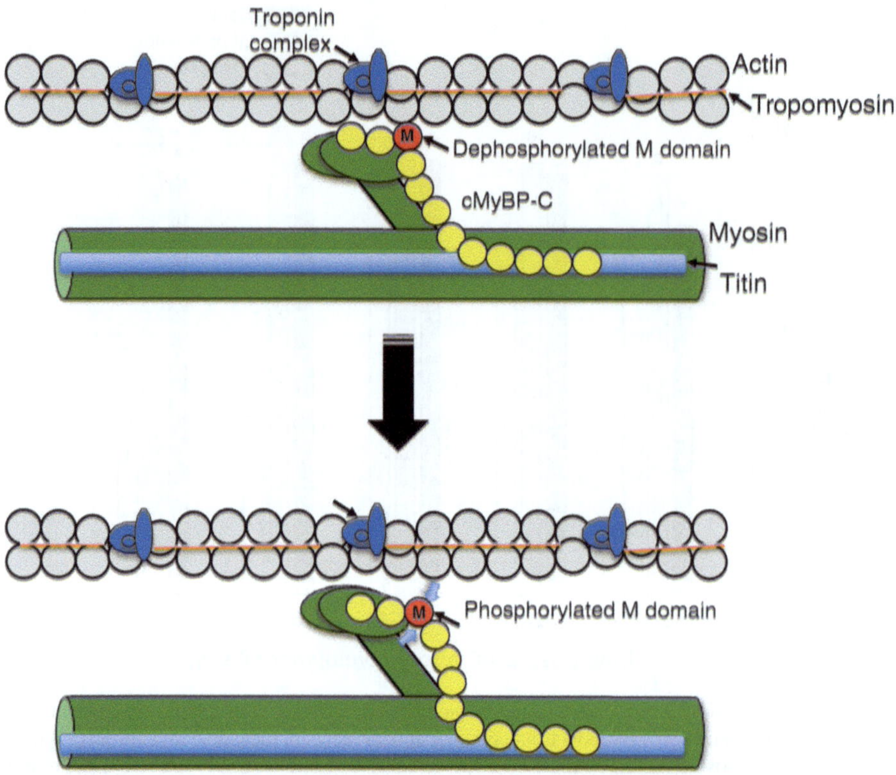

Fig. 6.4 The phosphorylation of cMyBP-C's M domain influences the interaction between myosin and actin within the contractile element. When not phosphorylated, the M domain of cMyBP-C in the N-terminus of the protein acts as a tether linking myosin and actin together. This interaction regulates the ability of the myosin head to bind to the actin filament during cross-bridge cycling. The C terminus of the cMyBP-C is anchored to titin. The phosphorylation of the M domain releases its interaction with myosin and actin. This removes the tether and enables a change in the orientation of the myosin head resulting in an increase in the rate of force production by the cardiac myocyte. Figure modified from Barefield and Sadayappan (2010)

(Fig. 6.6) (Fentzke et al. 1999). This effect of PKA activity translates into an increase in the rate at which the contractile element can be deactivated when Ca^{2+} is pumped back into the SR between cardiac contractions.

When cTnI is replaced with ssTnI within mouse (or murine) hearts using transgenic approaches, the effect of PKA activity on the Ca^{2+} sensitivity of force generation is eliminated (Fig. 6.6) (Fentzke et al. 1999). As described above, the phosphorylation of PLB increases the rate at which Ca^{2+} is pumped back into the SR via SERCA 2A following a contraction (Bers 2001). This causes an increase in the rate at which intercellular Ca^{2+} decreases between contractions. When this effect is coupled with the increased rate of Ca^{2+} dissociation from the Tn complex, caused by the phosphorylation of cTnI, the rate of muscle relaxation between contractions increases (Bers 2001).

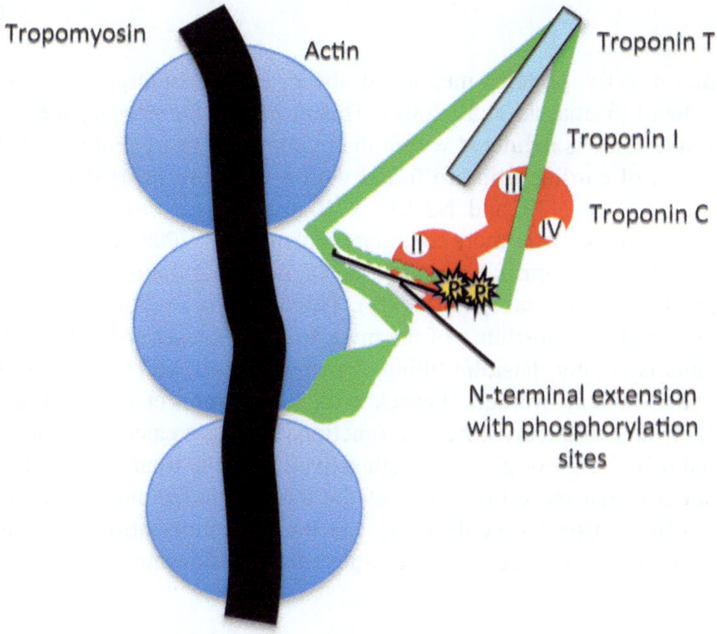

Fig. 6.5 Schematic of the interactions between the components of the troponin complex during diastole. Here the N-terminal extension of the mammalian isoform of cardiac troponin I (cTnI) is phosphorylated at serines 23 and 24, and as a result is interacting with the N-terminus of cardiac Troponin C (cTnC). This interaction regulates the ability of the cTnC to bind Ca^{2+}. cTnC is illustrated in the apo-state where site II in the regulatory domain is not bound by Ca^{2+}. Figure is modified from Solaro et al. (2008)

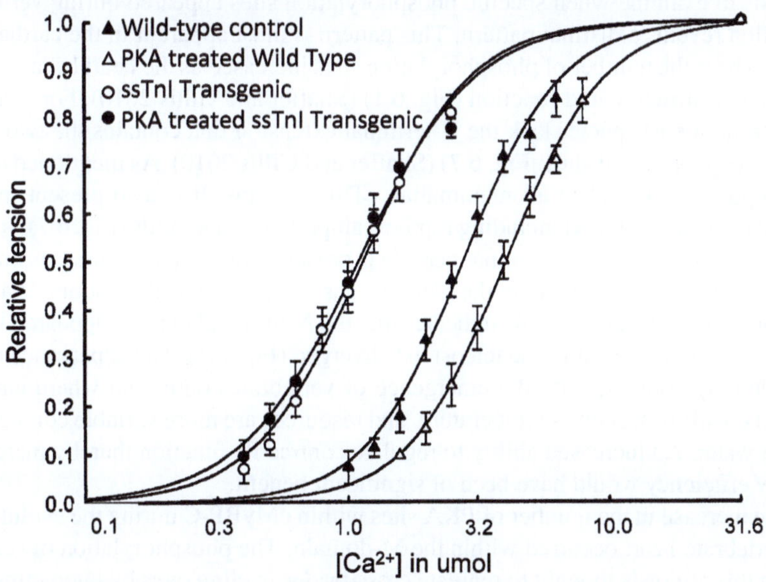

Fig. 6.6 The influence of PKA treatment on the Ca^{2+}-tension relationships of wild-type and transgenic cardiomyocytes expressing slow skeletal troponin I (ssTnI). PKA treatment caused the Ca^{2+} concentration at half maximum force, used as a measure of Ca^{2+} sensitivity, to increase by 1.5-fold. There was no influence of PKA treatment on the Ca^{2+} sensitivity of the transgenic preparations. Figure modified from Fentzke et al. (1999)

In addition to the proteins mentioned above, PKA also targets titin within the vertebrate heart (Yamasaki et al. 2002). This protein is the primary mechanism by which passive force is regulated within the mammalian heart (Fukuda et al. 2008). Two isoforms of cardiac titin are found within the adult heart that vary in their stiffness. These are N2B and N2BA. N2B is a shorter molecule with limited extensibility, while N2BA is more compliant (Fukuda et al. 2008). However, the extensibility of the N2B isoform can be increased via phosphorylation by PKA or cGMP-dependent kinase (Yamasaki et al. 2002). This leads to a reduction in the passive tension and greater extensibility of the myocardium (Kruger and Linke 2009). The result of this is greater diastolic filling in response to β-AS (Fukuda et al. 2008). Work by Shiels and co-workers (Patrick et al. 2010; Shiels et al. 2006) has demonstrated that trout cardiac myocytes are functional over a greater range of sarcomere lengths and this allows for greater length activation of the trout myocardium. These authors suggest that these properties of the trout myocardium are due in part to differences in the titin molecule as well as the phosphorylation of the molecule by PKA. Further work is required to establish this relationship.

The Evolution of TnI and MyBP-C in the Vertebrate Heart

The differences in the number and location of the phosphorylation sites within the tissue-specific isoforms of TnI and MyBP-C are an example of how the proteins have become specialized to function within a specific muscle type. The use of phylogenic analysis to examine when specific phosphorylation sites appeared during vertebrate evolution reveals a distinct pattern. This pattern is most apparent in the cardiac isoforms where the number of phosphorylation sites increased as the heart became more complex in structure and function (Fig. 6.1) (Shaffer and Gillis 2010). For example, cTnI from teleost species lack the N-terminal extension that contains the two PKA/PKC phosphorylation sites (Fig. 6.7) (Shaffer and Gillis 2010). As mentioned above, this peptide is present in all mammalian cTnI isoforms. It is also present in cTnI cloned from any tetrapod including reptiles, amphibians, and birds (Fig. 6.7). It is not known, however, if the N-terminal peptide is present within monotremes, as only the C-terminus of the cTnI gene in the platypus has been sequenced (residues 154–210) (Shaffer and Gillis 2010). This indicates that the N-terminal peptide appeared within the vertebrate lineage after the teleost fish diverged (Fig. 6.1). The appearance of this peptide may coincide with the emergence of vertebrates onto land where environmental conditions, such as temperature, and resources are more variable compared to that in water. An increased ability to regulate contractile function thereby increasing energy efficiency would have been of significant benefit.

The increase in the number of PKA sites within cMyBP-C during the evolution of the vertebrate heart occurred within the M-domain. The phosphorylation of residues within this region is thought to regulate cross-bridge cycling rates by interacting with actin and myosin (Gautel et al. 1995; Gruen et al. 1999; Shaffer et al. 2009). In mammalian cMyBP-C isoforms, this domain contains four PKA targets; in birds and

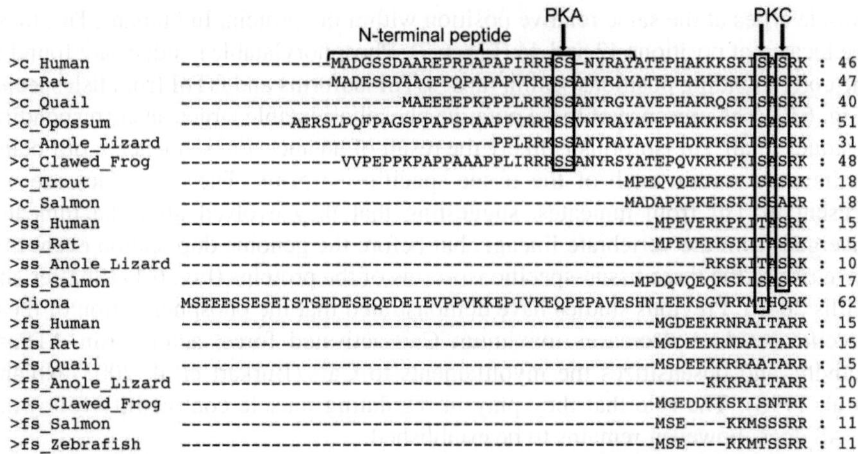

Fig. 6.7 Sequence alignment of the N-terminus of TnI sequences from cardiac, slow skeletal, and fast skeletal muscle from select vertebrate species. Human cTnI protein kinase A (PKA) and protein kinase C (PKC) targets are indicated as is the N-terminal extension in mammalian troponin I sequences. Figure modified from Shaffer and Gillis (2010)

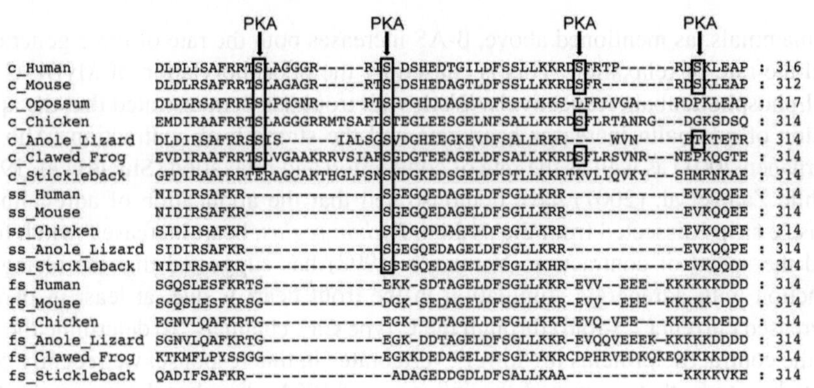

Fig. 6.8 Sequence alignment of the MyBP-C motif (M-domain) of cardiac, slow skeletal, and fast skeletal muscle from select vertebrate species. Human cMyBP-C PKA targets are indicated. Figure modified from Shaffer and Gillis (2010)

amphibians there are three within the same domain; and in fish there are two (Fig. 6.8). The additional sites present in birds and mammals appeared in cMyBP-C after fish diverged from the vertebrate lineage as they are not present in fast skeletal or slow skeletal isoforms of the protein. In fact there is no evidence that fast skeletal and slow skeletal isoforms of TnI and MyBP-C are phosphorylated by PKA. This analysis indicates that the regulation of contractile function via the phosphorylation of TnI and MyBP-C by PKA is only possible within cardiac myocytes.

While there are differences in the number and location of PKA sites within cTnI and cMyBP-C, there are two putative PKC sites that are found in TnI from all muscle types at the same relative position within the protein. In human cTnI, these are located at positions 42 and 44 (Fig. 6.7). Phosphorylatable residues are found at the corresponding positions within most ssTnI isoforms and fsTnI from fish species (Fig. 6.7). The presence of these putative phosphorylatable targets at corresponding residues in all TnI isoforms is likely the result of an ancestral sequence motif being maintained within each of the tissue-specific isoforms. These residues are not present in TnI from tunicates, suggesting that they evolved after the tunicates diverged from the vertebrate lineage, but before the genome duplication event that gave rise to the three tissue-specific isoforms of the proteins (Fig. 6.1) (Shaffer and Gillis 2010). Previous studies have demonstrated that the phosphorylation of these specific residues decreases maximum Ca^{2+}-activated force, actomyosin ATPase activity, and desensitizes the myofilaments to Ca^{2+} (Burkart et al. 2003; Noland et al. 1995). The role that they play in regulating muscle contractility following α-AS does, however, remains to be established.

The Role of PKA in Regulating the Functional Properties of Cardiac Muscle

In mammals, as mentioned above, β-AS increases both the rate of force generation and the rate of relaxation. This is caused by the phosphorylation of MyBP-C and TnI/phospholamban, respectively. Work with trout has demonstrated that the application of adrenalin increases heart rate and the strength of contraction (Aho and Vornanen 2001) as well as the rate of contraction and relaxation (Shiels et al. 1998). While Zaar et al. (2007) have demonstrated that the application of adrenaline to muscle strips dissected from the python, *Python regius*, heart increases twitch force and the speed of contraction. Vornanen (1998) has suggested that the change in function with adrenergic stimulation in the trout heart is due, at least in part, to increased current (2.3-fold) through the L-type Ca^{2+} channels. To determine the role of the contractile proteins in "lower" vertebrates in this response, we recently completed a study that examined the influence of PKA phosphorylation on the Ca^{2+} activation of chemically skinned trout cardiac trabeculae (Gillis and Klaiman 2011). Here, we measured the kinetics of Ca^{2+} activation, the Ca^{2+} sensitivity of force generation, and how this is influenced by PKA activation. As the preparations were skinned, the Ca^{2+} levels that the contractile elements were exposed to were controlled via the experimental solutions. There was no influence, therefore, of the Ca^{2+} handling proteins. Results demonstrated that PKA treatment does not influence the Ca^{2+} sensitivity of force generation, but it decreases maximum force generation by 25% (Fig. 6.9) and the rate of force re-development at maximal activation by 46% (Gillis and Klaiman 2011). These results are quite different to those of similar experiments completed with mammalian tissues where the Ca^{2+} sensitivity of force generation is reduced (Fig. 6.6), and there is no change in force generated at maximal activation (Fentzke et al. 1999; Kobayashi and Solaro 2005). Analysis of the PKA-treated cardiac tissue in the trout experiments for phospho-protein content revealed

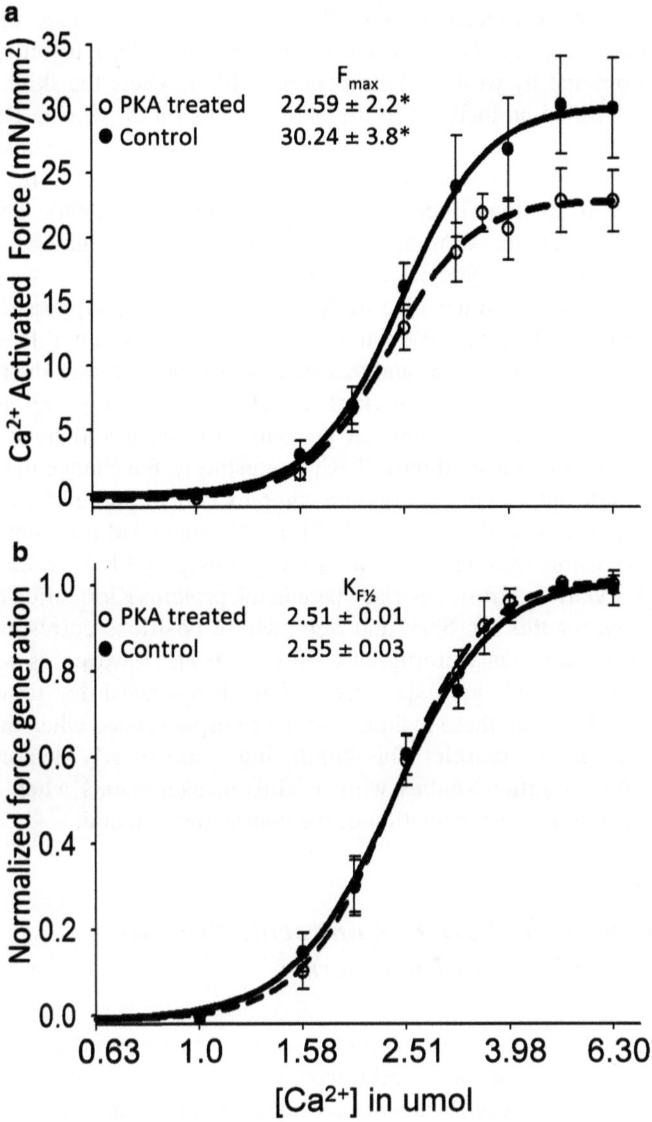

Fig. 6.9 Protein kinase A (PKA) treatment reduces the force generated by skinned trout cardiac trabeculae during maximum activation, but it does not change the Ca^{2+} sensitivity of force generation. (**a**) Actual force generated. (**b**) Relative force generated. Measurements made at 15°C and a sarcomere length of 2.2 μm. For the control preparations $n=10$, for the PKA-treated preparations $n=8$. The difference in maximum force generated (F_{max}) between the control and PKA-treated preparations is significantly different ($P<0.05$). This is indicated by the *superscript asterisk*. The difference between the Ca^{2+} concentration at half maximum tension ($K_{F1/2}$), used as a measure of Ca^{2+} sensitivity, is not significant. Figure modified from Gillis and Klaiman (2011)

that the only protein affected by the PKA treatment was myosin light chain 2 (MLC2) (Gillis and Klaiman 2011). The finding that MLC2 is a target for phosphorylation is supported by work by Patrick et al. (2010), where the skinning of trout cardiac tissue caused an increase in the phosphorylation of MLC2. There was no change in the level of phosphorylation of cTnI or cMyBP-C. The phosphorylation of MLC2 can influence the flexibility of the myosin head and therefore alter the rate of cross-bridge formation (Buck et al. 1999; Olsson et al. 2004). These results, therefore, demonstrate that cTnI and cMyBP-C do not play a significant role in the response of trout cardiac myocytes to β-AS.

The results of the trout trabeculae study are not completely supported by those of a recent solution study where we found that PKA does phosphorylate trout cTnI when it is in a troponin complex, and that such treatment decreases the Ca^{2+} affinity of the complex (Fig. 6.9) (Kirkpatrick et al. 2011). In this study Ca^{2+} activation of the complex, made from recombinant proteins, was measured by monitoring a fluorescent reporter engineered into cTnC. Interestingly, the change in Ca^{2+} affinity was not as significant as when a troponin complex containing rat cTnI was treated with PKA (Fig. 6.10) (Kirkpatrick et al. 2011). As trout cTnI does not contain the N-terminal extension, the change in Ca^{2+} affinity caused by PKA treatment was due to the phosphorylation of residues elsewhere in the protein (Kirkpatrick et al. 2011). Two candidates for this are Ser^{14} and Ser^{16} whose positions correspond to PKC targets within mammalian isoforms of cTnI (Fig. 6.7). However, as we found no change in the phosphorylation state of trout cTnI when trout cardiac tissue is treated with PKA it is likely that these residues are not phosphorylated when the protein is functioning within the muscle. This conflicting result clearly demonstrates the need to integrate solution studies with in vitro measurements when working to characterize the molecular regulation of the contractile reaction.

Modifications to the Cardiac Contractile Proteins During the Evolution of Endothermy

When the Ca^{2+} sensitivity of rat cardiac tissue is compared to that of a trout with each at their respective physiological temperature, the Ca^{2+} sensitivities are similar (Gillis et al. 2007). However, when the Ca^{2+} sensitivities of the two hearts are compared at the same temperature the Ca^{2+} sensitivity of rat cardiac tissue is one-tenth that of the trout. If the rat heart had the same Ca^{2+} sensitivity as the trout heart it would likely be stuck in contracture at 37°C as Ca^{2+} would remain bound to cTnC. The comparatively lower Ca^{2+} sensitivity of the mammalian heart is due at least in part to functional changes to the contractile element. For example, the Ca^{2+} affinity of trout cTnC is ~2.3-fold that of human cTnC when the function of the two molecules is compared at the same temperature (Gillis et al. 2000). This means that the protein is activated by Ca^{2+} at a lower concentration.

However, when the Ca^{2+} affinities of trout cTnC and human cTnC are compared at temperatures close to their respective physiological temperatures the Ca^{2+} affinities

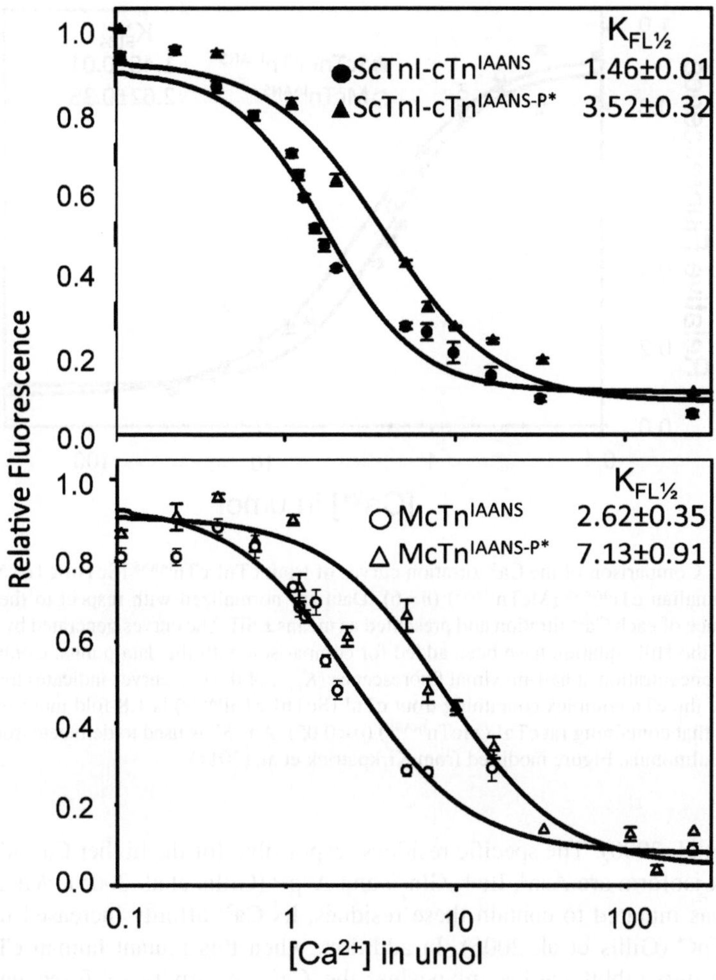

Fig. 6.10 Comparison of the effect of protein kinase A (PKA) phosphorylation on the Ca^{2+} titration curves of trout cTnI-cTn^{IAANS} (ScTnI-cTn^{IAANS}) and mammalian cTn^{IAANS} ($McTn^{IAANS}$). Data are normalized with respect to the maximal fluorescence of each Ca^{2+} titration and presented as means ± SE. The curves generated by fitting the data with the Hill equation have been added for comparison with the data points. (**a**) Titration of fluorescence of ScTnI-cTn^{IAANS} ($n=8$) and PKA phosphorylated ScTnI-$cTn^{IAANS*P}$ ($n=8$). (**b**) Titration of fluorescence of $McTn^{IAANS}$ ($n=6$) and PKA phosphorylated $McTn^{IAANS*P}$ ($n=6$). In both **a** and **b**, the phosphorylation of the cTn complex caused the Ca^{2+} concentration at half-maximal fluorescence, ($K_{FL1/2}$), to increase ($P<0.05$). This indicates that the Ca^{2+} affinity of the complex decreased. Figure modified from Kirkpatrick et al. (2011)

are more similar. This demonstrates that the Ca^{2+} affinity of the cTnC molecule increases with environmental temperature. In addition, if the pH of the experimental solutions used at the different temperatures (7°C trout cTnC, 30°C human cTnC) are buffered to mimic that which occur with α-stat regulation (pH 7.6 at 7°C, pH 7.0 at 30°C), the Ca^{2+} affinity of trout cTnC is significantly greater than that of human cTnC

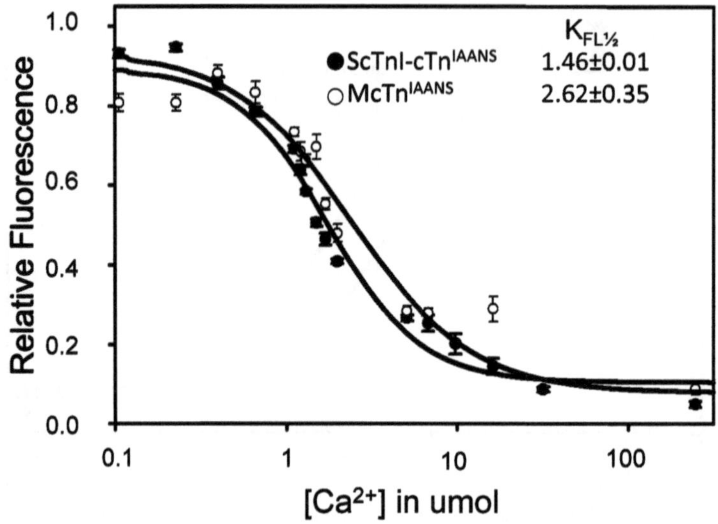

Fig. 6.11 Comparison of the Ca^{2+} titration curves of trout cTnI-cTn^{IAANS} (ScTnI-cTn^{IAANS}) ($n=8$) and mammalian cTn^{IAANS} (McTn^{IAANS}) ($n=6$). Data are normalized with respect to the maximal fluorescence of each Ca^{2+} titration and presented as means ± SE. The curves generated by fitting the data with the Hill equation have been added for comparison with the data points. Comparison of the Ca^{2+} concentration at half-maximal fluorescence ($K_{F1/2}$) of the two curves indicates that the Ca^{2+} affinity of the cTn complex containing trout cTnI (ScTnI-cTnI^{IAANS}) is 1.8-fold more sensitive to Ca^{2+} than that containing rat cTnI (McTn^{IAANS}) ($p<0.05$). An "S" is used to delineate trout cTnI as trout are salmonids. Figure modified from Kirkpatrick et al. (2011)

(Gillis et al. 2000). The specific residues responsible for the higher Ca^{2+} affinity of the trout isoform are Asn^{2}, Ile^{28}, Gln^{29}, and Asp^{30} (Gillis et al. 2005). When human cTnC was mutated to contain these residues, its Ca^{2+} affinity increased to that of trout cTnC (Gillis et al. 2005). In addition, when this mutant human cTnC was inserted into rabbit cardiac myocytes, the Ca^{2+} sensitivity of force generation increased by ~2-fold relative to controls containing recombinant human cTnC (Gillis et al. 2005). This means it is a reduction in the Ca^{2+} affinity of cTnC in mammalian cardiac tissues that is responsible, in part, for the comparatively lower Ca^{2+} sensitivity of the mammalian heart.

In a recent study, we demonstrated that the replacement of rat cTnI in a cTn complex with trout cTnI increases the Ca^{2+} affinity of the complex by ~1.8-fold (Fig. 6.11) (Kirkpatrick et al. 2011). The Ca^{2+} activation of the cTn complexes was monitored by measuring the signal of a fluorescent reporter engineered into cTnC. The increase in Ca^{2+} affinity caused by the replacement of rat cTnI with trout cTnI is thought to be due to the presence of His^{163}, Val^{165}, and Asn^{172} in trout cTnI (Kirkpatrick et al. 2011).

The corresponding residues are Ala, Glu, and His, respectively, in rat cTnI. Interestingly, His^{163}, Val^{165}, and Asn^{172} are present in all isoforms of fsTnI and ssTnI with few exceptions. These residues are also present in all cTnI isoforms until the lineage that led to mammals diverged from that which led to the birds and reptiles approximately 300 million years ago (Shaffer and Gillis 2010). Previous work has demonstrated that the insertion of any of these residues (Ala^{163}, Glu^{165}, and His^{172}) at their respective positions into rat ssTnI reduces the Ca^{2+} sensitivity of myocyte contraction (Westfall and Metzger 2007). This means that the presence of Ala^{163}, Glu^{165}, and His^{172} in mammalian cTnI translates into a reduction in the Ca^{2+} sensitivity of force generation. Together with the lower inherent Ca^{2+} affinity of mammalian cTnC, the lower Ca^{2+} affinity of the troponin complex containing mammalian cTnI would be responsible, at least in part, for the comparatively lower Ca^{2+} sensitivity of mammalian cardiac tissue. The reduction in Ca^{2+} sensitivity, caused by changes in the sequence of cTnI and cTnC, has likely evolved to offset the sensitizing effects of increasing temperature of contractile function.

Conclusions and Perspectives

The functional characteristics of the vertebrate heart have been tuned through changes in the functional properties of the contractile element. This includes changes in the Ca^{2+} binding characteristics to enable cardiac contraction at high temperatures as well as modifications to protein sequence to allow for targeting by PKA and greater regulatory control of contractile function. The appearance of PKA target residues in cTnI and cMyBP-C after fish diverged from the vertebrate lineage brings the contractile machinery under the regulatory influence of β-AS. The resulting changes in contractile function following β-AS neatly integrate with the changes already in place caused by the phosphorylation of the Ca^{2+} handling proteins. These changes in contractile function include the increased k_{off} of Ca^{2+} from the cardiac troponin complex, caused by the phosphorylation of cTnI, coupling with the increased rate of Ca^{2+} sequestration into the SR caused by the phosphorylation of PLB to enable a faster rate of muscle relation. This increased rate of relaxation and the increased rate of contraction integrate with the faster rate of myocyte stimulation caused by increased activity of the SA node. As a result, the functional coupling between the contractile stimulus and contractile machinery is increased. It is likely only through such coupling that heart rates of 600 bpm are attainable within the mouse heart. Coupled with the increased strength of contraction, the end result is a dramatic increase in cardiac output enabling extremely high rates of aerobic activity at high temperatures. Interestingly, there has been no study of how the Ca^{2+} handling proteins targeted by PKA following β-AS has evolved within striated muscle.

Future Work

To date, there has been limited examination of how myocyte contraction in reptilian and amphibian hearts is regulated by the specific contractile proteins and how this is influenced by β-AS. This represents a significant gap in our knowledge of how the control of the vertebrate heart evolved. The cTnI present in the hearts of these vertebrate groups posses the N-terminal extension that contains the two target residues of PKA. In addition, the cMyBP-C in amphibians and reptiles are also intermediate between teleosts and endotherms in regards to the number of PKA targets within the M-domain. Whether these are actually targeted is unknown. Functional data from these animals would be extremely interesting, as they are ectothermic but have some of the protein sequence motifs that are also present in mammalian species. Such studies would increase our understanding of the stepwise changes that occurred to the functional properties of the vertebrate heart with the evolution of endothermy. In addition, as significant changes occurred between the divergence of teleosts from the vertebrate lineage and when amphibians diverged, studying this transition would provide insight into how/if the regulation of the cardiovascular system was modified as vertebrates moved from the water onto land.

Studies of the regulation of contractile function in nonmammalian, nonmodel organisms are extremely useful for increasing our basic understanding of the molecular mechanisms that regulate contractility. The hearts of these species function under a wide range of physiological conditions such as low temperature, high temperature, or low oxygen. Understanding how the regulatory systems have evolved to enable cardiac function under such conditions provides insight into the adaptive capability of the vertebrate heart. Such knowledge is also potentially useful in the study and treatment of cardiac dysfunction caused by changes in the function of a number of the regulatory proteins, including cTnI and cMyPB-C. Such changes in function initiate compensatory changes in the heart that ultimately lead to pathological conditions and heart failure.

Acknowledgments This work was supported by operating grants from the Natural Sciences and Engineering Research Council (NSERC), and equipment grants from NSERC and the Canadian Foundation for Innovation. The author thanks Drs H.A. Shiels and S.L. Alderman for comments on an earlier version.

References

Ababou A, Rostkova E, Mistry S, Le Masurier C, Gautel M, Pfuhl M (2008) Myosin binding protein C positioned to play a key role in regulation of muscle contraction: structure and interactions of domain C1. J Mol Biol 384:615–630

Aho E, Vornanen M (2001) Cold acclimation increases basal heart rate but decreases its thermal tolerance in rainbow trout (Oncorhynchus mykiss). J Comp Physiol B 171:173–179

Barefield D, Sadayappan S (2010) Phosphorylation and function of cardiac myosin binding protein-C in health and disease. J Mol Cell Cardiol 48:866–875

Bers D (2001) Excitation-contraction coupling and cardiac contractile force. Kluwer Academic, London

Bishop C, Butler P (1995) Physiological modelling of oxygen consumption in birds during flight. J Exp Biol 198:2153–2163

Braz JC, Gregory K, Pathak A, Zhao W, Sahin B, Klevitsky R, Kimball TF, Lorenz JN, Nairn AC, Liggett SB et al. (2004) PKC-alpha regulates cardiac contractility and propensity toward heart failure. Nat Med 10:248–254

Buck SH, Konyn PJ, Palermo J, Robbins J, Moss RL (1999) Altered kinetics of contraction of mouse atrial myocytes expressing ventricular myosin regulatory light chain. Am J Physiol 276:H1167–H1171

Burggren W, Farrell AP, Lillywhite H (1997) Vertebrate cardiovascular systems. In: Dantzler WH (ed) Handbook of physiology, vol 1. Oxford University Press, New York, pp 215–308

Burkart EM, Sumandea MP, Kobayashi T, Nili M, Martin AF, Homsher E, Solaro RJ (2003) Phosphorylation or glutamic acid substitution at protein kinase C sites on cardiac troponin I differentially depress myofilament tension and shortening velocity. J Biol Chem 278:11265–11272

Cleto CL, Vandenberghe AE, MacLean DW, Pannunzio P, Tortorelli C, Meedel TH, Satou Y, Satoh N, Hastings KE (2003) Ascidian larva reveals ancient origin of vertebrate-skeletal-muscle troponin I characteristics in chordate locomotory muscle. Mol Biol Evol 20:2113–2122

DiFrancesco D (1993) Pacemaker mechanisms in cardiac tissue. Annu Rev Physiol 55:455–472

Dong WJ, Jayasundar JJ, An J, Xing J, Cheung HC (2007) Effects of PKA phosphorylation of cardiac troponin I and strong crossbridge on conformational transitions of the N-domain of cardiac troponin C in regulated thin filaments. Biochemistry 46:9752–9761

Farmer CG (1999) Evolution of the vertebrate cardio-pulmonary system. Annu Rev Physiol 61:573–592

Fentzke RC, Buck SH, Patel JR, Lin H, Wolska BM, Stojanovic MO, Martin AF, Solaro RJ, Moss RL, Leiden JM (1999) Impaired cardiomyocyte relaxation and diastolic function in transgenic mice expressing slow skeletal troponin I in the heart. J Physiol 517:143–157

Fukuda N, Granzier HL, Ishiwata S, Kurihara S (2008) Physiological functions of the giant elastic protein titin in mammalian striated muscle. J Physiol Sci 58:151–159

Gautel M, Zuffardi O, Freiburg A, Labeit S (1995) Phosphorylation switches specific for the cardiac isoform of myosin binding protein-C: a modulator of cardiac contraction? EMBO J 14:1952–1960

Gillis TE, Klaiman J (2011) The influence of PKA treatment on the Ca2+ activation of force generation by trout cardiac muscle. J Exp Biol 214:1989–1996

Gillis TE, Liang B, Chung F, Tibbits GF (2005) Increasing mammalian cardiomyocyte contractility with residues identified in trout troponin C. Physiol Genomics 22:1–7

Gillis TE, Marshall CR, Tibbits GF (2007) Functional and evolutionary relationships of troponin C. Physiol Genomics 32:16–27

Gillis TE, Marshall CR, Xue XH, Borgford TJ, Tibbits GF (2000) Ca^{2+} binding to cardiac troponin C: effects of temperature and pH on mammalian and salmonid isoforms. Am J Physiol Regul Integr Comp Physiol 279:R1707–R1715

Gordon AM, Homsher E, Regnier M (2000) Regulation of contraction in striated muscle. Physiol Rev 80:853–924

Gruen M, Prinz H, Gautel M (1999) cAPK-phosphorylation controls the interaction of the regulatory domain of cardiac myosin binding protein C with myosin-S2 in an on-off fashion. FEBS Lett 453:254–259

Harrison SM, Bers DM (1989) Influence of temperature on the calcium sensitivity of the myofilaments of skinned ventricular muscle from the rabbit. J Gen Physiol 93:411–428

Irisawa H, Brown HF, Giles W (1993) Cardiac pacemaking in the sinoatrial node. Physiol Rev 73:197–227

Kashimura T, Briston SJ, Trafford AW, Napolitano C, Priori SG, Eisner DA, Venetucci LA (2010) In the RyR2(R4496C) mouse model of CPVT, beta-adrenergic stimulation induces Ca waves by increasing SR Ca^{2+} content and not by decreasing the threshold for Ca waves. Circ Res 107:1483–1489

Kentish JC, McCloskey DT, Layland J, Palmer S, Leiden JM, Martin AF, Solaro RJ (2001) Phosphorylation of troponin I by protein kinase A accelerates relaxation and crossbridge cycle kinetics in mouse ventricular muscle. Circ Res 88:1059–1065

Kirkpatrick K, Robertson AS, Klaiman J, Gillis TE (2011) The influence of trout cardiac troponin I and PKA phosphorylation on the Ca^{2+} affinity of the cardiac troponin complex. J Exp Biol 214:1981–1988

Kobayashi T, Solaro RJ (2005) Calcium, thin filaments, and the integrative biology of cardiac contractility. Annu Rev Physiol 67:39–67

Kruger M, Linke WA (2009) Titin-based mechanical signalling in normal and failing myocardium. J Mol Cell Cardiol 46:490–498

Li MX, Wang X, Sykes BD (2004) Structural based insights into the role of troponin in cardiac muscle pathophysiology. J Muscle Res Cell Motil 25:559–579

McIvor ME, Orchard CH, Lakatta EG (1988) Dissociation of changes in apparent myofibrillar Ca2+ sensitivity and twitch relaxation induced by adrenergic and cholinergic stimulation in isolated ferret cardiac muscle. J Gen Physiol 92:509–529

Miller SC, Gillis TE, Wright PA (2011) The ontogeny of regulatory control of the rainbow trout (Oncorhynchus mykiss) heart and how this is influenced by chronic hypoxia exposure. J Exp Biol 214:2065–2072

Noland TA Jr, Guo X, Raynor RL, Jideama NM, Averyhart-Fullard V, Solaro RJ, Kuo JF (1995) Cardiac troponin I mutants. Phosphorylation by protein kinases C and A and regulation of Ca^{2+}-stimulated MgATPase of reconstituted actomyosin S-1. J Biol Chem 270:25445–25454

Noland TA Jr, Raynor RL, Jideama NM, Guo X, Kazanietz MG, Blumberg PM, Solaro RJ, Kuo JF (1996) Differential regulation of cardiac actomyosin S-1 MgATPase by protein kinase C isozyme-specific phosphorylation of specific sites in cardiac troponin I and its phosphorylation site mutants. Biochemistry 35:14923–14931

Olsson MC, Patel JR, Fitzsimons DP, Walker JW, Moss RL (2004) Basal myosin light chain phosphorylation is a determinant of Ca2+ sensitivity of force and activation dependence of the kinetics of myocardial force development. Am J Physiol Heart Circ Physiol 287:H2712–H2718

Ota S, Saitou N (1999) Phylogenetic relationship of muscle tissues deduced from superimposition of gene trees. Mol Biol Evol 16:856–867

Patrick SM, Hoskins AC, Kentish JC, White E, Shiels HA, Cazorla O (2010) Enhanced length-dependent Ca2+ activation in fish cardiomyocytes permits a large operating range of sarcomere lengths. J Mol Cell Cardiol 48:917–924

Priede IG (1974) The effect of swimming activity and section of the vagus nerves on heart rate in rainbow trout. J Exp Biol 60:305–319

Rarick HM, Tang HP, Guo XD, Martin AF, Solaro RJ (1999) Interactions at the NH2-terminal interface of cardiac troponin I modulate myofilament activation. J Mol Cell Cardiol 31:363–375

Shaffer JF, Gillis TE (2010) Evolution of the regulatory control of vertebrate striated muscle: the roles of troponin I and myosin binding protein-C. Physiol Genomics 42:406–419

Shaffer JF, Kensler RW, Harris SP (2009) The myosin-binding protein C motif binds to F-actin in a phosphorylation-sensitive manner. J Biol Chem 284:12318–12327

Shiels HA, Calaghan SC, White E (2006) The cellular basis for enhanced volume-modulated cardiac output in fish hearts. J Gen Physiol 128:37–44

Shiels HA, Stevens ED, Farrell AP (1998) Effects of temperature, adrenaline and ryanodine on power production in rainbow trout oncorhynchus mykiss ventricular trabeculae. J Exp Biol 201:2701–2710

Solaro RJ, Rarick HM (1998) Troponin and tropomyosin: proteins that switch on and tune in the activity of cardiac myofilaments. Circ Res 83:471–480

Solaro RJ, Rosevear P, Kobayashi T (2008) The unique functions of cardiac troponin I in the control of cardiac muscle contraction and relaxation. Biochem Biophys Res Commun 369:82–87

Stelzer JE, Patel JR, Walker JW, Moss RL (2007) Differential roles of cardiac myosin-binding protein C and cardiac troponin I in the myofibrillar force responses to protein kinase A phosphorylation. Circ Res 101:503–511

Suzuki M, Sasaki N, Miki T, Sakamoto N, Ohmoto-Sekine Y, Tamagawa M, Seino S, Marban E, Nakaya H (2002) Role of sarcolemmal K(ATP) channels in cardioprotection against ischemia/reperfusion injury in mice. J Clin Invest 109:509–516

Vornanen M (1998) L-type Ca^{2+} current in fish cardiac myocytes: effects of thermal acclimation and beta-adrenergic stimulation. J Exp Biol 201:533–547

Weisberg A, Winegrad S (1996) Alteration of myosin cross bridges by phosphorylation of myosin-binding protein C in cardiac muscle. PNAS 93:8999–9003

Westfall MV, Metzger JM (2007) Single amino acid substitutions define isoform-specific effects of troponin I on myofilament Ca2+ and pH sensitivity. J Mol Cell Cardiol 43:107–118

Yamasaki R, Wu Y, McNabb M, Greaser M, Labeit S, Granzier H (2002) Protein kinase A phosphorylates titin's cardiac-specific N2B domain and reduces passive tension in rat cardiac myocytes. Circ Res 90:1181–1188

Zaar M, Overgaard J, Gesser H, Wang T (2007) Contractile properties of the functionally divided python heart: two sides of the same matter. Comp Biochem Physiol A Mol Integr Physiol 146:163–173

Chapter 7
Ontogenesis of Myocardial Function

David Sedmera and Bohuslav Ostadal

Introduction

Unlike most other organ systems, function of the cardiovascular system is critical for embryonic survival and, as a result, the heart has to beat in order to support circulation well before its morphogenesis is complete. The function of the heart influences its morphology (Sedmera 2005), and vice versa. During development, complex changes in embryonic metabolism also influence the energy substrates available for heart pumping function and influence oxygen demands of the organ. We will cover in this chapter the developmental transitions in cardiac morphology in relation to mechanics of contraction, activation sequence of the heart, and accompanying metabolic switches. The description is organized into three prenatal stages with distinct morphology, contraction mechanics, and conduction parameters: tubular heart, trabeculated one, and fully septated heart with coronary perfusion (Fig. 7.1). Separately we treat the postnatal maturation that involves chiefly quantitative changes and appearance of aspects like ischemic preconditioning and gender differences in reactions to experimental perturbations.

D. Sedmera (✉)
Institute of Anatomy, First Faculty of Medicine, Charles University in Prague,
12800 Prague, Czech Republic

Institute of Physiology, Academy of Sciences of the Czech Republic,
14220 Prague, Czech Republic
e-mail: dsedmera@biomed.cas.cz

B. Ostadal
Institute of Physiology, Academy of Sciences of the Czech Republic,
Videnska 1083, 14220 Prague, Czech Republic

D. Sedmera and T. Wang (eds.), *Ontogeny and Phylogeny of the Vertebrate Heart*, 147
DOI 10.1007/978-1-4614-3387-3_7, © Springer Science+Business Media, LLC 2012

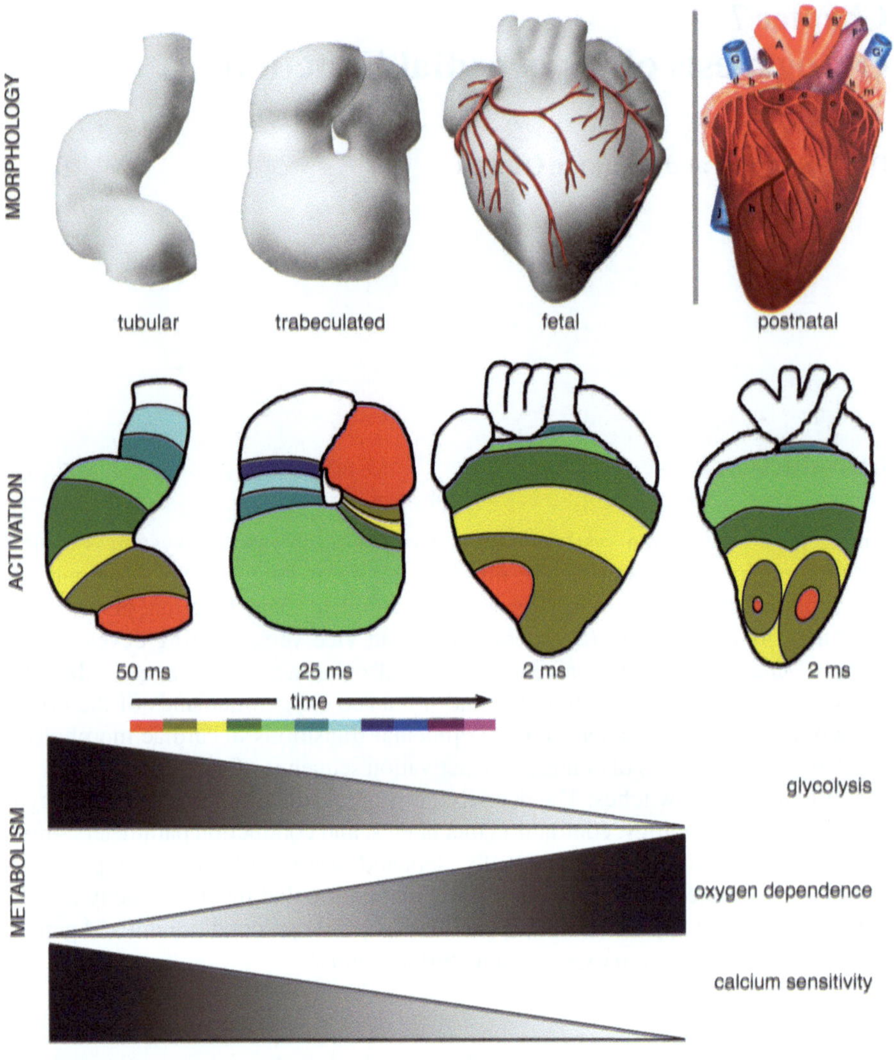

Fig. 7.1 Correlation with key phases of cardiac morphogenesis (*top row*) with activation patterns (*middle row*) and main developmental metabolic switches (*bottom*). Note that the metabolic switches are limited to the prenatal period, as postnatal development is not linear and presents frequently a biphasic pattern, Partially modified from (Sedmera 2011; Komarek et al. 1982)

Ontogenetic Approach to Cardiology

The importance of the developmental approach for experimental and clinical cardiology is indisputable. Clinical–epidemiological studies have clearly shown that the risk factors of serious cardiovascular diseases, such as atherosclerosis and ischemic heart disease, are already present during the early phases of ontogenetic development. Thus, experimental studies of the pathogenetic mechanisms of these disturbances must shift to the early developmental periods. Furthermore, congenital cardiovascular malformations remain the single largest cause of infant mortality from congenital defects in industrial countries. In spite of dramatic advances in surgical therapy, many patients' heart defects are not cured. The primary goal of medical progress must be prevention rather than more or better treatment alone.

There are, however, many controversies concerning the mechanisms of developmental variations in cardiovascular responsiveness; the obvious discrepancies may be partly due to different experimental procedures used for studying different animal species. Nevertheless, it is worth noting that ontogenetic conclusions are often based on observations from two or three developmental stages, such as fetal, newborn, and adult. It is therefore necessary to point out that the possible developmental changes between selected (and not exactly defined) points remain undetermined. The precise knowledge of individual ontogenetic periods, critical for cardiac ontogeny (prenatal, neonatal, suckling, weaning, adolescence, adulthood) is thus crucial for the prediction and explanation of cardiac reactions to various pathogenetic stimuli. Furthermore, experimental data on ontogenetic differences in cardiac responsiveness were obtained almost exclusively from healthy individuals. However, such an approach is far from clinical situations, where interventions applied in the critical developmental periods may have serious consequences for further maturation. Unfortunately, it is necessary to stress that the interest of developmental cardiologists concentrates almost exclusively on the prenatal development; postnatal changes remain, unfortunately neglected.

The advances in molecular and cell biology over the last few decades amount to a major revolution in our knowledge of the development of living systems. Development of in vitro systems has advanced our understanding of the ways in which gene expression is controlled in a cell-specific way. It is, however, necessary to mention that the integrative, synthetic view of different developmental problems is still indispensable. The multicellular or whole-organ problems with which cardiology was concerned before the massive switch to the cell and molecular level were often unresolved or inadequately worked out or even not yet defined. Furthermore, the systems, whose overall behavior we are trying to predict from mechanisms involving cells and molecules, are highly nonlinear. To meet the challenge of comprehending the process of developmental cardiology as a whole, we must combine both analytic and synthetic approach.

Tubular Heart

Morphology and Mechanics

The first cardiac contractions start at the straight tube stage (Fig. 7.1), approximately 2 days in the chick (Kamino et al. 1981), ED8.5 in the mouse (Chen et al. 2010) and 22 days in humans (Sadler 2011). The first heart beats are irregular and do not propel

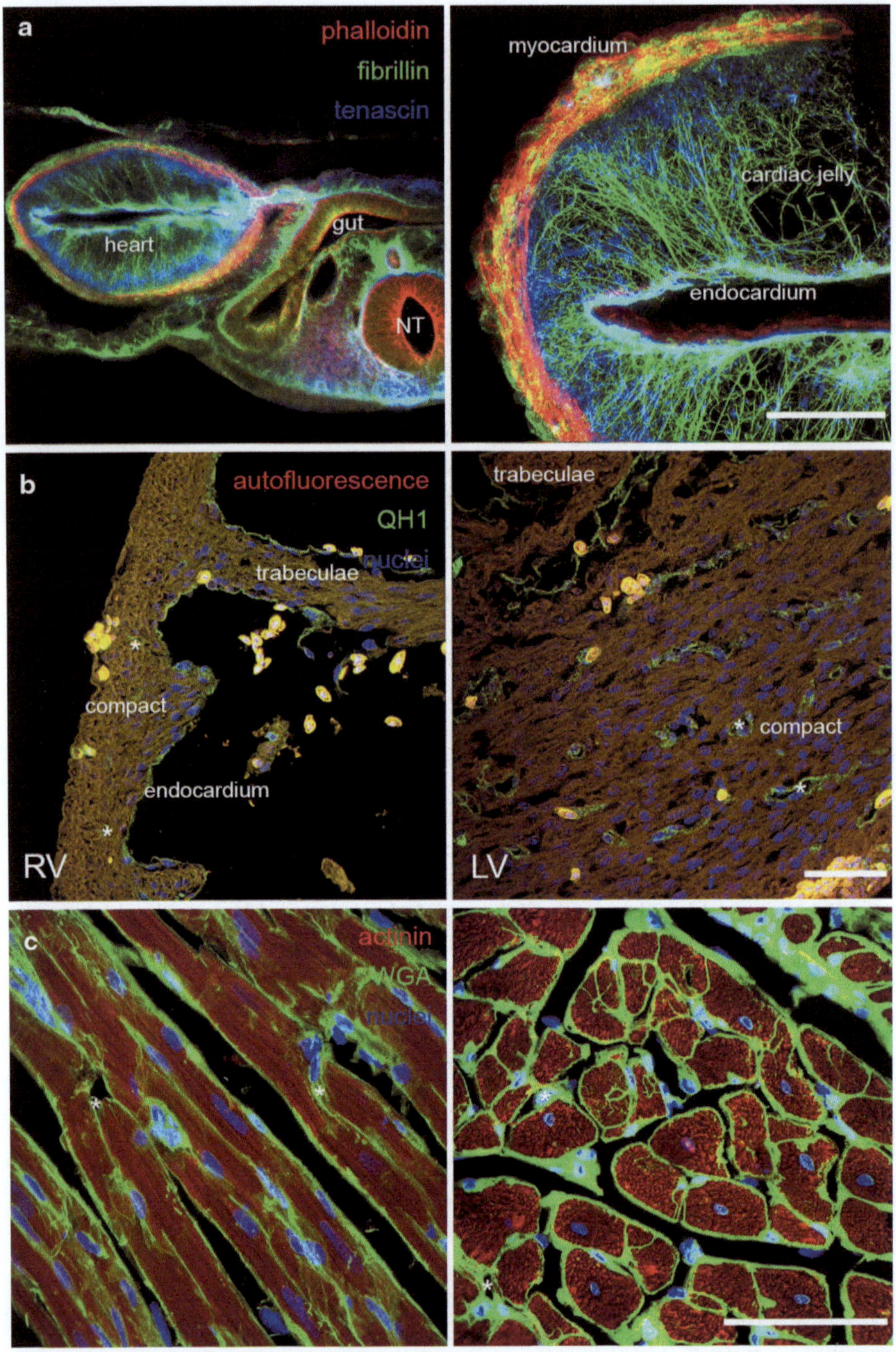

Fig. 7.2 Morphogenesis of ventricular myocardium. (**a**) Tubular chick embryonic heart at Stage 13 [after McQuinn et al. (2007)]. The myocardium, stained *red* with rhodamine-phalloidin, forms a mantle around an asymmetric layer of cardiac jelly and endocardium. There is no epicardium at this stage. *NT* neural tube. (**b**) Septated trabeculated heart of quail embryo (embryonic day 9)

blood through the circulation (Hirota et al. 1985), but activity quickly becomes coordinated and primitive blood starts to circulate through the embryo, thus enabling its further growth by decreasing diffusion distance for nutrients and oxygen.

Description of mechanics of the tubular heart (Xavier-Neto et al. 2010), initially described as peristaltic [e.g. de Jong et al. (1992); Sakai et al. (1996)], is now being gradually replaced by a more precise, albeit more complex, suction pump model (Forouhar et al. 2006). At this stage there is complete obstruction of the lumen during contraction by apposition of endocardium reinforced by cardiac jelly (Fig. 7.2), as demonstrated using high-resolution ultrasound (McQuinn et al. 2007) or optical coherence tomography (Yelbuz et al. 2002; Manner et al. 2009; Garita et al. 2010). Such tubular hearts have thus, by definition 100% ejection fraction, which is never achieved at later stages. Their geometry is at the extreme end between cylindrical, optimal for generating pressure, and spherical, optimal for volume displacement (Hutchins et al. 1978). In avian and mammalian embryos, pumping function of the tubular heart is required from this stage on for embryonic survival (Koushik et al. 2001).

Conduction Through a Tubular Heart

The conduction system at that stage has a well-defined pacemaker at the inflow (Kamino et al. 1981), but the general speed of conduction is slow (Hirota et al. 1985), correlating with the "peristaltoid" mode of contraction. With further development, conduction remains slow in the regions of the atrioventricular canal and ventricular outflow, which are considered to maintain the primary myocardium phenotype (Christoffels et al. 2010) while there is some acceleration in the prospective ventricular region [Fig. 7.1; de Jong et al. (1992); Reckova et al. (2003); reviewed in Gourdie et al. (2003)].

Early Heart Metabolism

Studies covering the early period of heart formation and function are, mostly for methodological reasons, sparse. Oxygen uptake during the early stages of heart formation was studied by Raddatz and associates in the chick embryo (Raddatz et al.

Fig. 7.2 (continued) shows different thickness of compact myocardium in left (LV) versus right ventricle (RV). Capillary bed has developed in the compact zone, and is already connected to coronary circulation, as evidenced by the presence of red blood cells (*yellow*) in the lumen. **(c)** Adult rat left ventricular myocardium shows strong directional preferences (*left panel* shows longitudinal, *right panel* transverse section). This well-organized myocardial syncytium is abundantly supplied by capillary bed (*asterisks*) and is supported by network of extracellular matrix, highlighted with wheat germ agglutinin staining. Scale bars 50 μm

1992). It was found that the uptake per unit area of the ventricular surface almost doubled between Hamburger–Hamilton (Hamburger and Hamilton 1951) stages 10 and 12, indicating that originally mostly anaerobic metabolism changes stepwise into the oxidative aerobic mode. There are significant variations in oxygen consumption along the heart tube already at these early stages, with lowest values reported from the conotruncus. Calculations made from various sources showed after normalization to protein amount further increase during later embryonic and fetal development (range from 0.41 nmol O_2 per hour and microgram of protein at Stage 10 to 2 at Stage 44).

The tubular heart shows high resistance to lack of oxygen (Sedmera et al. 2002). In comparison with trabeculated stages, the tubular heart at the second day of incubation in the chick embryo did not stop beating during 30 min of anoxia, the only reaction at Stage 14 was minute transitory tachycardia followed by slowing to about 80% of the normoxic rate. Similarly, there was no arrest of propagation of the excitation/contraction wave at the atrioventricular canal, observed at later stages, suggesting that not only automaticity, but also conduction is not dependent on oxygen availability at this stage.

The energetic metabolism of the tubular heart is based upon glycolysis and rate of glycogenolysis and oxygen uptake is proportional to mechanical load (Romano et al. 2001). Oxidative capacity of the entire Stage 11 heart is significantly lower than that of Stage 20 (1 day later). There is also higher reserve of oxidative capacity and high anaerobic glycolytic capacity (measured as lactate production). While the normoxic lactate production (normalized per microgram of protein) is similar to later stages, it can be increased almost three times under anoxia, a feat impossible in just slightly older hearts (Sedmera et al. 2002). Schematic representation of longitudinal development of key metabolic parameters is depicted in Fig. 7.1.

Trabeculated Heart

Morphology and Chamber Formation

Soon after the beginning of looping, morphological differentiation of myocardium along the cardiac tube becomes apparent (Knaapen et al. 1996). This includes disappearance of cardiac jelly in the atria and ventricles, coinciding with the process of development of trabeculae, and, soon afterwards, pectinate muscles in the atria (Fig. 7.3). In the chick embryo, this period occurs between embryonic days 2.5 (Stage 16) and 8 (Stage 34), when trabecular compaction starts and ventricular septation is completed; this corresponds to embryonic days 9.5–13.5 in the mouse and fourth to seventh week in humans (Sadler 2011). The morphological differences are accompanied by changes in gene expression patterns (Moorman and Christoffels 2003; Aanhaanen et al. 2010). This process is well conserved in vertebrates, no matter whether there is one or two atria or ventricles. Ventricular trabeculae provide a means of increasing myocardial mass in the absence of coronary circulation (Minot 1901). Detailed morphological description of trabeculation as well as its experimental perturbations was provided by numerous investigators in the chick (Ben-Shachar et al. 1985; Icardo and Fernandez-Teran 1987; Sedmera et al. 1997, 1999), mouse (Vuillemin and Pexieder 1989; Gruber et al. 1996), and humans (Moscoso and Pexieder 1990; Pexieder and Janecek 1984; Sedmera et al. 2000; Lamers et al. 1992).

Conduction System Differentiation

The conduction system starts to show some mature characteristics once the chambers are formed. This includes activation from a well-defined pacemaker (Leaf et al. 2008), spreading through the atrial tissues using preferential pathways (Sedmera et al. 2006) correlating with heterogeneities in atrial myoarchitecture, and atrioventricular delay based upon slow conduction through the myocardium of the atrioventricular canal (Chen et al. 2010; de Jong et al. 1992; Arguello et al. 1986). In the ventricle, the trabecular network serves as a nascent specialized ventricular conduction system by the virtue of its faster conduction due both to its anisotropic arrangement (de Jong et al. 1992; Reckova et al. 2003) as well as differential gene expression of specific channel proteins (Moorman and Christoffels 2003; Dominguez et al. 2008) or gap junction protein connexin40 (van Kempen et al. 1995; Miquerol et al. 2004). Even prior to functional maturation of bundle branches, there exists a preferential conduction pathway along the "primary ring" (Wessels et al. 1992), localized approximately in the position of future interventricular septum. This preferential activation pathway was documented using optical mapping in the chick (Sedmera et al. 2004; Sankova et al. 2010), rat (Sedmera et al. 2005a), and mouse [(Rentschler et al. 2002) and our unpublished data]. The outflow valve function is exercised, rather efficiently (McQuinn et al. 2007), by a combination of close apposition of the outflow tract cushions and contraction of the slowly conducting myocardium of the outflow tract [(de Jong et al. 1992; Reckova et al. 2003; Sedmera et al. 2003a), Fig. 7.2].

Contraction Mechanics of a Trabeculated Heart

At this stage, the embryonic heart is big enough for imaging and acquisition of functional parameters. It was shown that many functional aspects of the trabeculated heart are comparable with the adult heart, and the same parameters can be meaningfully determined after some modification of the equipment. Video recordings of chick hearts (Keller et al. 1990), together with direct servo-null pressure measurements and Doppler flow waveforms (Clark et al. 1986) have taught us that the trabeculated embryonic heart observes the Frank–Starling relationship, dorsal aortic blood flow scales with embryo weight and that the increase in heart rate results in decreased cardiac output (Dunnigan et al. 1987). The reason why the trabeculated heart poorly tolerates wider variations in heart rate lies in its inability to distend more during diastole due to relatively low compliance of its trabeculated wall to compensate for decreased heart rate, as well as its dependence on rate-limited passive filling (Hu et al. 1991) that prevents maintaining sufficient output at increased rate (Grobety et al. 1999). Video recordings in mice followed shortly (Keller et al. 1996), and ultrasound biomicroscopy studies even allowed longitudinal follow-up of individual embryos (Phoon 2006). Such systematic studies provided functional insights into the demise of some prenatally lethal mouse mutants such as NFATc1 nulls that lack outflow valves (Phoon et al. 2004). Likewise, insufficient valve function of the cushions in the outflow tract in ED11.5 mouse embryos leads to diastolic flow reversal and death at ED12.5 (Nomura-Kitabayashi et al. 2009). Recently, data from early human embryos have been reported using advances in gynecological ultrasound (Rein et al. 2009).

Metabolic Aspects of the Trabeculated Heart

Energy metabolism during this period is characterized by transition from mainly anaerobic to oxidative (Raddatz et al. 1992; Sedmera et al. 2002; Romano et al. 2001). The energy producing systems gradually mature resulting in increased efficiency. Calcium handling in the trabeculated heart is significantly different from the adult myocardium (Tenthorey et al. 1998). Despite its morphological immaturity, sarcoplasmic reticulum seems to be partly functional at this stage of development. Both L-type calcium channels as well as ryanodine receptor channels are involved in anoxia–reoxygenation-induced dysfunction. Regulation of pH is critical for function of the embryonic heart, and predominantly depends on bicarbonate transport (Meiltz et al. 1998), as inhibition of this system significantly worsened both anoxic and reoxygenation arrhythmias. Availability of energy substrates is also important, as removal of glucose or blockade of glycolysis was protective against arrhythmias during experimental anoxia and reoxygenation in vitro (Tran et al. 1996).

Transition to the Fetal Heart

Transformation of Myocardial Architecture

While the interest of developmental cardiologists wanes after completion of septation, the opposite is true in the clinical settings, since the earliest stages under 12 weeks are seldom studied apart from confirming the presence of a beating heart. Many interesting things happen between ventricular septation and birth, especially in humans where this period lasts 6 months, rather than six days for the mouse. One critical step in myocardial morphogenesis and prerequisite for its maturation is the connection of the coronary plumbing that occurs in all higher vertebrates. Its importance for homeotherms is clearly shown by lethality of mouse mutants with impaired compaction [for review, see Wessels and Sedmera (2003)] or hypoxic avian embryos with absent coronary arteries and thin compact myocardium (Tintu et al. 2009; Nanka et al. 2008). Studying the development of coronary arteries (Tomanek 1996) helps to explain some of the coronary artery anomalies. Critical players are the epicardium, providing most of the cellular material, and myocardial signaling that provides the guidance (Lavine and Ornitz 2009). The fetal heart not only grows in size, but further develops spiral architecture of the compact layer (Jouk et al. 2000). This morphogenetic step was shown to depend on hemodynamic loading, being accelerated by increased pressure load and delayed by decreased preload (Tobita et al. 2005). Abnormal hemodynamics can lead to development of some forms of severe congenital heart disease, such as aortic atresia in hearts that were perfectly normal at the routine scan at 19 weeks of human gestation (Hornberger et al. 1995). From clinical perspective, the critical period is between 20 and 28 weeks of gestation is the time when fetal intervention can be attempted for some severe lesions with predictably devastating consequences such as severe aortic stenosis (Tworetzky et al. 2004). Impaired performance of malformed hearts also affects the other fetal systems, most critically the central nervous system (Limperopoulos et al. 2010). It is thus likely that fetal interventions improving cardiac output, such as balloon dilation of the stenotic aortic (Marshall et al. 2005) or pulmonary valve (Tworetzky et al. 2009), could have potential benefits exceeding the simple restoration of normal cardiac anatomy and circulation. Several fetal models were developed to test the effects of prenatal interventions (deAlmeida et al. 2007) or to study the growth response of the fetal heart to hemodynamic challenge (Saiki et al. 1997; Rudolph 2000). It is important to know the limits of adaptation of the fetal circulatory system, since some procedures that might work well in theory may not be tolerated by the whole organism [e.g. requirement of 3-stage palliation of the hypoplastic left heart syndrome ending up in Fontan circulation rather than “all at once” procedure; reviewed in Sedmera et al. (2005b)]. Interventions in prenatal life have the potential to take advantage of the growth capacity of fetal myocardium based upon myocyte proliferation rather than hypertrophy (deAlmeida et al. 2007). Such restoration of myocardial mass with appropriate vascularization might also result in superior performance in the long term.

Maturation of Impulse Conduction in the Atria and Ventricles

The heart with formed cardiac chambers is from the conduction system point of view directly comparable to the mature one. The pacemaking activity is localized to the sino-atrial node, which starts to acquire its autonomic innervation pattern for neural modulation (Hildreth et al. 2009). The function of delay generation between the activation of atria and ventricles begins to shift from the slowly conducting myocardium of the atrioventricular canal (Chen et al. 2010; de Jong et al.

1992; Arguello et al. 1986) that is similar to arrangement in the adult hearts of lower vertebrates (Sedmera et al. 2003b) to the atrioventricular node (Viragh and Challice 1977a). The transition from primitive base-to-apex activation of the ventricles to mature apex-to-base sequence correlates with ventricular septation (Chuck et al. 1997). This is rather logical, as the proximal part of the fast component of the ventricular conduction system, His bundle and bundle branches, is located in the interventricular septum (Miquerol et al. 2004; Vassall-Adams 1982). However, the actual transition occurs before the completion of septation, and is structurally supported by trabecular sheets connected directly to the atrioventricular canal myocardium (Sedmera et al. 2004). Interestingly, the apex to base activation of the ventricle is first heralded by an epicardial breakthrough localized to the right of the interventricular septum, near the apex of the forming right ventricle. This was documented in the mouse (Rentschler et al. 2001), chick (Reckova et al. 2003), rat (Sedmera et al. 2005a), and rabbit (Rothenberg et al. 2005) and is indicative of earlier right bundle branch functionality. This precocious functionality suggests that geometry of this bundle—narrow with fewer side branches compared with left bundle branch—plays an important role in conduction speed. Earlier functionality of the right bundle is conserved among species (Sedmera et al. 2005a) despite molecularly more advanced differentiation of the left ventricle evidenced by earlier and higher level expression of connexin40, shown both in the chick (Gourdie et al. 1993; Hall et al. 2004) and mouse (Miquerol et al. 2004). Development of ventricular activation patterns is depicted in Fig. 7.1. Conservation and potential utility of this arrangement could be explained by the close proximity of the ventricular inflow and outflow at the base of the heart, which favors beginning of contraction from the opposite end—the apex, and is present already in fishes and amphibians (Sedmera et al. 2003b).

Remodeling of Atrioventricular Myocardium

The myocardium of the atrioventricular canal initially fulfills, together with the atrioventricular cushions, valve function in the preseptation heart. In higher vertebrates, the atrioventricular canal myocardium gradually regresses, at least in part through apoptosis (Cheng et al. 2002), and is replaced by fibrous insulating tissue derived from the epicardium (Kolditz et al. 2008). The function of delay generator between the atrium and ventricle is then localized to the atrioventricular node (Viragh and Challice 1977a, 1977b). This structure is prominent and well recognized in mammals, and has recently been characterized at the molecular level as a rather complex structure with several functional subdomains with differential gene expression (Aanhaanen et al. 2010). It is well recognized that the adult node is derived from a slowly conducting embryonic myocardial segment expressing, among other genes, Tbx2 (Aanhaanen et al. 2009) that is down-regulated in the developing chambers. A molecular model of conduction system differentiation was recently presented by the Amsterdam group (Christoffels et al. 2010). Detailed understanding of the normal development of atrioventricular canal myocardium is essential for our ability to explain the developmental origins of cardiac arrhythmias, such as ventricular preexcitation, characterized by the presence of rapidly conducting anomalous myocardial connections between the atria and ventricles. Abnormal atrioventricular canal remodeling in the quail model of epicardial ablation can lead to persistence of accessory pathways that can lead in turn to ventricular preexcitation (Kolditz et al. 2008) or form a substrate of atrioventricular reentry circuit resulting in reciprocating tachycardia. This situation is observed in the early embryonic mouse heart (Valderrabano et al. 2006), where these connections are still normally present and functional.

Problems can arise also from the opposite situation, where conduction from the atrioventricular node is slowed or even completely blocked. The atrioventricular node is one of the targets of the autoantibodies present in mothers with lupus. These patients stand up to 5% chance (Friedman et al. 2002) of having a child with congenital heart block. Third-degree atrioventricular block results often in profound bradycardia, leading to circulatory failure with accumulation of fluid in fetal tissues and cavities (hydrops) and rapid demise. Fortunately, diagnosis is nowadays possible in utero, so serial monitoring of at risk pregnancies is possible (Friedman et al. 2008) and treatment can be initiated as soon as the condition develops; however, no firm therapeutic guidelines are currently established.

Effect of Chronic Hypoxia on Prenatal Cardiac Development

Human epidemiological studies have shown a clear association of adverse intrauterine environment and an increased risk of ischemic heart disease in later adult life (Barker et al. 1993, 1989). Of all these stresses to which the fetus is subjected, perhaps the most important and clinically relevant is that of hypoxia (Xue and Zhang 2009). The fetus may experience prolonged hypoxic stress under many different conditions, including pregnancy at high altitude, pregnancy with anemia, placental insufficiency, cord compression, and heart, lung, and kidney disease. There is clear evidence of a link between hypoxia and fetal intrauterine growth restriction. Human studies at high altitude suggest that hypoxia per se, independent of maternal nutrition, causes fetal growth restriction, resulting in low birth weight and altered body shape at birth (Giussani et al. 2001).

Experimental studies have demonstrated that reduced fetal oxygen or nutrient supply influence heart development. In fetal sheep long-term hypoxemia (natural altitude 3,820 m, 110 days) reduced cardiac output and contractility (Gilbert 1998), increased lactate dehydrogenase and citrate synthase (Ohtsuka and Gilbert 1995) and resulted in cardiomegaly (Martin et al. 1998; Murotsuki et al. 1997). In this connection it may be of interest to mention that undernutrition increased cardiac expression of several genes that regulate the development of cardiomegaly and remodeling by midgestation (Han et al. 2004). Bae and colleagues (Bae et al. 2003) have observed that maternal chronic hypoxia (CH) (10.5% oxygen, 7 days) increased hypoxia-inducible factor (HIF) 1α and increased apoptotic cell death in fetal rat hearts. The increased cell death may lead to asymmetric cardiac enlargement. Although the mechanisms underlying hypoxia-induced apoptosis are not clear and are likely to be multiplex, the above study demonstrated that apoptosis was associated with an increase in Fas receptors and decrease in Bcl-2 proteins. In addition, CH significantly suppressed expression of heat shock protein (Hsp) 70 and differentially regulated β-adrenoreceptor (AR) subtypes in the fetal heart. There was no difference in β2-AR protein levels between the control and hypoxic heart, but β1-AR protein levels were significantly increased in the hypoxic group. Recently, Patterson and associates (Patterson et al. 2010) have observed that hypoxic treatment of pregnant rats from days 15 to 21 of gestation resulted in significant decreases in PKCε protein and mRNA in fetal hearts. The finding demonstrates a direct effect of hypoxia on cardiac PKCε gene repression, linking fetal hypoxia to possible pathophysiological consequences in the heart of adult offspring.

The population-based, prospective cohort study has studied fetal hemodynamic adaptive changes related to intrauterine growth retardation (Verburg et al. 2008). Fetal growth characteristics and fetal circulation variables were assessed with ultrasound and Doppler examinations in 1,215 healthy women. It has been observed that decreased fetal growth is associated with cardiac remodeling and changes of cardiac output, consistent with gradual increase in afterload and compromised arterial compliance. These changes have already begun to occur before the stage of clinically apparent fetal growth restriction.

These data imply that limitation of either oxygen or nutrient supply to the fetus produces functional and structural changes in the neonatal heart. However, less is known regarding the persistence and long-term consequences of these changes and their possible contribution to the increased risk of cardiovascular disease in later life.

Postnatal Development

Cardiac Adaptations During the Early Postnatal Period

Changes in circulatory patterns after birth are well known from embryology and obstetrics textbooks. The early postnatal period differs from subsequent growth in childhood, as there is a time-window during which cardiomyocytes are still capable of proliferation. In rodents, this privileged period lasts for about 3 days after birth (Li et al. 1996), while in humans, it is believed to last about 6 months, and surgical interventions (Foker et al. 1999) such as biventricular repair for hypoplastic left heart syndrome are most successful during this window (Foker et al. 1999; Tchervenkov et al. 2000).

Perinatal period represents a typical example of developmentally decisive critical period. The fetus makes numerous adaptations to ensure normal development in spite of its low arterial pO_2, which corresponds to an altitude of 8,000 m

["Mt. Everest in utero," (Eastman 1954)]. Among these adaptations are polycythemia, leftward shift of the dissociation curve, fetal hemoglobin, a high cardiac output, and low arterial blood pressure, which help provide adequate oxygen and nutrient delivery to the developing tissues. Fetal heart operates at near maximum levels and is very sensitive to increase in afterload.

The major changes can be observed during delivery. Within the short period of time mammalian fetus (and its heart) comes from the hypoxic environment with low pO_2 and low oxygen saturation into the normal atmosphere, arterial saturation increases more than 3 times. As a consequence of these dramatic changes at birth, mammalian hearts meet suddenly with an extremely high reactive oxygen species (ROS) concentration. The delivery is, furthermore, accompanied by the transition from the amniotic fluid to the air, by the marked decrease of the ambient temperature, and by the termination of placental nutrition. This transition requires appropriate physiological adaptations: onset of pulmonary respiration, transition from fetal to neonatal circulation, switching-on of thermoregulation and increase of basal metabolic rate. Very important is the change in nutrition: trans-placental supply is interrupted and mother's milk nutrition is not yet fully developed (Kuma et al. 2004). The neonatal heart can probably use ROS for the increase of protein degradation, which permits utilization of amino acids, necessary for the maintenance of energy homeostasis during neonatal starvation (Muhlfeld et al. 2005; Ostadalova et al. 2010).

The early postnatal period is characterized by further adaptation to postnatal living conditions: relative heart weight significantly increases (Ostadalova et al. 1993), proliferation of myocytes (hyperplasia) is replaced by hypertrophic growth (Li et al. 1996), rapid rate of coronary capillary formation occurs, cardiac mitochondrial function and calcium handling change substantially (Skarka et al. 2003), and contractility increases; cardiac sensitivity to the increasing concentration of calcium is the highest (Ostadalova et al. 1993) [for review see Ostadal et al. (1999)]. Hemodynamic parameters are stepwise approaching the adult values (blood pressure, heart rate). The changes are very fast: e.g. the contractile response to low extracellular sodium differs in rats even day-by-day during the first week of life (Ostadalova et al. 1995). The whole period is characterized by intensive body growth (Babicky et al. 1973). It is clear that the early postnatal period is critical for normal cardiac development and different interventions imposed on the heart may significantly influence myocardial structure and function.

Postnatal Development of Cardiac Tolerance to Oxygen Deprivation

Cardiac tolerance of the immature heart to oxygen deficiency is significantly higher as compared with the adult myocardium; myocardial infarction is extremely rare during early ontogenetic development. Riva and Hearse (2009) observed that age-dependent changes in resistance to global ischemia in the isolated rat heart showed a biphasic pattern with increasing tolerance from 5 to 23 days of age, followed by a decline to adulthood. Detailed analysis of the tolerance of the isolated rat heart to global ischemia during the first week of life has revealed a significant decrease from day 1 to 7 (Ostadalova et al. 1998), suggesting a possible triphasic pattern of the ontogenetic development of cardiac sensitivity to ischemia. The sensitivity of neonatal myocardium may be species dependent; Baker et al. (1995) have shown that the neonatal pig heart is more susceptible to ischemia than the neonatal rabbit heart.

The mechanisms of the higher resistance of the immature heart to oxygen deprivation have not yet been satisfactorily clarified. For the explanation of this fact, the above-mentioned physiological changes during perinatal period should be taken into consideration, particularly the transition from the "Mt. Everest" to the sea level. The reason of the higher tolerance during further development can be still only hypothetical. It may be speculated that an explanation of the phenomenon lies in the greater anaerobic glycolytic capacity, higher glycogen reserves of the immature heart (Hoerter 1976) or amino acid utilization by transamination (Julia et al. 1990). Metabolism of ATP also changes during development: the immature heart is better equipped to ATP synthesis than to its breakdown, the situation that might be advantageous in conditions of low substrate availability. The immature heart thus suffers less ischemic injury after the same ischemic insult. Another factor that may contribute to the increased tolerance of the immature heart is the age-related change in calcium handling. The relative contribution of transsarcolemmal calcium influx and calcium release from the sarcoplasmic reticulum varies significantly during development. The contraction of the immature myocardium, where the sarcoplasmic reticulum is not fully developed, depends to a large extent on the fluxes of calcium across sarcolemma (Vetter et al. 1995). Calcium overload, common in the adult myocardium, was not described in the immature heart.

Still unclear is the role of mitochondria in the developmental changes of cardiac tolerance to oxygen deprivation, in spite of the fact that mitochondria are responsible for cellular oxygen handling. Mitochondrial oxidative phosphorylation is not completely developed in the rat heart at birth; cardiac maturation during the first postnatal week is characterized by increasing content and specific activity of cytochrome *c* oxidase and enhanced flux of adenine nucleotides across the inner mitochondrial membrane (Drahota et al. 2004; Schagger et al. 1995). We have shown previously (Skarka et al. 2003) that the content of cytochromes in cardiac mitochondria increased twofold between birth and day 30, similarly as the expression of adenine nucleotide translocase 1. Furthermore, in newborn animals, a single population of mitochondria with relatively high mitochondrial membrane potential (MMP) was observed. Starting with the weaning period, a second population of mitochondria with significantly lower MMP occurs. The collapse of MMP due to the opening of a high-conductance mitochondrial permeability transition pore (MPTP) has been implicated in the molecular mechanisms associated with ischemia/reperfusion (I/R) injury of the adult heart (Di Lisa and Bernardi 1998, 2005). We have observed, however, significant ontogenetic differences in the role of MPTP in the I/R injury.

Whereas the blockade of MPTP by sanglifehrin in perfused rat heart had a protective effect on I/R-induced damage in the adult myocardium as already demonstrated (Di Lisa et al. 2001), it had no effect in the neonatal heart (Milerova et al. 2010). For the explanation of this difference, a modified amount of cyclophilin receptors or lower sensitivity of MPTP in the neonatal heart to pore opening factors has to be taken into consideration. Furthermore, we found (Milerova et al. 2010) that in cardiac mitochondria isolated from neonatal rats Ca^{2+}-dependent and cyclosporine-sensitive MPTP is less sensitive to Ca^{2+} ions as compared with adults. We can only speculate that its lower sensitivity to the calcium-induced swelling may be related to the higher ischemic tolerance of the neonatal heart. All these results support the view that cardiac mitochondria are deeply involved in the regulation of cardiac tolerance to oxygen deprivation during ontogenetic development.

Developmental Pathologies of Conduction System

Some pathologies of the conduction system develop in adulthood, but their genetic origin points to an earlier trigger resulting in disease. For example, children of mothers with systemic lupus and anti-Ro antibodies stand a 5% chance of developing complete atrioventricular block (Friedman et al. 2002). Atrioventricular block is also part of phenotype associated with Nkx2.5 mutations (Benson et al. 1999), and morphological and functional anomalies of the conduction axis were also reported in the mouse model of haploinsufficiency (Jay et al. 2004). Current trends are trying to explain developmental origins of atrial arrhythmias through second heart field derivatives around the pulmonary veins (Gittenberger-de Groot et al. 2007; Hoogaars et al. 2007; Mommersteeg et al. 2007). Abnormal patterns of atrial conduction were also reported in mouse model of connexin40 deficiency (Leaf et al. 2008), and right bundle branch block was shown in the adults (Tamaddon et al. 2000).

Regeneration Potential from Ontogenetic Perspective

The ability of the adult mammalian myocardium to regenerate is limited. Nevertheless, because of the potential significance of regeneration to counter the current human plague of infarction and resulting chronic heart failure in survivors, exploration of regenerative potential is a subject of intensive research. The identification of resident cardiac stem cells (Beltrami et al. 2003) generated much hope as well as much controversy, and the current state of the field is nicely summarized in Rosenthal and Harvey (2010). It was confirmed recently that adult cardiomyocytes also possess proliferative capacity, but it is mostly used to compensate for physiological cellular turnover, while the repair of injury results in recruitment of resident stem cells (Hsieh et al. 2007). This (low) proliferative capacity maintaining myocardial homeostasis decreases during aging process (Bergmann et al. 2009). A further insight into normal developmental processes is necessary to define optimal therapeutic strategies for regeneration and repair of the diseased adult heart.

Protective Strategies for the Immature Myocardium

As mentioned above, cardiac tolerance of the immature heart to oxygen deficiency is significantly higher as compared with the adult myocardium. Thus the question arises whether we can further increase the already high resistance of the immature heart. To the most effective experimental protective mechanisms belongs long-lasting adaptation to CH [for review see Ostadal and Kolar (2007)] and brief adaptation, so-called ischemic preconditioning [IP, for reviews see Bolli (2007); Yellon and Downey (2003)]. Data on the effects of these two phenomena on the immature cardiac muscle are, however, only sporadic. The protective effect of postconditioning (Vinten-Johansen 2007) is comparable with IP and was observed in many species including humans (Laskey 2005). Data on the possible effect of postconditioning on the immature heart are, however, still lacking.

Ischemic Preconditioning

It is now 23 years since the phenomenon termed "ischemic preconditioning" has been formally recognized. In their seminal study, Murry and colleagues (1986) reported that infarct size produced by 40 min of sustained coronary occlusion could be markedly reduced if they first "preconditioned" the heart with four 5-min episodes of brief ischemia. Whereas extensive data are available on IP in the adult myocardium, information on whether this protective phenomenon also occurs in immature hearts is inadequate. We have shown (Ostadalova et al. 1998) that classical IP, at least in rats, is not present at birth and that the enhanced postischemic recovery of contractile function can be observed only at the end of the first postnatal week. The decreasing tolerance of the neonatal heart to ischemia is thus counteracted by the development of the inducible endogenous protective mechanism. It was shown that the absence of IP in immature isolated hearts cannot be overcome by increasing the preconditioning stimulus (Awad et al. 1998). In addition, a pharmacological stimulus (phenylephrine) also failed to protect the immature heart.

Molecular mechanisms of IP in the adult myocardium are still unclear and the same is true for the immature heart. Signaling cascade obviously starts by the release of some "protective molecules" [for review see Yellon and Downey (2003)]. These molecules are able to bind to the receptors of cardiac cells coupled with G-proteins, thereby starting the activation of one or more signaling pathways; this finally leads to the phosphorylation of the target proteins [for review see Bolli (2007)]. The end-effector of this cascade might be e.g. the mitochondrial ATP dependent potassium (KATP) channels (Garlid et al. 2003). IP is associated with activation of mitochondrial KATP channels even in the immature rabbit heart (Baker et al. 1999). On the other hand, in the immature rat heart the administration of 5-hydroxydecanoate, a selective blocker of the mitochondrial KATP channels, had no effect on the protection by IP (Ostadalova et al. 2002). It seems, therefore, that in the mechanism of IP in the immature heart may exist significant differences between species. It remains unanswered, whether the mechanisms of IP, participating in the immature heart, differ from the mechanisms, described for the adult myocardium.

Adaptation to Chronic Hypoxia

CH is the main pathophysiological feature of hypoxemic congenital heart disease. Understanding the mechanisms by which cyanotic congenital heart malformations modify the myocardium and how the modifications impact on the cardiac tolerance to oxygen deprivation may provide insight into developing treatments for limiting myocardial damage. Unfortunately, no existing model adequately reproduces chronic myocardial perfusion with the hypoxic blood caused by congenital cyanotic defects. Therefore, similar experimental models as in adults, i.e. CH simulated in the normobaric or hypobaric chambers are used in studies performed during early stages of ontogenetic development.

Already in the late 1950s, the first observation (Hurtado 1960), showing that the incidence of myocardial infarction is lower in people living at high altitude, appeared. These epidemiological studies on the protective effect of high altitude were confirmed in experimental studies using simulated hypoxia [for review see Ostadal and Kolar (2007)]. However, only a few authors have compared tolerance to oxygen deprivation in chronically hypoxic versus normoxic immature myocardium. We have observed (Ostadal et al. 1995) that CH, simulated in the barochamber, results in similarly enhanced cardiac resistance in rats exposed to CH either from the fourth day of postnatal life or in adulthood. Similarly, Baker et al. (1995) demonstrated that adaptation to CH increased the tolerance of the developing rabbit heart. However, it follows from our results [(Ostadalova et al. 2002), Fig. 7.3] that the protective effect of CH is absent in newborn rats and the protective effect of adaptation develops—similarly as that of IP—only during the first postnatal week. These results suggest that we might be dealing with the more general biological phenomenon: the already high resistance of the cardiac muscle cannot be further increased by different protective mechanisms. A similar situation as in the immature mammalian heart can be observed also in the highly tolerant hearts of poikilotherms or in the myocardium of young–adult females. In addition to protective effects, adaptation to CH also induces other adaptive responses including polycythemia, hypoxic pulmonary hypertension, and right ventricular hypertrophy. The degree of these changes is comparable in animals exposed to CH after birth or in adulthood (Ostadal et al. 1995; Kolar et al. 1989).

Overwhelming majority of studies, analyzing the possible mechanisms of cardiac adaptation to CH, deals exclusively with the adult myocardium. We have shown

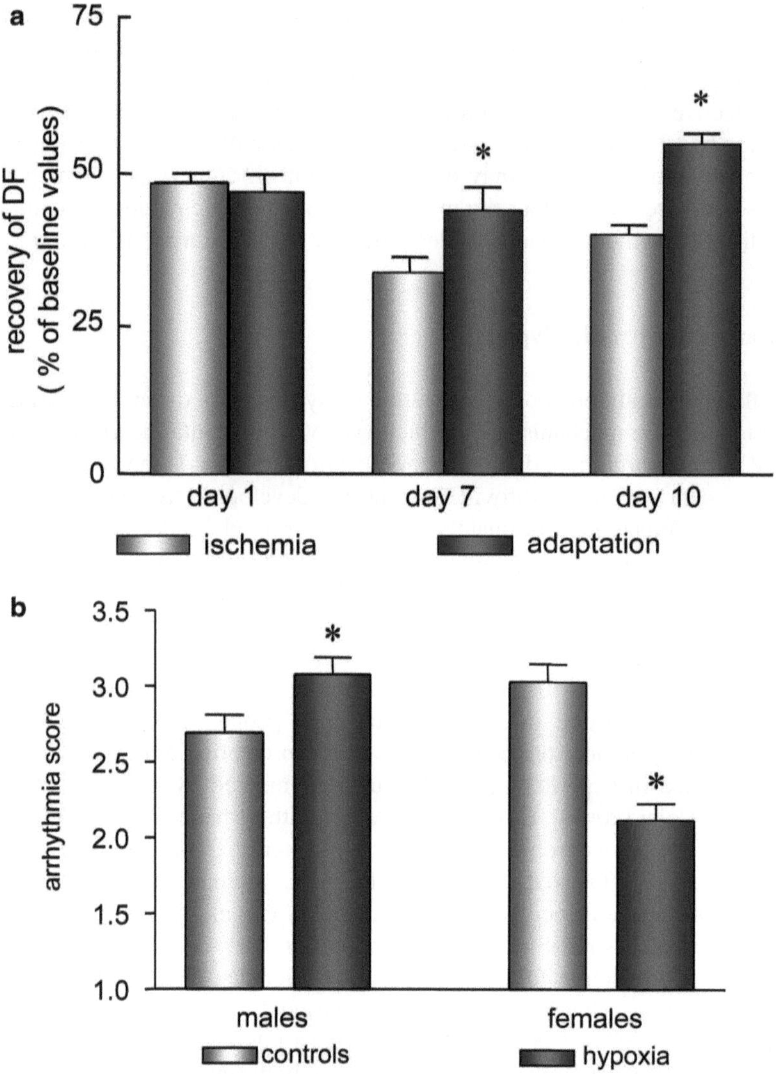

Fig. 7.3 Examples of physiological features occurring in postnatal development. (**a**) Protective effect of adaptation to chronic hypoxia is only apparent from postnatal day 7, since at earlier stages, the capacity appears already at maximum due to immaturity of the myocardium. (**b**) Gender differences in adult rats exposed perinatally (1 week before and 1 week after birth) to high-altitude hypoxia equivalent of 7,000 m above sea level. While incidence of ischemic arrhythmias is significantly increased in the adapted males, the same exposure is protective in females. Adapted from Ostadal et al. (2009)

previously (Ostadalova et al. 2002) that blockade of mitochondrial KATP channels with 5-hydroxydecanoate completely abolished the cardioprotective effect of adaptation to CH also in neonatal rats; a similar effect had the blockade of NO by using L-NAME. It seems, therefore, that—unlike in IP—both mitochondrial KATP channels and NO may play an important role in the mechanism of adaptation of the neonatal heart to CH. Recently, it has been shown that also angiotensin II is involved in the mechanisms of adaptation of the immature heart to CH: the chronic blockade of angiotensin II type 1 receptors (AT1) by irbesartan completely abolished the cardioprotective effect of CH (Rakusan et al. 2007). The involvement of AT1 receptor pathway in the adaptive responses of the immature hearts to CH should be taken into consideration in the treatment of children suffering from cyanotic congenital heart disease.

Effect of Chronic Hypoxia on Postnatal Cardiac Development

CH is the main pathophysiological feature of hypoxemic congenital heart disease. In the chronically hypoxic newborn mammals, body growth is blunted (Chvojkova et al. 2005). Neonatal growth retardation during moderate (15% O_2) or severe (10% O_2) hypoxic exposure can be almost entirely attributed to the effects of hypoxia on the newborn, and is not mediated by the maternal response (Mortola et al. 1990). The faulty maternal lactation and limited food availability to the suckling are not the primary mechanisms of the neonatal growth retardation in CH, as was suggested by the observations that the cellular responses to hypoxia (Naeye 1966) differ from those of experimental starvation. In this connection it is necessary to mention that nutritional status markedly influences cardiac development. Slow-growing pups confer smaller cardiomyocyte length and volume (Bai et al. 1990), accompanied by qualitative changes of the subcellular structures. The development of α1- and β-adrenergic receptors is retarded and the resulting receptor deficit probably contributes to reduced responsiveness to adrenergic stimulation (Bell and Slotkin 1988). These changes are connected with alterations of cardiac ornithine decarboxylase activity (Bell et al. 1987). Furthermore, we have observed (Brodsky et al. 1992) that the number of cardiomyocytes in both the right and left ventricular myocardium was significantly lower in slow-growing weanling rats. Early postnatal nutritional modification altered also protein remodeling in the immature myocardium: concentration of collagenous proteins in slow-growing rats significantly decreased (Pelouch et al. 1997). Furthermore, undernutrition markedly decreased the basal values of left ventricular pressure and contractility in 3-week-old rats (Dowell and Martin 1984); data on younger animals are, unfortunately, lacking.

Mild (19% O_2) or moderate levels of 1-week hypoxic exposure significantly increased cardiac weight and DNA synthesis in neonatal rats (Mortola et al. 1990). This suggests that hypoxia can truly stimulate cardiac muscle cell multiplication as has been demonstrated in neonatal rats exposed to sideropenic anemia (Neffgen and Korecky 1972; Rakusan and Poupa 1966) or to low oxygen atmosphere (Hollenberg et al. 1976; Wachtlova et al. 1977). In neonatal animals exposed to high altitude cardiac enlargement and activation of DNA synthesis was significantly more expressed in the right ventricular myocardium. The mechanisms behind the hypoxia-induced cardiac hyperplasia are unclear but it is possible that they relate to the greater cardiac work caused by the higher cardiac output, blood viscosity, and pulmonary vascular resistance.

Timing of corrective surgery for hypoxemic congenital heart disease is critically important, with early surgery desirable to promote more normal development. Many children undergoing cardiac surgery in the first year of life exhibit varying degrees of cyanotic heart disease where the myocardium is chronically perfused with hypoxic blood. We have observed (Samanek et al. 1989) metabolic adaptation to CH in the myocardium of children with cyanotic congenital cardiac malformations. The capacity of the aerobic energy metabolism was significantly reduced in hypoxic hearts as compared with normoxic patients. Understanding the mechanisms by which cyanotic congenital heart disease modifies the myocardium and how the modifications impact on the cardiac tolerance to ischemia may provide insight into the treatments for limiting myocardial damage during cardiac surgery (Fitzpatrick et al. 2005). Unfortunately, so far the clinical data are missing. To date, no experimental model is available that adequately mimics chronic perfusion of immature hearts with hypoxic blood, caused by congenital cyanotic defects. Therefore, similar models as in adults, i.e. CH simulated in the normobaric or hypobaric chamber, are being used in experimental studies performed during early stages of ontogenetic development (Ostadal et al. 2009). This last study showed that gender differences in cardiac responsiveness to ischemic insult, not present in the newborn animals, appear during later life. We have found that the same "preconditioning" protocol, namely perinatal hypoxia, was clearly protecting adult females against arrhythmias induced by subsequent experimental ischemia, but was deleterious in males (Fig. 7.3). While potentially useful in clinical management and risk stratification, further research is needed to uncover the underlying mechanisms (estrogens?) and precise the critical exposure period.

Conclusions and Perspectives

Developmental cardiology provides sometimes unique perspective on pathogenetic mechanisms of some cardiovascular diseases, and can help to explain them from developmental perspective. The structure and function of the developing heart are intrinsically linked, so studying both morphology and physiology provides added value and helps with data interpretation. The differences in heart metabolism and function between stages are often more profound than differences between species, so translation of data obtained in animal studies to humans needs to be primarily considered in developmental terms (for example, newborn rodents are more similar to third semester human fetuses than to neonates). Recent advances in our understanding of reactions of the immature heart to experimental hypoxia could help in devising optimal protection strategies for the immature heart, which is subject to increasingly complex surgical procedures aimed at correcting hitherto untractable forms of congenital heart disease.

Acknowledgments We would like to thank to all our past and present collaborators who were instrumental in generating results discussed in this paper and helped to shape our thinking about the developing heart.

Funding: Supported by Ministry of Education VZ 0021620806 and 1 M0510, Academy of Sciences Purkinje Fellowship to D.S., and institutional AV0Z50110509. Further support comes from Grant Agency of the Czech Republic 304/08/0615 and P302/11/1308.

Conflict of Interest None

References

Aanhaanen WT, Brons JF, Dominguez JN, Rana MS, Norden J, Airik R, Wakker V, de Gier-de VC, Brown NA, Kispert A, Moorman AF, Christoffels VM (2009) The Tbx2+ primary myocardium of the atrioventricular canal forms the atrioventricular node and the base of the left ventricle. Circ Res 104(11):1267–1274

Aanhaanen WT, Mommersteeg MT, Norden J, Wakker V, de Gier-de VC, Anderson RH, Kispert A, Moorman AF, Christoffels VM (2010) Developmental origin, growth, and three-dimensional architecture of the atrioventricular conduction axis of the mouse heart. Circ Res 107(6): 728–736

Arguello C, Alanis J, Pantoja O, Valenzuela B (1986) Electrophysiological and ultrastructural study of the atrioventricular canal during the development of the chick embryo. J Mol Cell Cardiol 18(5):499–510

Awad WI, Shattock MJ, Chambers DJ (1998) Ischemic preconditioning in immature myocardium. Circulation 98(19 Suppl):II206–II213

Babicky A, Parizek J, Ostadalova I, Kolar J (1973) Initial solid food intake and growth of young rats in nests of different sizes. Physiol Bohemoslov 22(6):557–566

Bae S, Xiao Y, Li G, Casiano CA, Zhang L (2003) Effect of maternal chronic hypoxic exposure during gestation on apoptosis in fetal rat heart. Am J Physiol Heart Circ Physiol 285(3): H983–990

Bai SL, Campbell SE, Moore JA, Morales MC, Gerdes AM (1990) Influence of age, growth, and sex on cardiac myocyte size and number in rats. Anat Rec 226(2):207–212

Baker EJ, Boerboom LE, Olinger GN, Baker JE (1995) Tolerance of the developing heart to ischemia: impact of hypoxemia from birth. Am J Physiol 268(3 Pt 2):H1165–1173

Baker JE, Holman P, Gross GJ (1999) Preconditioning in immature rabbit hearts: role of KATP channels. Circulation 99(9):1249–1254

Barker DJ, Osmond C, Golding J, Kuh D, Wadsworth ME (1989) Growth in utero, blood pressure in childhood and adult life, and mortality from cardiovascular disease. BMJ 298(6673): 564–567

Barker DJ, Martyn CN, Osmond C, Hales CN, Fall CH (1993) Growth in utero and serum cholesterol concentrations in adult life. BMJ 307(6918):1524–1527

Bell JM, Slotkin TA (1988) Postnatal nutritional status influences development of cardiac adrenergic receptor binding sites. Brain Res Bull 21(6):893–896

Bell JM, Whitmore WL, Queen KL, Orband-Miller L, Slotkin TA (1987) Biochemical determinants of growth sparing during neonatal nutritional deprivation or enhancement: ornithine decarboxylase, polyamines, and macromolecules in brain regions and heart. Pediatr Res 22(5): 599–604

Beltrami AP, Barlucchi L, Torella D, Baker M, Limana F, Chimenti S, Kasahara H, Rota M, Musso E, Urbanek K, Leri A, Kajstura J, Nadal-Ginard B, Anversa P (2003) Adult cardiac stem cells are multipotent and support myocardial regeneration. Cell 114(6):763–776

Ben-Shachar G, Arcilla RA, Lucas RV, Manasek FJ (1985) Ventricular trabeculations in the chick embryo heart and their contribution to ventricular and muscular septal development. Circ Res 57(5):759–766

Benson DW, Silberbach GM, Kavanaugh-McHugh A, Cottrill C, Zhang Y, Riggs S, Smalls O, Johnson MC, Watson MS, Seidman JG, Seidman CE, Plowden J, Kugler JD (1999) Mutations in the cardiac transcription factor NKX2.5 affect diverse cardiac developmental pathways [see comments]. J Clin Invest 104(11):1567–1573

Bergmann O, Bhardwaj RD, Bernard S, Zdunek S, Barnabe-Heider F, Walsh S, Zupicich J, Alkass K, Buchholz BA, Druid H, Jovinge S, Frisen J (2009) Evidence for cardiomyocyte renewal in humans. Science 324(5923):98–102

Bolli R (2007) Preconditioning: a paradigm shift in the biology of myocardial ischemia. Am J Physiol Heart Circ Physiol 292(1):H19–27

Brodsky VY, Pelouch V, Arefyeva AM, Milerova M, Ostadal B (1992) Lack of proportionality between gene dosage and total muscle protein content in the rat heart. Int J Dev Biol 36(2):339–342

Chen F, De Diego C, Chang MG, McHarg JL, John S, Klitzner TS, Weiss JN (2010) Atrioventricular conduction and arrhythmias at the initiation of beating in embryonic mouse hearts. Dev Dyn 239(7):1941–1949

Cheng G, Wessels A, Gourdie RG, Thompson RP (2002) Spatiotemporal distribution of apoptosis in embryonic chicken heart. Dev Dyn 223:119–133

Christoffels VM, Hoogaars WM, Moorman AF (2010) Patterning and development of the conduction system of the heart: origins of the conduction system in development. In: Rosenthal N, Harvey RP (eds) Heart development and regeneration, vol 1. Elsevier, London, pp 171–194

Chuck ET, Freeman DM, Watanabe M, Rosenbaum DS (1997) Changing activation sequence in the embryonic chick heart. Implications for the development of the His-Purkinje system. Circ Res 81(4):470–476

Chvojkova Z, Ostadalova I, Ostadal B (2005) Low body weight and cardiac tolerance to ischemia in neonatal rats. Physiol Res 54(4):357–362

Clark EB, Hu N, Dummett JL, Vandekieft GK, Olson C, Tomanek R (1986) Ventricular function and morphology in chick embryo from stages 18 to 29. Am J Physiol 250(3 Pt 2):H407–413

de Jong F, Opthof T, Wilde AA, Janse MJ, Charles R, Lamers WH, Moorman AF (1992) Persisting zones of slow impulse conduction in developing chicken hearts. Circ Res 71(2):240–250

deAlmeida A, McQuinn T, Sedmera D (2007) Increased ventricular preload is compensated by myocyte proliferation in normal and hypoplastic fetal chick left ventricle. Circ Res 100(9): 1363–1370

Di Lisa F, Bernardi P (1998) Mitochondrial function as a determinant of recovery or death in cell response to injury. Mol Cell Biochem 184(1–2):379–391

Di Lisa F, Bernardi P (2005) Mitochondrial function and myocardial aging. A critical analysis of the role of permeability transition. Cardiovasc Res 66(2):222–232

Di Lisa F, Menabo R, Canton M, Barile M, Bernardi P (2001) Opening of the mitochondrial permeability transition pore causes depletion of mitochondrial and cytosolic NAD+ and is a causative event in the death of myocytes in postischemic reperfusion of the heart. J Biol Chem 276(4):2571–2575

Dominguez JN, de la Rosa A, Navarro F, Franco D, Aranega AE (2008) Tissue distribution and subcellular localization of the cardiac sodium channel during mouse heart development. Cardiovasc Res 78(1):45–52

Dowell RT, Martin AF (1984) Perinatal nutritional modification of weanling rat heart contractile protein. Am J Physiol 247(6 Pt 2):H967–972

Drahota Z, Milerova M, Stieglerova A, Houstek J, Ostadal B (2004) Developmental changes of cytochrome c oxidase and citrate synthase in rat heart homogenate. Physiol Res 53(1):119–122

Dunnigan A, Hu N, Benson DW Jr, Clark EB (1987) Effect of heart rate increase on dorsal aortic flow in the stage 24 chick embryo. Pediatr Res 22(4):442–444

Eastman NJ (1954) Mount Everest in utero. Am J Obstet Gynecol 67(4):701–711

Fitzpatrick CM, Shi Y, Hutchins WC, Su J, Gross GJ, Ostadal B, Tweddell JS, Baker JE (2005) Cardioprotection in chronically hypoxic rabbits persists on exposure to normoxia: role of NOS and KATP channels. Am J Physiol Heart Circ Physiol 288(1):H62–68

Foker JE, Berry J, Steinberger J (1999) Ventricular growth stimulation to achieve two-ventricular repair in unbalanced common atrioventricular canal. Prog Pediatr Cardiol 10:173–186

Forouhar AS, Liebling M, Hickerson A, Nasiraei-Moghaddam A, Tsai HJ, Hove JR, Fraser SE, Dickinson ME, Gharib M (2006) The embryonic vertebrate heart tube is a dynamic suction pump. Science 312(5774):751–753

Friedman DM, Rupel A, Glickstein J, Buyon JP (2002) Congenital heart block in neonatal lupus: the pediatric cardiologist's perspective. Indian J Pediatr 69(6):517–522

Friedman DM, Kim MY, Copel JA, Davis C, Phoon CK, Glickstein JS, Buyon JP (2008) Utility of cardiac monitoring in fetuses at risk for congenital heart block: the PR Interval and Dexamethasone Evaluation (PRIDE) prospective study. Circulation 117(4):485–493

Garita B, Jenkins MW, Han M, Zhou C, Vanauker M, Rollins AM, Watanabe M, Fujimoto JG, Linask KK (2010) Blood flow dynamics of one cardiac cycle and relationship to mechanotransduction and trabeculation during heart looping. Am J Physiol Heart Circ Physiol 300(3): H879–891

Garlid KD, Dos Santos P, Xie ZJ, Costa AD, Paucek P (2003) Mitochondrial potassium transport: the role of the mitochondrial ATP-sensitive K(+) channel in cardiac function and cardioprotection. Biochim Biophys Acta 1606(1–3):1–21

Gilbert RD (1998) Fetal myocardial responses to long-term hypoxemia. Comp Biochem Physiol A Mol Integr Physiol 119(3):669–674

Gittenberger-de Groot AC, Mahtab EA, Hahurij ND, Wisse LJ, Deruiter MC, Wijffels MC, Poelmann RE (2007) Nkx2.5-negative myocardium of the posterior heart field and its correlation with podoplanin expression in cells from the developing cardiac pacemaking and conduction system. Anat Rec (Hoboken) 290(1):115–22

Giussani DA, Phillips PS, Anstee S, Barker DJ (2001) Effects of altitude versus economic status on birth weight and body shape at birth. Pediatr Res 49(4):490–494

Gourdie RG, Green CR, Severs NJ, Anderson RH, Thompson RP (1993) Evidence for a distinct gap-junctional phenotype in ventricular conduction tissues of the developing and mature avian heart. Circ Res 72(2):278–289

Gourdie RG, Harris BS, Bond J, Justus C, Hewett KW, O'Brien TX, Thompson RP, Sedmera D (2003) Development of the cardiac pacemaking and conduction system. Birth Defects Res 69C:46–57

Grobety M, Sedmera D, Kappenberger L (1999) The chick embryo heart as an experimental setup for the assessment of myocardial remodeling induced by pacing. Pacing Clin Electrophysiol 22(5):776–782

Gruber PJ, Kubalak SW, Pexieder T, Sucov HM, Evans RM, Chien KR (1996) RXR alpha deficiency confers genetic susceptibility for aortic sac, conotruncal, atrioventricular cushion, and ventricular muscle defects in mice. J Clin Invest 98(6):1332–1343

Hall CE, Hurtado R, Hewett KW, Shulimovich M, Poma CP, Reckova M, Justus C, Pennisi DJ, Tobita K, Sedmera D, Gourdie RG, Mikawa T (2004) Hemodynamic-dependent patterning of endothelin converting enzyme 1 expression and differentiation of impulse-conducting Purkinje fibers in the embryonic heart. Development 131(3):581–592

Hamburger V, Hamilton HL (1951) A series of normal stages in the development of the chick embryo. J Morphol 88(1):49–92

Han HC, Austin KJ, Nathanielsz PW, Ford SP, Nijland MJ, Hansen TR (2004) Maternal nutrient restriction alters gene expression in the ovine fetal heart. J Physiol 558(Pt 1):111–121

Hildreth V, Anderson RH, Henderson DJ (2009) Autonomic innervation of the developing heart: origins and function. Clin Anat 22(1):36–46

Hirota A, Kamino K, Komuro H, Sakai T, Yada T (1985) Early events in development of electrical activity and contraction in embryonic rat heart assessed by optical recording. J Physiol 369:209–227

Hoerter J (1976) Changes in the sensitivity to hypoxia and glucose deprivation in the isolated perfused rabbit heart during perinatal development. Pflugers Arch 363(1):1–6

Hollenberg M, Honbo N, Samorodin AJ (1976) Effects of hypoxia on cardiac growth in neonatal rat. Am J Physiol 231(5 Pt. 1):1445–1450

Hoogaars WM, Engel A, Brons JF, Verkerk AO, de Lange FJ, Wong LY, Bakker ML, Clout DE, Wakker V, Barnett P, Ravesloot JH, Moorman AF, Verheijck EE, Christoffels VM (2007) Tbx3

controls the sinoatrial node gene program and imposes pacemaker function on the atria. Genes Dev 21(9):1098–1112
Hornberger LK, Sanders SP, Rein AJ, Spevak PJ, Parness IA, Colan SD (1995) Left heart obstructive lesions and left ventricular growth in the midtrimester fetus. A longitudinal study. Circulation 92(6):1531–1538
Hsieh PC, Segers VF, Davis ME, MacGillivray C, Gannon J, Molkentin JD, Robbins J, Lee RT (2007) Evidence from a genetic fate-mapping study that stem cells refresh adult mammalian cardiomyocytes after injury. Nat Med 13(8):970–974
Hu N, Connuck DM, Keller BB, Clark EB (1991) Diastolic filling characteristics in the stage 12 to 27 chick embryo ventricle. Pediatr Res 29(4 Pt 1):334–337
Hurtado A (1960) Some clinical aspects of life at high altitudes. Ann Intern Med 53:247–258
Hutchins GM, Bulkley BH, Moore GW, Piasio MA, Lohr FT (1978) Shape of the human cardiac ventricles. Am J Cardiol 41(4):646–654
Icardo JM, Fernandez-Teran A (1987) Morphologic study of ventricular trabeculation in the embryonic chick heart. Acta Anat 130(3):264–274
Jay PY, Harris BS, Maguire CT, Buerger A, Wakimoto H, Tanaka M, Kupershmidt S, Roden DM, Schultheiss TM, O'Brien TX, Gourdie RG, Berul CI, Izumo S (2004) Nkx2-5 mutation causes anatomic hypoplasia of the cardiac conduction system. J Clin Invest 113(8):1130–1137
Jouk PS, Usson Y, Michalowicz G, Grossi L (2000) Three-dimensional cartography of the pattern of the myofibres in the second trimester fetal human heart. Anat Embryol (Berl) 202(2):103–118
Julia P, Young HH, Buckberg GD, Kofsky ER, Bugyi HI (1990) Studies of myocardial protection in the immature heart. II. Evidence for importance of amino acid metabolism in tolerance to ischemia. J Thorac Cardiovasc Surg 100(6):888–895
Kamino K, Hirota A, Fujii S (1981) Localization of pacemaking activity in early embryonic heart monitored using voltage-sensitive dye. Nature 290(5807):595–597
Keller BB, Hu N, Clark EB (1990) Correlation of ventricular area, perimeter, and conotruncal diameter with ventricular mass and function in the chick embryo from stages 12 to 24. Circ Res 66(1):109–114
Keller BB, MacLennan MJ, Tinney JP, Yoshigi M (1996) In vivo assessment of embryonic cardiovascular dimensions and function in day-10.5 to -14.5 mouse embryos. Circ Res 79(2): 247–255
Knaapen MW, Vrolijk BC, Wenink AC (1996) Nuclear and cellular size of myocytes in different segments of the developing rat heart. Anat Rec 244(1):118–125
Kolar F, Ostadal B, Prochazka J, Pelouch V, Widimsky J (1989) Comparison of cardiopulmonary response to intermittent high-altitude hypoxia in young and adult rats. Respiration 56(1–2):57–62
Kolditz DP, Wijffels MC, Blom NA, van der Laarse A, Hahurij ND, Lie-Venema H, Markwald RR, Poelmann RE, Schalij MJ, Gittenberger-de Groot AC (2008) Epicardium-derived cells in development of annulus fibrosis and persistence of accessory pathways. Circulation 117(12): 1508–1517
Komarek V, Malinovsky L, Lemez L (1982) Anatomia avium domesticarum et embryologia galli. Priroda, Bratislava, http://www.priroda.sk
Koushik SV, Wang J, Rogers R, Moskophidis D, Lambert NA, Creazzo TL, Conway SJ (2001) Targeted inactivation of the sodium-calcium exchanger (Ncx1) results in the lack of a heartbeat and abnormal myofibrillar organization. FASEB J 15(7):1209–1211
Kuma A, Hatano M, Matsui M, Yamamoto A, Nakaya H, Yoshimori T, Ohsumi Y, Tokuhisa T, Mizushima N (2004) The role of autophagy during the early neonatal starvation period. Nature 432(7020):1032–1036
Lamers WH, Wessels A, Verbeek FJ, Moorman AF, Viragh S, Wenink AC, Gittenberger-de Groot AC, Anderson RH (1992) New findings concerning ventricular septation in the human heart. Implications for maldevelopment. Circulation 86(4):1194–1205

Laskey WK (2005) Brief repetitive balloon occlusions enhance reperfusion during percutaneous coronary intervention for acute myocardial infarction: a pilot study. Catheter Cardiovasc Interv 65(3):361–367

Lavine KJ, Ornitz DM (2009) Shared circuitry: developmental signaling cascades regulate both embryonic and adult coronary vasculature. Circ Res 104(2):159–169

Leaf DE, Feig JE, Vasquez C, Riva PL, Yu C, Lader JM, Kontogeorgis A, Baron EL, Peters NS, Fisher EA, Gutstein DE, Morley GE (2008) Connexin40 imparts conduction heterogeneity to atrial tissue. Circ Res 103(9):1001–1008

Li F, Wang X, Capasso JM, Gerdes AM (1996) Rapid transition of cardiac myocytes from hyperplasia to hypertrophy during postnatal development. J Mol Cell Cardiol 28(8):1737–1746

Limperopoulos C, Tworetzky W, McElhinney DB, Newburger JW, Brown DW, Robertson RL Jr, Guizard N, McGrath E, Geva J, Annese D, Dunbar-Masterson C, Trainor B, Laussen PC, du Plessis AJ (2010) Brain volume and metabolism in fetuses with congenital heart disease: evaluation with quantitative magnetic resonance imaging and spectroscopy. Circulation 121(1):26–33

Manner J, Thrane L, Norozi K, Yelbuz TM (2009) In vivo imaging of the cyclic changes in cross-sectional shape of the ventricular segment of pulsating embryonic chick hearts at stages 14 to 17: a contribution to the understanding of the ontogenesis of cardiac pumping function. Dev Dyn 238(12):3273–3284

Marshall AC, Tworetzky W, Bergersen L, McElhinney DB, Benson CB, Jennings RW, Wilkins-Haug LE, Marx GR, Lock JE (2005) Aortic valvuloplasty in the fetus: technical characteristics of successful balloon dilation. J Pediatr 147(4):535–539

Martin C, Yu AY, Jiang BH, Davis L, Kimberly D, Hohimer AR, Semenza GL (1998) Cardiac hypertrophy in chronically anemic fetal sheep: increased vascularization is associated with increased myocardial expression of vascular endothelial growth factor and hypoxia-inducible factor 1. Am J Obstet Gynecol 178(3):527–534

McQuinn TC, Bratoeva M, Dealmeida A, Remond M, Thompson RP, Sedmera D (2007) High-frequency ultrasonographic imaging of avian cardiovascular development. Dev Dyn 236(12): 3503–3513

Meiltz A, Kucera P, de Ribaupierre Y, Raddatz E (1998) Inhibition of bicarbonate transport protects embryonic heart against reoxygenation-induced dysfunction. J Mol Cell Cardiol 30(2): 327–335

Milerova M, Charvatova Z, Skarka L, Ostadalova I, Drahota Z, Fialova M, Ostadal B (2010) Neonatal cardiac mitochondria and ischemia/reperfusion injury. Mol Cell Biochem 335(1–2): 147–153

Minot CS (1901) On a hitherto unrecognised circulation without capillaries in the organs of Vertebrata. Proc Boston Soc Nat Hist 29:185–215

Miquerol L, Meysen S, Mangoni M, Bois P, van Rijen HV, Abran P, Jongsma H, Nargeot J, Gros D (2004) Architectural and functional asymmetry of the His-Purkinje system of the murine heart. Cardiovasc Res 63(1):77–86

Mommersteeg MT, Brown NA, Prall OW, de Gier-de Vries C, Harvey RP, Moorman AF, Christoffels VM (2007) Pitx2c and Nkx2-5 are required for the formation and identity of the pulmonary myocardium. Circ Res 101(9):902–909

Moorman AF, Christoffels VM (2003) Cardiac chamber formation: development, genes, and evolution. Physiol Rev 83(4):1223–1267

Mortola JP, Xu LJ, Lauzon AM (1990) Body growth, lung and heart weight, and DNA content in newborn rats exposed to different levels of chronic hypoxia. Can J Physiol Pharmacol 68(12):1590–1594

Moscoso G, Pexieder T (1990) Variations in microscopic anatomy and ultrastructure of human embryonic hearts subjected to three different modes of fixation. Pathol Res Pract 186(6): 768–774

Muhlfeld C, Singer D, Engelhardt N, Richter J, Schmiedl A (2005) Electron microscopy and microcalorimetry of the postnatal rat heart (Rattus norvegicus). Comp Biochem Physiol A Mol Integr Physiol 141(3):310–318

Murotsuki J, Challis JR, Han VK, Fraher LJ, Gagnon R (1997) Chronic fetal placental embolization and hypoxemia cause hypertension and myocardial hypertrophy in fetal sheep. Am J Physiol 272(1 Pt 2):R201–207

Murry CE, Jennings RB, Reimer KA (1986) Preconditioning with ischemia: a delay of lethal cell injury in ischemic myocardium. Circulation 74(5):1124–1136

Naeye RL (1966) Organ and cellular development in mice growing at simulated high altitude. Lab Invest 15(4):700–706

Nanka O, Krizova P, Fikrle M, Tuma M, Blaha M, Grim M, Sedmera D (2008) Abnormal myocardial and coronary vasculature development in experimental hypoxia. Anat Rec (Hoboken) 291(10):1187–1199

Neffgen JF, Korecky B (1972) Cellular hyperplasia and hypertrophy in cardiomegalies induced by anemia in young and adult rats. Circ Res 30(1):104–113

Nomura-Kitabayashi A, Phoon CKL, Kishigami S, Rosenthal J, Yamauchi Ÿ, Abe K, Yamamura K, Samtani R, Lo CW, Mishina Y (2009) Outflow tract cushions perform a critical valve-like function in the early embryonic heart requiring BMPRIA-mediated signaling in cardiac neural crest. Am J Physiol Heart Circ Physiol 297(5):H1617–1628

Ohtsuka T, Gilbert RD (1995) Cardiac enzyme activities in fetal and adult pregnant and nonpregnant sheep exposed to high-altitude hypoxemia. J Appl Physiol 79(4):1286–1289

Ostadal B, Kolar F (2007) Cardiac adaptation to chronic high-altitude hypoxia: beneficial and adverse effects. Respir Physiol Neurobiol 158(2–3):224–236

Ostadal B, Kolar F, Pelouch V, Widimsky J (1995) Ontogenetic differences in cardiopulmonary adaptation to chronic hypoxia. Physiol Res 44(1):45–51

Ostadal B, Ostadalova I, Dhalla NS (1999) Development of cardiac sensitivity to oxygen deficiency: comparative and ontogenetic aspects. Physiol Rev 79(3):635–659

Ostadal B, Netuka I, Maly J, Besik J, Ostadalova I (2009) Gender differences in cardiac ischemic injury and protection–experimental aspects. Exp Biol Med (Maywood) 234(9):1011–1019

Ostadalova I, Kolar F, Ostadal B, Rohlicek V, Rohlicek J, Prochazka J (1993) Early postnatal development of contractile performance and responsiveness to Ca2+, verapamil and ryanodine in the isolated rat heart. J Mol Cell Cardiol 25(6):733–740

Ostadalova I, Ostadal B, Kolar F (1995) Effect of prenatal hypoxia on contractile performance and responsiveness to Ca2+ in the isolated perinatal rat heart. Physiol Res 44(2):135–137

Ostadalova I, Ostadal B, Kolar F, Parratt JR, Wilson S (1998) Tolerance to ischaemia and ischaemic preconditioning in neonatal rat heart. J Mol Cell Cardiol 30(4):857–865

Ostadalova I, Ostadal B, Jarkovska D, Kolar F (2002) Ischemic preconditioning in chronically hypoxic neonatal rat heart. Pediatr Res 52(4):561–567

Ostadalova I, Charvatova Z, Wilhelm J (2010) Lipofuscin-like pigments in the rat heart during early postnatal development: effect of selenium supplementation. Physiol Res 59(6):881–886

Patterson AJ, Chen M, Xue Q, Xiao D, Zhang L (2010) Chronic prenatal hypoxia induces epigenetic programming of PKC{epsilon} gene repression in rat hearts. Circ Res 107(3):365–373

Pelouch V, Kolar F, Milerova M, Ostadal B (1997) Effect of the preweaning nutritional state on the cardiac protein profile and functional performance of the rat heart. Mol Cell Biochem 177(1–2):221–228

Pexieder T, Janecek P (1984) Organogenesis of the human embryonic and early fetal heart as studied by microdissection and SEM. In: Nora JJ, Takao A (eds) Congenital heart disease: causes and processes. Futura, New York, pp 401–421

Phoon CK (2006) Imaging tools for the developmental biologist: ultrasound biomicroscopy of mouse embryonic development. Pediatr Res 60(1):14–21

Phoon CK, Ji RP, Aristizabal O, Worrad DM, Zhou B, Baldwin HS, Turnbull DH (2004) Embryonic heart failure in NFATc1-/- mice: novel mechanistic insights from in utero ultrasound biomicroscopy. Circ Res 95(1):92–99

Raddatz E, Servin M, Kucera P (1992) Oxygen uptake during early cardiogenesis of the chick. Am J Physiol 262(4 Pt 2):H1224–1230

Rakusan K, Poupa O (1966) Differences in capillary supply of hypertrophic and hyperplastic hearts. Cardiologia 49(5):293–298

Rakusan K, Chvojkova Z, Oliviero P, Ostadalova I, Kolar F, Chassagne C, Samuel JL, Ostadal B (2007) ANG II type 1 receptor antagonist irbesartan inhibits coronary angiogenesis stimulated by chronic intermittent hypoxia in neonatal rats. Am J Physiol Heart Circ Physiol 292(3): H1237–1244

Reckova M, Rosengarten C, deAlmeida A, Stanley CP, Wessels A, Gourdie RG, Thompson RP, Sedmera D (2003) Hemodynamics is a key epigenetic factor in development of the cardiac conduction system. Circ Res 93(1):77–85

Rein AJ, Mevorach D, Perles Z, Gavri S, Nadjari M, Nir A, Elchalal U (2009) Early diagnosis and treatment of atrioventricular block in the fetus exposed to maternal anti-SSA/Ro-SSB/La antibodies: a prospective, observational, fetal kinetocardiogram-based study. Circulation 119(14): 1867–1872

Rentschler S, Vaidya DM, Tamaddon H, Degenhardt K, Sassoon D, Morley GE, Jalife J, Fishman GI (2001) Visualization and functional characterization of the developing murine cardiac conduction system. Development 128:1785–1792

Rentschler S, Zander J, Meyers K, France D, Levine R, Porter G, Rivkees SA, Morley GE, Fishman GI (2002) Neuregulin-1 promotes formation of the murine cardiac conduction system. Proc Natl Acad Sci USA 99(16):10464–10469

Riva A, Hearse D (2009) Age-dependent changes in myocardial susceptibility to ischemic injury. Cardioscience 4:58–92

Romano R, Rochat AC, Kucera P, De Ribaupierre Y, Raddatz E (2001) Oxidative and glycogenolytic capacities within the developing chick heart. Pediatr Res 49(3):363–372

Rosenthal N, Harvey RP (2010) Heart development and regeneration, vol 2, vol 2. Elsevier, London

Rothenberg F, Nikolski V, Watanabe M, Efimov I (2005) Electrophysiology and anatomy of embryonic rabbit hearts before and after septation. Am J Physiol Heart Circ Physiol 288(1): H344–351

Rudolph AM (2000) Myocardial growth before and after birth: clinical implications. Acta Paediatr 89(2):129–133

Sadler TW (2011) Langman's medical embryology, 11th edn. Lippincot Williams & Wilkins, Baltimore

Saiki Y, Konig A, Waddell J, Rebeyka IM (1997) Hemodynamic alteration by fetal surgery accelerates myocyte proliferation in fetal guinea pig hearts. Surgery 122(2):412–419

Sakai T, Hirota A, Kamino K (1996) Video-imaging assessment of initial beating patterns of the early embryonic chick heart. Jpn J Physiol 46(6):465–472

Samanek M, Bass A, Ostadal B, Hucin B, Stejskalova M (1989) Effect of hypoxaemia on enzymes supplying myocardial energy in children with congenital heart disease. Int J Cardiol 25(3): 265–269

Sankova B, Machalek J, Sedmera D (2010) Effects of mechanical loading on early conduction system differentiation in the chick. Am J Physiol Heart Circ Physiol 298(5):H1571–1576

Schagger H, Noack H, Halangk W, Brandt U, von Jagow G (1995) Cytochrome-c oxidase in developing rat heart. Enzymic properties and amino-terminal sequences suggest identity of the fetal heart and the adult liver isoform. Eur J Biochem 230(1):235–241

Sedmera D (2005) Form follows function: developmental and physiological view on ventricular myocardial architecture. Eur J Cardiothorac Surg 28(4):526–528

Sedmera D (2011) Function and form in the developing cardiovascular system. Cardiovasc Res 91(2):252–259

Sedmera D, Pexieder T, Hu N, Clark EB (1997) Developmental changes in the myocardial architecture of the chick. Anat Rec 248(3):421–432

Sedmera D, Pexieder T, Rychterova V, Hu N, Clark EB (1999) Remodeling of chick embryonic ventricular myoarchitecture under experimentally changed loading conditions. Anat Rec 254(2):238–252

Sedmera D, Pexieder T, Vuillemin M, Thompson RP, Anderson RH (2000) Developmental patterning of the myocardium. Anat Rec 258(4):319–337

Sedmera D, Kucera P, Raddatz E (2002) Developmental changes in cardiac recovery from anoxia-reoxygenation. Am J Physiol Regul Integr Comp Physiol 283(2):R379–388

Sedmera D, Reckova M, DeAlmeida A, Coppen SR, Kubalak SW, Gourdie RG, Thompson RP (2003a) Spatiotemporal pattern of commitment to slowed proliferation in the embryonic mouse heart indicates progressive differentiation of the cardiac conduction system. Anat Rec 274A(1):773–777

Sedmera D, Reckova M, DeAlmeida A, Sedmerova M, Biermann M, Volejnik J, Sarre A, Raddatz E, McCarthy RA, Gourdie RG, Thompson RP (2003b) Functional and morphological evidence for a ventricular conduction system in the zebrafish and Xenopus heart. Am J Physiol Heart Circ Physiol 284:H1152–H1160

Sedmera D, Reckova M, Bigelow MR, DeAlmeida A, Stanley CP, Mikawa T, Gourdie RG, Thompson RP (2004) Developmental transitions in electrical activation patterns in chick embryonic heart. Anat Rec 280A(2):1001–1009

Sedmera D, Reckova M, Rosengarten C, Torres MI, Gourdie RG, Thompson RP (2005a) Optical mapping of electrical activation in developing heart. Microsc Microanal 11(3):209–215

Sedmera D, Cook AC, Shirali G, McQuinn TC (2005b) Current issues and perspectives in hypoplasia of the left heart. Cardiol Young 15(1):56–72

Sedmera D, Wessels A, Trusk TC, Thompson RP, Hewett KW, Gourdie RG (2006) Changes in activation sequence of embryonic chick atria correlate with developing myocardial architecture. Am J Physiol Heart Circ Physiol 291(4):H1646–1652

Skarka L, Bardova K, Brauner P, Flachs P, Jarkovska D, Kopecky J, Ostadal B (2003) Expression of mitochondrial uncoupling protein 3 and adenine nucleotide translocase 1 genes in developing rat heart: putative involvement in control of mitochondrial membrane potential. J Mol Cell Cardiol 35(3):321–330

Tamaddon HS, Vaidya D, Simon AM, Paul DL, Jalife J, Morley GE (2000) High-resolution optical mapping of the right bundle branch in connexin40 knockout mice reveals slow conduction in the specialized conduction system. Circ Res 87(10):929–936

Tchervenkov CI, Jacobs ML, Tahta SA (2000) Congenital Heart Surgery Nomenclature and Database Project: hypoplastic left heart syndrome. Ann Thorac Surg 69(4 Suppl): S170–179

Tenthorey D, de Ribaupierre Y, Kucera P, Raddatz E (1998) Effects of verapamil and ryanodine on activity of the embryonic chick heart during anoxia and reoxygenation. J Cardiovasc Pharmacol 31(2):195–202

Tintu A, Rouwet E, Verlohren S, Brinkmann J, Ahmad S, Crispi F, van Bilsen M, Carmeliet P, Staff AC, Tjwa M, Cetin I, Gratacos E, Hernandez-Andrade E, Hofstra L, Jacobs M, Lamers WH, Morano I, Safak E, Ahmed A, le Noble F (2009) Hypoxia induces dilated cardiomyopathy in the chick embryo: mechanism, intervention, and long-term consequences. PLoS One 4(4):e5155

Tobita K, Garrison JB, Li JJ, Tinney JP, Keller BB (2005) Three-dimensional myofiber architecture of the embryonic left ventricle during normal development and altered mechanical loads. Anat Rec A Discov Mol Cell Evol Biol 283(1):193–201

Tomanek RJ (1996) Formation of the coronary vasculature: a brief review. Cardiovasc Res 31(Spec No):E46–E51

Tran L, Kucera P, de Ribaupierre Y, Rochat AC, Raddatz E (1996) Glucose is arrhythmogenic in the anoxic-reoxygenated embryonic chick heart. Pediatr Res 39(5):766–773

Tworetzky W, Wilkins-Haug L, Jennings RW, van der Velde ME, Marshall AC, Marx GR, Colan SD, Benson CB, Lock JE, Perry SB (2004) Balloon dilation of severe aortic stenosis in the fetus: potential for prevention of hypoplastic left heart syndrome: candidate selection, technique, and results of successful intervention. Circulation 110(15):2125–2131

Tworetzky W, McElhinney DB, Marx GR, Benson CB, Brusseau R, Morash D, Wilkins-Haug LE, Lock JE, Marshall AC (2009) In utero valvuloplasty for pulmonary atresia with hypoplastic right ventricle: techniques and outcomes. Pediatrics 124(3):e510–518

Valderrabano M, Chen F, Dave AS, Lamp ST, Klitzner TS, Weiss JN (2006) Atrioventricular ring reentry in embryonic mouse hearts. Circulation 114(6):543–549 van Kempen MJ, ten Velde I, Wessels A, Oosthoek PW, Gros D, Jongsma HJ, Moorman AF, Lamers WH (1995) Differential connexin distribution accommodates cardiac function in different species. Microsc Res Tech 31(5):420–436

Vassall-Adams PR (1982) The development of the atrioventricular bundle and its branches in the avian heart. J Anat 134(Pt 1):169–183

Verburg BO, Jaddoe VW, Wladimiroff JW, Hofman A, Witteman JC, Steegers EA (2008) Fetal hemodynamic adaptive changes related to intrauterine growth: the Generation R Study. Circulation 117(5):649–659

Vetter R, Studer R, Reinecke H, Kolar F, Ostadalova I, Drexler H (1995) Reciprocal changes in the postnatal expression of the sarcolemmal Na+-Ca(2+)-exchanger and SERCA2 in rat heart. J Mol Cell Cardiol 27(8):1689–1701

Vinten-Johansen J (2007) Postconditioning: a mechanical maneuver that triggers biological and molecular cardioprotective responses to reperfusion. Heart Fail Rev 12(3–4):235–244

Viragh S, Challice CE (1977a) The development of the conduction system in the mouse embryo heart. II. Histogenesis of the atrioventricular node and bundle. Dev Biol 56(2):397–411

Viragh S, Challice CE (1977b) The development of the conduction system in the mouse embryo heart. I. The first embryonic A-V conduction pathway. Dev Biol 56(2):382–396

Vuillemin M, Pexieder T (1989) Normal stages of cardiac organogenesis in the mouse: II. Development of the internal relief of the heart. Am J Anat 184(2):114–128

Wachtlova M, Mares V, Ostadal B (1977) DNA synthesis in the ventricular myocardium of young rats exposed to intermittent high altitude (IHA) hypoxia. An autoradiographic study. Virchows Arch B Cell Pathol 24(4):335–342

Wessels A, Sedmera D (2003) Developmental anatomy of the heart: a tale of mice and man. Physiol Genomics 15(3):165–176

Wessels A, Vermeulen JL, Verbeek FJ, Viragh S, Kalman F, Lamers WH, Moorman AF (1992) Spatial distribution of "tissue-specific" antigens in the developing human heart and skeletal muscle. III. An immunohistochemical analysis of the distribution of the neural tissue antigen G1N2 in the embryonic heart; implications for the development of the atrioventricular conduction system. Anat Rec 232(1):97–111

Xavier-Neto J, Davidson B, Simoes-Costa MS, Castro RA, Castillo HA, Sampaio AC, Azambuja AP (2010) Evolutionary origins of hearts. In: Rosenthal N, Harvey RP (eds) Heart development and regeneration, vol 1. Elsevier, London, pp 3–46

Xue Q, Zhang L (2009) Prenatal hypoxia causes a sex-dependent increase in heart susceptibility to ischemia and reperfusion injury in adult male offspring: role of protein kinase C epsilon. J Pharmacol Exp Ther 330(2):624–632

Yelbuz TM, Choma MA, Thrane L, Kirby ML, Izatt JA (2002) Optical coherence tomography: a new high-resolution imaging technology to study cardiac development in chick embryos. Circulation 106(22):2771–2774

Yellon DM, Downey JM (2003) Preconditioning the myocardium: from cellular physiology to clinical cardiology. Physiol Rev 83(4):1113–1151

Chapter 8
Basic Cardiac Development: The Heart and Its Electrical Components

L.Y.E. Wong, A.F. Moorman, and P. Barnett

Basic Embryology

In humans, fertilization marks the start of one of the most critical phases of life, embryogenesis, a short but extremely eventful period in terms of development. This proceeds until approximately the eighth week of development, after which point the embryo enters the fetal period. After fertilization the zygote, derived as the fusion product of two haploid, meiotic cells, undergoes several rounds of cell division, rapidly giving rise to a ball-like cluster of around 16 cells known as the morula (Fig. 8.1a), the outermost cells of which become distinguishable from the cells at the center. Development has now entered the blastocyte stage, a phase which has been of particular interest to cellular research, providing the most very primitive, pluripotent form of stem cells. The blastocyst consists of a blastocyst cavity, which is filled with fluid, and the inner cell mass or embryoblast, which will eventually form the embryo itself. The blastocyst cavity and embryoblast are surrounded by a layer of outer cells, the trophoblast, which will give rise to the extra embryonic tissues of the placenta. The cells of embryoblast now undergo an early differentiation phase that gives rise to the germinal disc, a bilaminar layer of cells, each layer with its own specific function and role during further development (Fig. 8.1b).

The embryo, now completely embedded in the endometrium of its mother, enters approximately its third week of development, during which gastrulation occurs. This important phenomenon in embryology culminates in the embryonic disc becoming trilaminar, setting up the organizational plan for organ and tissue development (Fig. 8.2). During gastrulation, a localized movement and division of cells give rise to three germinal layers known as the ectoderm, mesoderm and endoderm. From each of these three layers specific tissues and organs will arise. The ectoderm

L.Y.E. Wong • A.F. Moorman • P. Barnett (✉)
Department of Anatomy, Embryology and Physiology,
Academic Medical Center, Amsterdam, The Netherlands
e-mail: p.barnett@amc.uva.nl

D. Sedmera and T. Wang (eds.), *Ontogeny and Phylogeny of the Vertebrate Heart*,
DOI 10.1007/978-1-4614-3387-3_8, © Springer Science+Business Media, LLC 2012

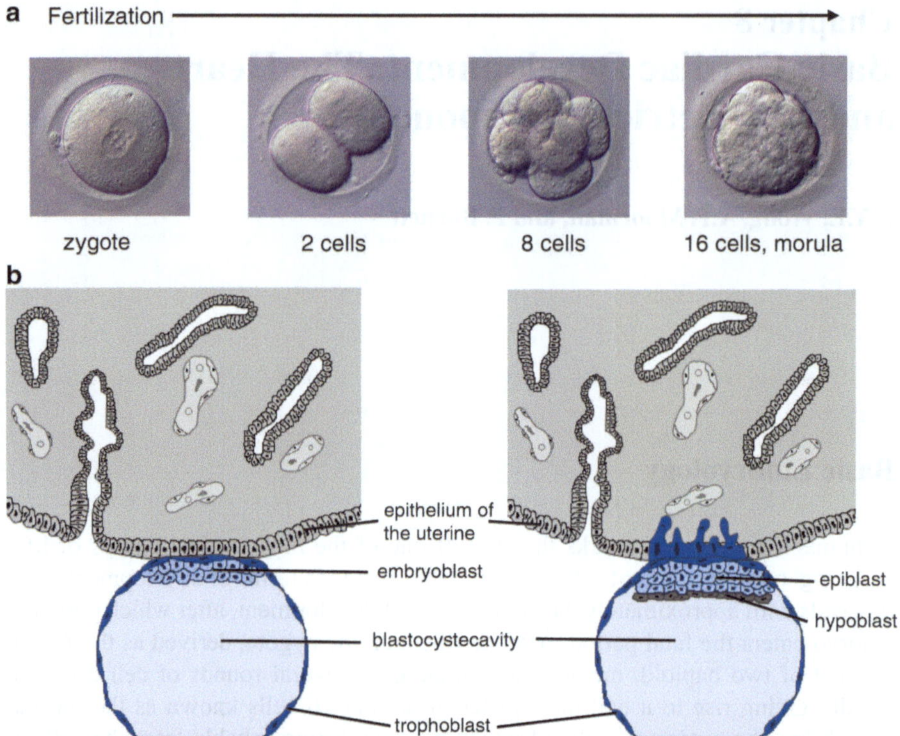

Fig. 8.1 (**a**) Cell divisions after fertilization. (**b**) The blastocyst comes into contact with the epithelium of the uterus (*left panel*). Subsequently, the blastocyst embeds itself deeper in the uterine wall. The cells of the embryoblast that are adjacent to the blastocyst cavity differentiate into the hypoblast and the cells that have not differentiated make up the epiblast. This process gives rise to a double-layered germinal disc (*right panel*)

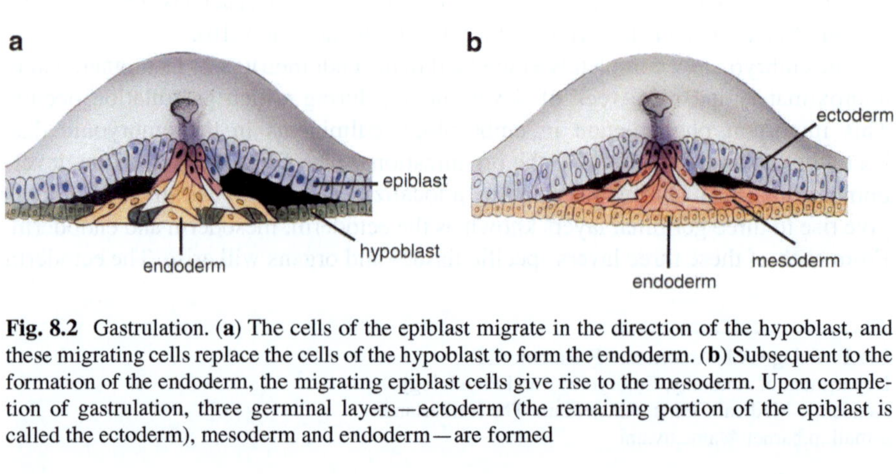

Fig. 8.2 Gastrulation. (**a**) The cells of the epiblast migrate in the direction of the hypoblast, and these migrating cells replace the cells of the hypoblast to form the endoderm. (**b**) Subsequent to the formation of the endoderm, the migrating epiblast cells give rise to the mesoderm. Upon completion of gastrulation, three germinal layers—ectoderm (the remaining portion of the epiblast is called the ectoderm), mesoderm and endoderm—are formed

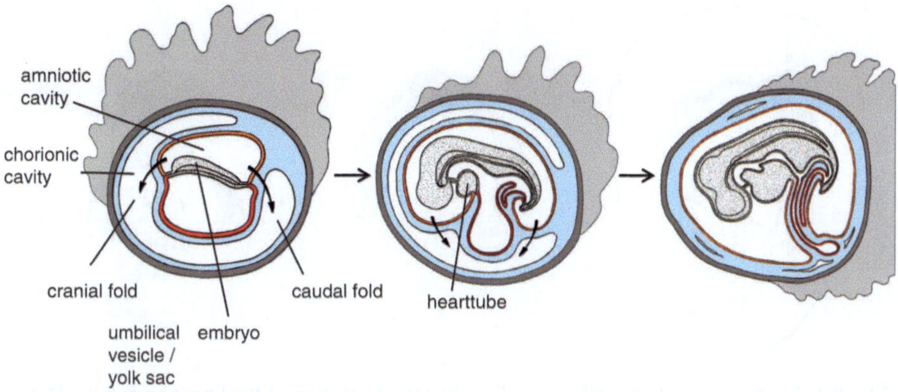

Fig. 8.3 Folding of the embryo. The embryo folds at both the cranial and caudal ends, and at the same time folds also laterally. After cranial folding is completed, the developing brain is positioned most cranially, followed by the facial, pharyngeal and then heart regions that have reached their final locations

mainly contributes to the structures that maintain the communication with the outside world like the central nervous system and epidermis. The mesoderm derives the fibrous tissues, skeleton and muscles, including the cardiovascular system that consists of the heart and major vessels, with, lastly, the endoderm giving rise to the epithelial linings of the intestinal tract and its derivatives.

Organogenesis starts as soon as gastrulation is complete. The first phase of organogenesis is the period of neurulation, a process that takes place within the ectoderm, being characterized by the formation of the neural tube which will eventually give rise to the nervous system. After the formation of the neural tube, cells located on the lateral sides of the neural tube, the neural crest, migrate to other parts of the embryo to contribute to the formation of various tissues including some structures of the heart.

Until the end of the third week, the trilaminar embryo has retained a flat disc-like structure. However, as the embryo develops further a complex three-dimensional folding process ensues, which starts to give the embryo a more cylindrical structure (Fig. 8.3). The embryo bends at the cranial (head) and caudal ("tail") end by as much as 180°, while at the same time folding laterally. These folding processes lead to the enclosure of the mesoderm and endoderm by ectoderm. Because of the head folding, the mouth area bends to the lower front and the cardiac plate moves to the thoracic region where the heart will finally be positioned in the adult (Fig. 8.3).

From Heart Tube to Four-Chambered Heart

The heart is the first organ to form and function during embryonic development. It develops from the cardiac plate that is formed by cells from the mesoderm in the most anterior part of the flat embryonic disc. During the process of gastrulation,

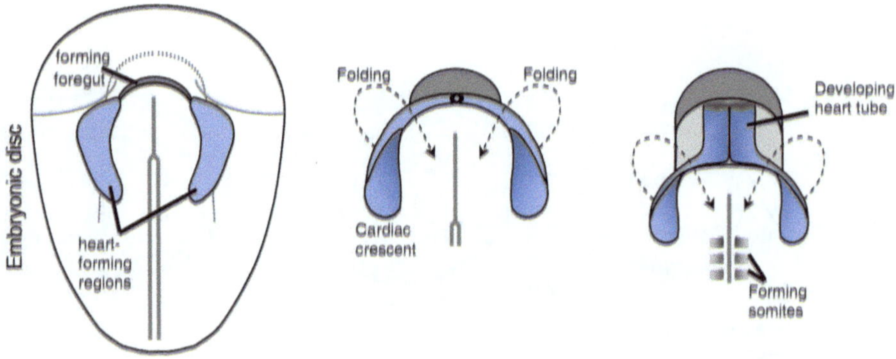

Fig. 8.4 Ventral view of the embryonic disc showing formation of the linear heart tube. During embryonic folding, the heart-forming regions migrate across the embryonic midline and fuse to form the heart tube. Adapted from van den Berg and Moorman (2009)

early cardiac progenitor cells are located in two symmetrical, bilaterally positioned regions on either side of the embryonic midline. These regions are often referred to as the heart-forming regions. The developing embryo undergoes what is perhaps one of the most difficult descriptive morphological transformations during development that of embryonic folding. By the lateral folding of the embryo the heart-forming regions expand across the midline of the embryonic disc, deriving a horse-shoe like region of cells aptly referred to as the cardiac crescent (Fig. 8.4). Subsequently, the lateral walls of the cardiac crescent bend toward the central region and fuse at the midline to form the primitive heart tube. At the same time as the lateral folding, the head folding of the embryo drives the cardiac crescent to move from the upper part to the central region of the embryonic disc where the heart takes up its position in what will become the thoracic region.

The heart tube consists of an outer myocardial layer and an inner endocardial layer, and these two layers are separated by cardiac jelly produced by the myocardium. The linear heart tube, which is bilaterally symmetrical, has an inverted "Y" shape, the legs of which form the venous pole of the heart, while the stem forms the arterial pole, respectively referred to as the inflow and outflow regions of the heart tube (Fig. 8.5). Because of the process of embryonic folding, the heart tube becomes progressively pushed into the pericardial sac. At this stage, the myocardial mantel of the heart tube is not yet closed and is attached along its length via a tissue bridge, termed the dorsal mesocardium, to the mesoderm of the body. As the heart tube subsequently begins to close and looping ensues, the dorsal mesocardium bridge is broken, releasing the now sealed tubular heart into the pericardial sac. The heart will remain connected to the body, but only via its venous and arterial poles.

The heart, as mentioned earlier, is one of the first organs to form and function. The primary myocardium of the early heart tube possesses a phenotype that resembles that of the nodal tissues in the adult heart, having slow contraction, slow transmission of electrical signals, but high automaticity. These characteristics give initial heart

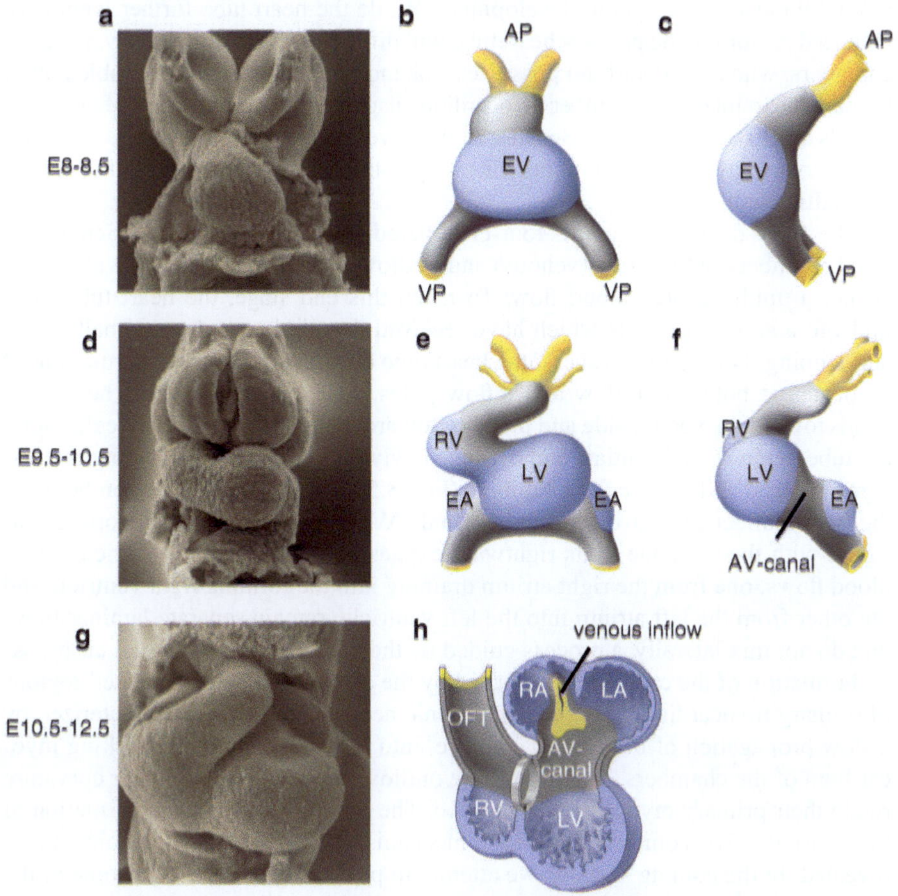

Fig. 8.5 Ballooning and looping of the heart tube in the mouse embryo. (**a**, **d**, **g**) Heart tube at different stages captured by scanning electron microscope. (**a–c**) The heart tube of an E8-8.5 embryo, showing the inverted "Y" shape with the venous pole located at the legs and arterial pole at the stem of the Y. The first signs of local expansion of the heart tube giving rise to the embryonic ventricle are seen here (*blue*). (**a**, **b**) Ventral view; (**c**) left lateral view. (**d–f**) Approximately at stage E8.5-10.5, the heart tube loops rightward and the chambers balloon out of the heart tube; the developing atria can be recognized. (**d**, **e**) Ventral view; (**f**) left lateral view. (**g**, **h**) The heart at E10.5-12.5, showing the four chambers and the opening of the venous inflow into the right atrium. (**g**, **h**) Ventral view, and (**h**) the ventral wall is removed to show the inside of the heart. Primary myocardium is in *gray*; myocardium of the chambers is in *blue*; nonmyocardial components of the inflow and outflow region in *yellow*. *Abbreviations*: *AP* arterial pole, *VP* venous pole, *EV* embryonic ventricle, *LV* left ventricle, *RV* right ventricle, *EA* embryonic atrium, *AV-canal* atrioventricular canal, *OFT* outflow tract. Figures are reproduced from Moorman et al. (2003)

pump function the form of waves of peristaltic-like, rhythmic contractions, generating a unidirectional blood flow from the venous to the arterial pole. Dominant pacemaker activity develops in the venous pole of the heart tube. In humans, the heart begins to contract around 22–23 days of embryonic development and in the mouse around 9 days of embryonic development. While the heart tube further lengthens, localized regions in the primary heart tube will differentiate into atrial and ventricular chambers, whose myocardium possesses a phenotype that does not resemble that of the early heart tube. The chamber myocardium displays rapid transmission of electrical signals and low automaticity. Because of the development of the chambers, the heart begins to display more synchronous-like, sequential contractions instead of the initial peristaltic contractions.

The fully developed, adult, four-chambered heart consists of ventricular and atrial chambers with inflow (venous) and outflow (arterial) regions and valves that ensure a unidirectional blood flow. To reach this end stage, the heart tube must undergo a series of events which have received descriptive analogy of ballooning and looping. During this process, the heart tube elongates due to the recruitment of myocytes at both its outflow and inflow poles. At the same time, the heart tube bends toward the ventral side and loops rightward, and localized areas of cells along the tube begin to differentiate and divide, slowly ballooning out, providing the first signs of atrial and ventricular chambers (Fig. 8.5). The intervening region between the two chambers, the atrioventricular canal (AVC), connects both developing ventricles with the atria due to its rightward expansion relative to the ventricles. Two blood flows, one from the right atrium draining into the forming right ventricle and the other from the left atrium into the left ventricle, remain separate laminar flows that do not mix laterally, a process guided by the developing endocardial cushions.

Formation of the chambers is marked by the differentiation of localized regions of primary myocardium along the embryonic heart tube, which is characterized by a slow propagation of the cardiac impulse, into the fast-conducting working myocardium of the chambers. The inflow, the outflow, the AVC and the inner curvature retain their primary myocardial phenotype. The looping and chamber formation of the heart involve complex molecular mechanisms that are just beginning to be revealed. In the coming sections we attempt to provide an overview of some of the major events that are driven by these processes and provide a basic insight into the early stage of cardiac development.

Early Cardiac Growth

Primary Myocardium Versus Working Myocardium

Cardiac muscle cells, termed cardiomyocytes, possess a number of functional features that distinguish them from the nonheart-related muscle cells. They display automaticity, meaning that they are capable of generating an intrinsic cycle of electrical activity which results in contraction. Cardiac progenitor cells undergo varying degrees of differentiation leading to the different phenotypes of working (chamber), nodal and ventricular conducting myocardium (Table 8.1). The early heart tube, as described earlier, consists of a primary form of myocardium, albeit the cardiomyocytes show a level of dissimilarity owing to a cranio–caudal polarity over the heart

Table 8.1 Characteristics of different types of myocardium present during heart development

	Type of myocardium			
Phenotype	Primary	Nodal	Bundle and branches	Chamber
Automaticity	High	High	High	Low
Conduction velocity	Low	Low	High	High
Contractility	Low	Low	Low	High
SR activity	Low	Low	Low	High
Proliferation	Low	Low	Low	High

Primary and nodal myocardia have the same phenotypic and molecular features. The phenotype of the cells of the chamber or working myocardium is developed in such a way that these cells are specialized for contraction and conduction of the electrical signals. See the text section for a more extensive description of this table. *SR* sarcoplasmic reticulum. The data are based on Canale et al. (1986), Kleber et al. (2001), Sissman (1966), Soufan et al. (2006), Thompson et al. (1990)

tube. Cardiomyocytes in the primary myocardium are small, have poorly developed sarcomeres and sarcoplasmic reticular structure and possess slow conduction properties. This form of myocardium resembles the nodal myocardium of the sinuatrial and atrioventricular nodes in many phenotypic and molecular aspects. The common pacemaker channel, Hcn4, is expressed in both primary and nodal myocardium which coincides with the high automaticity found in both these types of myocardium (Satin et al. 1988; van Mierop 1967). The cardiomyocytes of the chambers possess a phenotype more attuned to the high contractility and conductivity required of a working myocardial cell. They have well-developed sarcomeres and sarcoplasmic reticuli, and a high abundance of fast conducting gap junction proteins Cx40 and Cx43, which ensures fast electrical conduction and synchronous contraction of the chambers.

Growth of the Chambers

The Ballooning Model of Chamber Formation

During looping of the heart tube numerous compartments become visible: the outflow tract (arterial pole), embryonic right ventricle, embryonic left ventricle, atria and sinus venosus (venous pole) (Anderson et al. 1978; Moorman and Lamers 1994). Although these regions will continue to develop to derive their functional counterparts in the mature heart, the underlying process proceeds in a dynamic fashion. Differentiating cells that constitute a part of the outflow tract at an early stage may finally be located in the adult right ventricle. According to the classical view of heart development, the heart tube already consists of all components that will be found in the mature heart, the components being arranged in segments along the anteroposterior, or cranio–caudal axis, hence the term segmental model (Anderson et al. 1978; De la Cruz et al. 1989; Stalsberg 1969). The segmental model of development was accepted for a long time until another, better fitting and accurate model was put

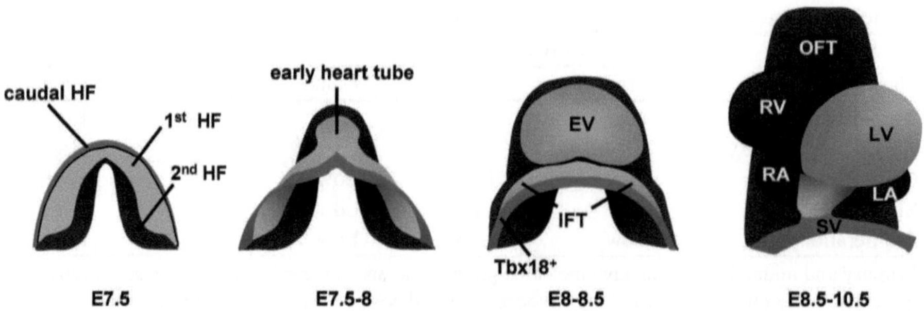

Fig. 8.6 Schematic overview of the different cardiac precursor cell populations. The first heart field (HF), cells lying in the heart tube region prior to rupture of the mesocardium, gives rise mainly to the left ventricle. The right ventricle, outflow tract and a part of the atria are derived from the second heart field, i.e. cells that must enter the heart via either the venous pole or outflow tract regions due to mesocardium rupture. The sinus venosus (SV) of the venous heart pole is derived from a *Tbx18*-positive progenitor cell population, distinct form the second heart field, called the caudal heart field. *Abbreviations*: *EV* embryonic ventricle, *IFT* inflow tract, *OFT* outflow tract, *RV* right ventricle, *LV* left ventricle, *RA* right atrium, *LA* left atrium

forward (Christoffels et al. 2000; Moorman et al. 2000). With the advent of molecular biology, many of the historical descriptive anatomical teachings of developmental processes have been resolved to a level of better understanding and no more so than for the heart. Studies making use of gene expression patterns have provided crucial insights into the early processes of embryonic heart development, providing clear and evidenced views of differing developmental stages. These findings demonstrate that chamber formation actually occurs predominantly in two phases. First, the heart tube shows specification in gene expression and morphology not only along the anteroposterior axis as described in the segmental model, but also along the dorsoventral axis. Second, subsequent to the formation and looping of the heart tube, the atrial and ventricular chambers start their development locally, forming at the outer curvatures (Fig. 8.5). The observed expression patterns of transcription factors associated with chamber formation demonstrate that the inner curvature does not differentiate into chamber working myocardium, but retains a more primary phenotype. Cardiomyocytes located within the chamber-forming area proliferate fast, causing the chambers "balloon" out from the heart tube.

Other evidence that also strengthened the developmental mechanisms underlying the "ballooning" model was the discovery of what is commonly referred to as the second heart field. The second heart field is a cardiac precursor cell population that is located within the pharyngeal mesenchyme outside the bounds of the early heart tube (Fig. 8.6). The precursor cells that make up the linear heart tube are descriptively derived from the first heart field, and give rise to the embryonic ventricle and inflow tract region that go to make up the early heart tube as we have described (Fig. 8.6). Perhaps one of the surprising findings, although historically a process described in the 1960s (DeHaan 1967), is that large part of the observed growth of the heart tube is due to the addition of cells from the second heart field, cells which are particularly key to the development of the right ventricle and outflow tract as well as making significant contributions to other regions of the heart (Kelly and Buckingham 2002; Mjaatvedt et al. 2001; van den Berg et al. 2009; Waldo et al. 2001). This finding shows that precursor cells needed for the building of the heart are not already present in the heart tube. The characterization of heart development in terms of the spatiotemporal expression of key markers of cardiac development has then clearly provided some of the most profound insights into this anatomically complex developmental process.

Proliferation of the Heart Tube

The early heart tube of chicken embryos increases in length at a rate of 70 μm per hour (van den Berg et al. 2009). Quantitative reconstructions of embryonic chicken hearts show that the growth of the early heart tube is due to the recruitment of precursor cells from a pool of precursor cells at the venous pole of the heart tube, which has been referred to as the caudal precursor pool (Soufan et al. 2006) (Fig. 8.6). These reconstructions show that the caudal pool of precursor cells proliferates rapidly, whereas the primary cardiac precursor cells (primary myocardium) of the heart tube proliferate very slowly, if at all, reinforcing the idea that actual growth of the heart tube is mainly due to the recruitment of cells from a cardiac precursor pool located outside the heart tube, and not to large-scale proliferation of the primary myocardium itself. Once the highly proliferative caudal precursor pool of cells enters the heart tube, it becomes less proliferative, concomitant with overt differentiation into cardiomyocytes. Subsequent to the quiescent phase of the cell-cycle, proliferation becomes re-initiated at the outer curvature of the heart tube, i.e. the region of chamber development. In contrast to the inversely related proliferation and differentiation rate observed in the caudal precursor pool, the ballooning chambers undergo simultaneous differentiation and proliferation, as marked by the expression of working myocardial genes such as *Nppa* (Christoffels et al. 2000; Delorme et al. 1997; Houweling et al. 2002; van Kempen et al. 1996) within the high proliferative zones of the ballooning chambers. Furthermore, the increase in cell size has been shown to contribute to chamber formation (Soufan et al. 2006). The mechanisms underlying the growth of the heart tube and formation of the chambers, put forward mainly using mouse and chicken studies, have recently also been shown to be conserved in humans (Sizarov et al. 2011).

A cell-tracing study in chicken shows that the caudally defined precursor cells described above, contributing to the growth of the heart tube, also migrate cranially and enter the heart via the dorsal pericardial wall into the arterial pole of the heart tube and subsequently differentiate into cardiomyocytes. This, although seeming to redefine thoughts regarding separate fields of heart cells each contributing to the heart, in fact ties the heart fields together in terms of their proliferative characteristics by simply defining one single growth center contributing cells to both poles of the heart. On the basis of all these studies, then it is clear that the spatiotemporal location of cardiac precursor cells during embryogenesis will ultimately play a chief role in determining their terminal location and functional differentiation state (Rochais et al. 2009; van den Berg et al. 2009; Xu et al. 2004).

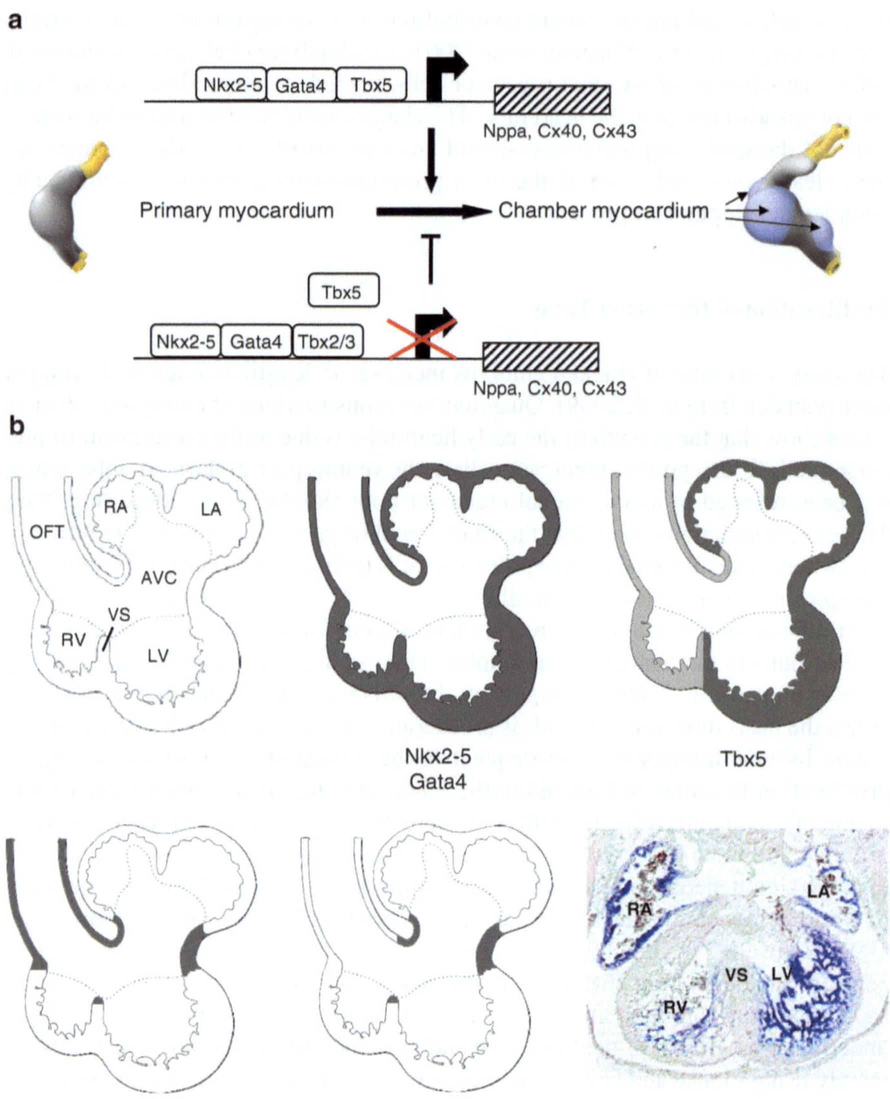

Fig. 8.7 Roles of T-box-dependant networks of transcription factors in the regulation of chamber formation. (**a**) Model of a regulatory network for chamber formation. Nkx2-5, Gata4 and Tbx5 physically interact to activate the expression of chamber-specific genes and thereby stimulating the differentiation of primary into chamber myocardium. In the atrioventricular canal, Tbx2 and Tbx3 compete with Tbx5 and interact with Nkx2-5 and Gata4 in order to repress the chamber-specific genes, thereby maintaining the primary myocardial phenotype in the atrioventricular canal. (**b**) Schematic representation of an E10.5-12.5 four-chambered heart showing the myocardial expression patterns of genes important for the regulation of heart development. The ventral wall is removed to show the myocardial layer of the heart. The regions of expression are shown in *gray*. *Nkx2-5* and *Gata4* are expressed throughout the myocardium. *Tbx5* shows a gradient in expression,

Differentiation of the Chambers

The heart tube initially consists of cardiomyocytes with primary myocardial features (Table 8.1). Cardiac chamber development requires the localized differentiation and proliferation of primary myocardial cells into working myocardial cells. The remaining portion of the tube, constituting the inflow tract, AVC, inner curvature and outflow tract, will largely retain its primary "nodal" phenotype and does not differentiate into the working myocardium phenotype.

The primary and working myocardium can be distinguished from each other by the expression and repression patterns of numerous genes. Differentiation of primary myocardium into working myocardium is tightly regulated by conserved transcription factors, ensuring local initiation and restriction of a chamber-specific gene program. Natriuretic precursor peptide type A (*Nppa*, also known as ANF), Connexin 40 (*Cx40*) and Connexin 43 (*Cx43*) are genes expressed in cells of the working myocardium (Christoffels et al. 2004a). Expression and regulation of *Nppa* have been extensively studied, and its expression can be linked to the differentiation of the working myocardium of the atria and ventricles (Houweling et al. 2005), making it suitable and thus commonly used marker for this process. The transcription factors Nkx2-5, Gata4 and Tbx5 have been shown to be able to activate the expression of *Nppa* in the working myocardium of the chambers, thereby indicating that these factors play a role in chamber differentiation (Fig. 8.7a) (Bruneau et al. 2001; Garg et al. 2003; Hiroi et al. 2001; Small and Krieg 2003). When the heart-forming regions fuse in the midline to form the cardiac crescent, the expression of *Nkx2-5*, *Gata4* and *Tbx5* can be detected in the cardiac progenitor cells, making these factors suitable as early markers for cardiogenesis.

Nkx2-5, GATA4 and Tbx5 Activate Chamber Differentiation

Nkx2-5 belongs to the evolutionarily conserved homeodomain proteins of the tinman family, which is known to be an important class of transcription factors involved in heart development. Initial experiments making use of Drosophila identified a mutant phenotype lacking a heart, aptly referred to as the Tinman phenotype after the Wizard of Oz fairy tale. Investigations showed that this phenotype was caused by mutations in the gene Nkx2-5 (Bodmer 1993). Nkx2-5 is a transcription factor that binds DNA through a conserved 60 amino acid DNA-binding domain called the

Fig. 8.7 (continued) which is higher in the atria, the left ventricle and the left aspect of the ventricular septum than the right ventricle, the right aspect of the ventricular septum and the outflow tract (*light gray*). *Tbx2* and *Tbx3* are expressed in the nonchamber myocardium. *Nppa* expression, visualized by in situ hybridization, overlaps the expressions of *Nkx2-5*, *Gata4* and *Tbx5* in the chamber myocardium, and is absent in the *Tbx2* and *Tbx3* expression-regions, indicating the roles of these factors in the up- and down-regulation of *Nppa*. *Abbreviations*: *OFT* outflow tract, *RA* right atrium, *LA* left atrium, *AVC* atrioventricular canal, *LV* left ventricle, *VS* ventricular septum, *RV* right ventricle. The in situ hybridization figure is adapted from Moorman and Christoffels (2003)

homeodomain. Nkx2-5 is an early marker of cardiac precursor cells and is expressed throughout the early heart tube and within the second heart field, and plays a crucial role in the early steps of heart development. During heart development, Nkx2-5 expression can be found in the outflow tract, AVC, ventricles and atria (Fig. 8.7b) (Komuro and Izumo 1993; Lints et al. 1993). The crucial role of Nkx2-5 in early heart development was evidenced by the observation that functional ablation of Nkx2-5 in mice leads to early embryonic death (at embryonic day E9.5) mainly due to cardiac insufficiency caused by phenotypes including looping defects and the lack of chamber formation. Furthermore, the expression of several cardiac genes in Nkx2-5-deficient embryos, including *Nppa*, is reduced, suggesting the pivotal role of Nkx2-5 in the transcriptional regulation of cardiac genes that are important for chamber differentiation after heart looping (Lyons et al. 1995; Tanaka et al. 1999). *NKX2-5* mutations in humans lead to congenital heart diseases including atrial and ventricular septal defects, which underlines that Nkx2-5 is indispensable for normal chamber formation (Akazawa and Komuro 2005; Reamon-Buettner and Borlak 2006). In addition to chamber formation, Nkx2-5 also plays an important role in the formation of the conduction system as evidenced by the presence of atrioventricular conduction abnormalities in humans with a defective copy of *NKX2-5* (Benson et al. 1999; Schott et al. 1998). The role of Nkx2-5 in the development of the conduction system is discussed in detail in the section on "The Cardiac Conduction System."

Gata4 belongs to a large evolutionarily conserved family of zinc-finger transcription factors. The presence of zinc ions is a major component of their structure. The Gata family shares DNA-binding ability through the C-terminal conserved zinc-finger domain (Garg et al. 2003). Although Gata4 is not the only Gata member important for heart development, it is expressed in the entire heart throughout embryonic development (Fig. 8.7b) and in adult life (Laverriere et al. 1994). It has been identified as a causative factor of human congenital heart defects, most often being linked to atrial septal defects (Chen et al. 2010; Garg et al. 2003; Okubo et al. 2004; Rajagopal et al. 2007). Nonconditional Gata4 knock-out mice die during the early stages of embryogenesis (between E8.5 and E10.5) due to defects in embryonic folding and heart tube formation (Kuo et al. 1997). Conditional deletion of Gata4 in the cardiomyocyte population specifically results in reduced proliferation of the right ventricle, leading to right ventricular hypoplasia (Zeisberg et al. 2005), a form of congenital heart disease observed in humans with *GATA4* mutations (Rajagopal et al. 2007).

Tbx5 belongs to the evolutionarily conserved T-box family of transcription factors. T-box factors bind DNA through a highly conserved 180 amino acid DNA-binding domain called the T-box. To date more than 20 T-box family members have been identified, all of which regulate various aspects of embryonic development. T-box proteins have been shown to bind consensus DNA motifs, T-box-binding elements (TBE), associated with gene promoters and enhancer elements, forming complexes with other transcription factors thereby regulating gene expression (Boogerd et al. 2009; Kispert and Herrmann 1993; Smith 1999). Tbx5 expression displays a posterior–anterior gradient along the heart tube, being most intense at the inflow tract of the heart. During heart development, Tbx5 expression is found strongly up-regulated in the atria, the left ventricle and the left aspect of the ventricular septum, and to a lesser extent in the right ventricle and the outflow tract (Fig. 8.7b) (Bruneau et al. 1999; Christoffels et al. 2000), its presence coinciding with the up-regulation of markers of chamber differentiation like *Nppa*. In humans, *TBX5* haploin sufficiency has been linked to Holt-Oram syndrome, manifest by congenital heart defects, conduction system abnormalities and upper limb deformities (Basson et al. 1997; 1999; Boogerd et al. 2010; Li et al. 1997). Heterozygous Tbx5 knockout mice recapitulate many of the phenotypic abnormalities observed in Holt-Oram syndrome patients (Bruneau et al. 2001). Expression of both *Nppa* and *Cx40* was also found to be reduced in these mice during development. Complete deletion of Tbx5 in mice leads to embryonic death (around E10.5) due to failure of heart tube looping and underdeveloped atria (Bruneau et al. 2001). Furthermore, Tbx5 overexpression in an embryonic cell line was shown to result in significant upregulation of *Nppa* and *Cx40*, these cells displaying a premature beating phenotype, indicative of the early onset of myocardial differentiation (Fijnvandraat et al. 2003; Hiroi et al. 2001).

Cooperation of Nkx2-5, Gata4 and Tbx5 in Gene Transcription

Tbx5, Nkx2.5 and Gata4 are critical factors involved in chamber differentiation. Nkx2-5, Gata4 and Tbx5 have been shown to bind regulatory DNA elements embedded in promoters and enhancers of key cardiac genes. Genes including *Nppa*, *Cx40*, *Cx43*, *Cx30.2* and *Pitx2* have been shown to respond to the up-regulation and loss of these factors (Bruneau et al. 2001; Chen et al. 2004; Habets et al. 2002; Munshi et al. 2009; Nowotschin et al. 2006; Plageman and Yutzey 2004). With the advent of modern high-throughput techniques such as microarrays, RNA-seq and ChIP-seq (Park 2009; Pepke et al. 2009; Ramsay 1998; Wang et al. 2009), new targets and gene packets of these factors are being identified. What is also becoming clearer is how these factors function at the molecular level, and how they cooperate with each other and other proteins to ensure correct development of the heart. In vitro studies have revealed a direct interaction between Nkx2-5, Gata4 and Tbx5 leading to the activation of cardiac gene transcription. Co-transfection experiments of Nkx2-5, Gata4 and Tbx5 in cardiomyocytes and various cell lines have demonstrated the synergistic activation of *Nppa* (Durocher et al. 1997; Hiroi et al. 2001; Lee et al. 1998; Sepulveda et al. 1998), and interaction studies have shown the physical interactions of these factors (Bruneau et al. 2001; Durocher et al. 1997; Garg et al. 2003; Hiroi et al. 2001; Lee et al. 1998). Furthermore, it has been shown that mutations in Tbx5 known to underlie Holt-Oram syndrome lead to a reduced ability of Tbx5 to bind DNA and Nkx2-5 (Basson et al. 1999; Boogerd et al. 2010; Fan et al. 2003; Mori and Bruneau 2004) and even an increased ability (Postma et al. 2008). The interaction of Gata4 and Tbx5 was discovered in a genetic linkage study aiming at finding gene mutations underlying isolated cardiac septal defects. A causal mutation in Gata4 was identified. The underlying mechanism leading to the associated cardiac phenotype was shown to be a loss of interaction with Tbx5 (Garg et al. 2003).

Physical interactions between Nkx2-5, Gata4 and Tbx5 or between transcription factors in general appear to be an important mechanism underlying gene-target specificity and the regulatory processes they induce.

Tbx2 and Tbx3 Repress Chamber Differentiation in the Atrioventricular Canal

The afore-mentioned transcription factors Nkx2-5, Gata4 and Tbx5 play an important role in the differentiation of the working myocardial phenotype. As stated earlier, these factors are expressed both in working myocardial cells and the more nodal type primary myocardium. Therefore their presence alone cannot be considered as dominant, as these factors are also expressed in areas that do not differentiate into working myocardium. Notably, *Nppa* expression is absent in the primary myocardium. Therefore one can hypothesize that other factors must be present that can repress the differentiation of primary myocardium. The causal linkage of Tbx5 to Holt-Oram syndrome was in part responsible for the discovery of two of these key repressing factors, which are themselves T-box proteins. These T-box factors, Tbx2 and Tbx3, are both expressed in primary myocardium, act as transcriptional repressors and display high homology in both structure and function (Christoffels et al. 2004b; Habets et al. 2002; He et al. 1999; Hoogaars et al. 2004; Paxton et al. 2002) (Fig. 8.7a). During heart development, *Tbx2* and *Tbx3* show similar patterns of expression (Fig. 8.7b). *Tbx2* is expressed in the myocardium of the AVC, inflow tract and outflow tract in mouse (Chapman et al. 1996; Christoffels et al. 2004b). *Tbx3* expression overlaps *Tbx2* expression in the inflow region and AVC (Christoffels et al. 2004b; Hoogaars et al. 2004). Later in development, *Tbx2* expression gradually fades, whereas *Tbx3* expression persists in the heart into adult life, becoming restricted to the proximal components of the cardiac conduction system: the sinus node, the atrioventricular node and the proximal atrioventricular bundle (Bakker et al. 2008; Christoffels et al. 2004b; Hoogaars et al. 2004).

In mouse, *Tbx2* expression and expression of many chamber-specific genes have been shown to be mutually exclusive. Markers of working myocardium such as Cx43, Cx40 and Nppa have all been shown to be absent from the AVC region and other regions expressing Tbx2. This led to the hypothesis that Tbx2 could function as an antagonist of myocardial differentiation, thus suppressing the working myocardial phenotype (Christoffels et al. 2004b; Moorman et al. 2004). Subsequent studies indeed showed this to be the case. Tbx2 homozygous knock-out mice show ectopic expression of *Nppa*, *Cx40* and *Cx43* in the primary myocardium of the AVC, suggesting that Tbx2 is required to repress chamber differentiation in the AVC. These mice die early in development due to severe cardiovascular defects (Aanhaanen et al. 2009; Harrelson et al. 2004). Perhaps the most conclusive observation underlying the theory of Tbx2 function comes from studies which used a promoter of a ubiquitously myocardial expressed protein to drive Tbx2 expression in the entire heart tube. This resulted in the loss of normal expression of *Nppa* and *Cx40* and a loss of chamber formation (Christoffels et al. 2004b). It has now been firmly established that both Tbx3 and Tbx2 play important roles in the maintenance of the nodal-like phenotype of primary myocardium that will eventually derive the nodal components of the cardiac conduction system.

The Cardiac Conduction System

Components of the Conduction System

The cardiac conduction system is responsible for the initiation and coordination of the electrical signals that cause the rhythmic and synchronized contractions of the heart. Well-coordinated contractions of the heart are extremely important for efficient blood perfusion through the lungs and the systemic circulation. It is also crucial to point out that the cardiac conduction system is non-neuronal, being derived from mesodermally derived myocardial tissue and not ectodermal neural cells (Virágh and Challice 1980; Virágh and Challice 1982). The conduction system can be divided into two component parts (1) the nodal conduction system comprising of the sinuatrial node (SAN), the atrioventricular node (AVN), the atrioventricular bundle (AVB) and the proximal part of the bundle branches (BBs); and (2) the peripheral ventricular conduction system which comprises the distal part of the BBs and Purkinje fiber network (Fig. 8.8). In the mature heart, the SAN is located at the junction of the superior caval vein and right atrium where it serves as the primary pacemaker, generating paced electrical signals to activate the atrial myocardium (Fig. 8.8). The SAN is innervated by the parasympathetic as well as the sympathetic nervous systems, stimulation of which can cause decrease or increase of the impulse-generating rate of the SAN, respectively. The electrical signals propagate rapidly through the myocardium of both atria and converge at the AVN, the only conductive bridge between the atria and ventricles. Here the electrical impulse is slightly delayed, a delay brought about by the different composition of conducting proteins such as connexins and ion channels as well as its interwoven myocardial architecture. The delay in transmitting electrical signals to the ventricles is important because it gives time for the ventricles to be sufficiently filled with blood after atrial contraction.

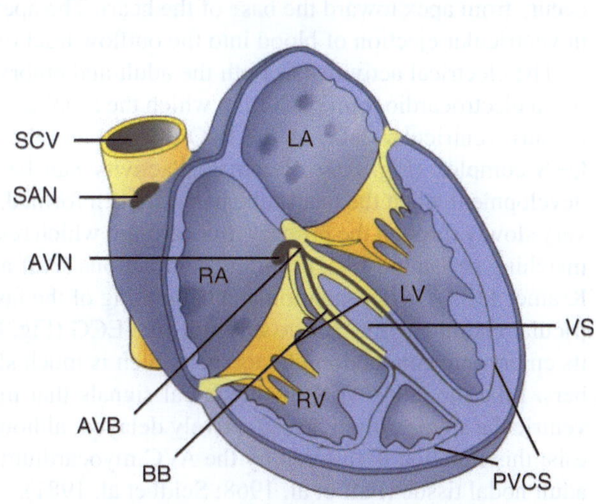

Fig. 8.8 Schematic overview of the adult mammalian heart. The cardiac conduction system consists of the pacemaker sinuatrial node (SAN), the atrioventricular node (AVN) and the fast-conducting atrioventricular bundle (AVB), bundle branches (BB) at both sides of the ventricular septum (VS) and peripheral ventricular conduction system (PVCS). *SCV* superior caval vein

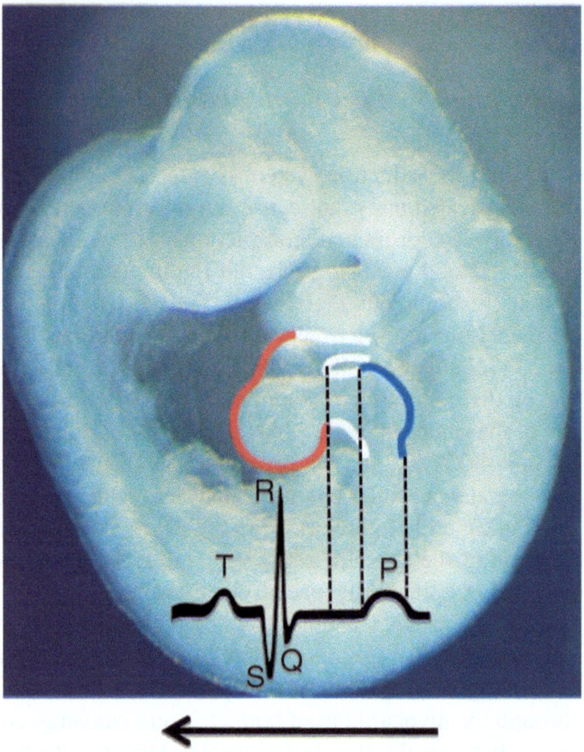

Fig. 8.9 Schematic overview of an E9.5 mouse embryo showing how the segments of an ECG correlate with the different components of the heart. An adult-like ECG can be recorded upon chamber differentiation. *Blue line* depicts the contour of the atria, *red line* the ventricles. The *arrow* indicates the direction of the action potential

Subsequently, the electrical signal enters the fast-conducting AVB and BBs at both sides of the ventricular septum, moving rapidly into the Purkinje fiber network, from which the ventricular working myocardium is activated, ensuring contraction occurs from apex toward the base of the heart. The apex-to-base contraction results in ventricular ejection of blood into the outflow tract of the heart.

The electrical activities of both the adult and embryonic heart can be registered by an electrocardiogram (ECG), in which the atrial activation shows as the P-wave, the atrioventricular delay as the P–Q interval and the ventricular activation as the QRS complex (Fig. 8.9). Pacemaker activity can be detected early during heart development when the heart tube has just been formed. The electrical signals travel very slowly through the primary myocardium which results in a sinusoidal ECG and matching peristaltic-like, rhythmic contractions (Paff and Boucek 1962; Patten and Kramer 1933). Differentiation and ballooning of the fast-conducting atrial and ventricular chambers give rise to an adult-like ECG (Fig. 8.9), with the AVC retaining its embryonic conduction phenotype, which is much slower than that of the chambers. Like the adult AVN the electrical signals that migrate from the atrial to the ventricular myocardium are effectively delayed, although during early embryogenesis; this function is fulfilled by the AVC myocardium itself and not differentiated adult nodal tissue (Paff et al. 1968; Seidl et al. 1981).

Gene Markers of the Developing Conduction System

A set of gene markers have been used to visualize the developmental processes of the conduction system. Among these gene markers gap-junction encoding genes provide functional markers for studying the development of the conduction system, contributing to conductivity, their isoform expression being related to the type and rate of conduction. The fast conducting gap-junction gene *Cx40* is expressed in the atrial and ventricular working myocardium, and later, after E14.5, also in parts of the ventricular conduction system (the AVB, BBs and the Purkinje fiber network). *Cx43* also encodes a fast-conducting gap junction, although unlike Cx40, its expression is restricted to the atrial and ventricular working myocardium, making it an excellent marker to discriminate these structures (Coppen et al. 2003; Delorme et al. 1997; van Kempen et al. 1996). The nodal components of the conduction system, the SAN and AVN, are slow conducting, and are characterized by the expression of the funny current channel *Hcn4* (Garcia-Frigola et al. 2003; Mangoni and Nargeot 2008; Mommersteeg et al. 2007; Stieber et al. 2003), slow-conducting gap junction genes *Cx30*, *Cx30.2* and *Cx45* (Aanhaanen et al. 2010; Boyett et al. 2006; Kreuzberg et al. 2006b) and lack of expression of the chamber-specific genes *Cx40*, *Cx43* and *Nppa* (Coppen et al. 2001; Gros et al. 2009; Kreuzberg et al. 2005, 2006a).

Tbx3 Represses the Working Myocardial Phenotype in the Conduction System

The transcriptional repressor Tbx3 plays a key role in the differentiation of the SAN and AVN. Tbx3 is specifically expressed in the nodal components of the conduction system throughout embryogenesis, making it a marker of choice for studying development of these elements (Hoogaars et al. 2004, 2007). Tbx3 has been found to define the SAN region by repressing the atrial gene program in the sinuatrial nodal cells (Hoogaars et al. 2007; Mommersteeg et al. 2007). In addition to the role of Tbx3 in the formation of the SAN, it has also been reported that Tbx3 defines the atrioventricular conduction system by repressing differentiation of the working myocardial phenotype (Bakker et al. 2008). Mutations in Tbx3 that result in its haploinsufficiency are associated with the human disorder ulnar-mammary syndrome, characterized by malformation of the upper limbs, apocrine-glands, hair, genitals and teeth, indicating the importance of Tbx3 dose levels during development (Bamshad et al. 1997, 1999). However, cardiac defects associated with ulnar-mammary syndrome patients have only very recently been reported, with the clinical description of an ulnar-mammary patient with defects in TBX3 and an associated ventricular septal defect and cardiac conduction defects (Linden et al. 2009). Heterozygous Tbx3 mutant mice show no apparent phenotype, whereas homozygous Tbx3 mutant mice show phenotypes very similar to those of ulnar-mammary syndrome patients (Davenport et al. 2003). Tbx3 represses chamber formation in the conduction system of the heart and thereby maintaining the nodal-like phenotype in these cells. The role of Tbx3 in specification and formation of the nodal conduction system is discussed in more detail in the next sections.

Formation of the SAN

The SAN develops within the sinus venosus myocardium, a structure comprising the left and right sinus horns and venous valves. During the early stages of heart development (<E9.5) the entire sinus venosus, of which the phenotype closely resembles that of the primary myocardium, displays pacemaker activity. After E9.5, noncardiac *Nkx2-5* negative, but *Tbx18*-positive, mesenchymal precursor cells located at the periphery of the inflow tract myocardium are recruited to form the myocardium of the sinus venosus (Christoffels et al. 2006) (Fig. 8.6). These cells are regarded as the caudal precursor cells, discussed in the section on "Early Cardiac Growth." These precursor cells also show expression of *Isl1* (Christoffels et al. 2006; Mommersteeg et al. 2007; Moretti et al. 2006; Sun et al. 2006), a marker that has often been used to define the subpopulation of mesenchyme, the second heart field, which is added to the heart after the definitive heart tube has formed (Buckingham et al. 2005; Cai et al. 2003). Logically then, the first heart field represents the myocardial population that gives rise to the definitive heart tube. After the dorsal mesocardium breaks along its length, the entry of additional cells is physically restricted to the outflow and inflow poles only. At E10, the *Tbx18*-positive precursor cells form a subdomain of the sinus venosus that will differentiate into a *Tbx3*/*Hcn4*-positive SAN primordium (Christoffels et al. 2006), located in the right sinus horn at the junction with the atrium (van Mierop and Gessner 1970; Virágh and Challice 1980). The remainder of the sinus venosus initiates the expression of working myocardial genes (Mommersteeg et al. 2007). In the SAN, the expression of *Hcn4* is retained and the initiation of working myocardial genes is prevented by Tbx3, thereby confining pacemaker activity to the definitive SAN structure (Hoogaars et al. 2004, 2007).

The SAN consists of a "head" region located within the right caval vein myocardium and a "tail" that runs along the right venous valve. These individual structures are to a degree structurally and functionally different from each other. The SAN "head" displays expression of *Tbx3* and *Tbx18* whereas the SAN "tail" loses its *Tbx18* expression (Mommersteeg et al. 2007; Wiese et al. 2009). *Tbx18*-deficient mice fail to develop the *Tbx3*-positive SAN "head" region, suggesting that *Tbx18* plays a primary role in the formation of this region from mesenchymal precursors (Christoffels et al. 2006; Wiese et al. 2009).

The Molecular Pathways of SAN Development

The SAN has a gene program that is distinct from that of the bordering atrial myocardial cells. A strict coordinated network of molecular pathways is involved in the restriction of the specific SAN gene program to the SAN. The SAN displays expression of *Hcn4* and *Tbx3*, and lacks expression of working myocardial genes like *Cx40*, *Cx43* and *Nppa*.

As discussed in section "Early Cardiac Growth," SAN myocardial cells show phenotypic resemblance to cells of the primary myocardium. Recently, Tbx3 has been shown to be a key factor in establishing the SAN gene program by inhibiting the working atrial phenotype in the SAN region. *Tbx3* is expressed in the SAN from E9.5 onward in the mouse (Hoogaars et al. 2004; Mommersteeg et al. 2007). Although *Tbx3*-deficient mice show expansion of expression of working myocardial genes like *Cx40* and *Scn5a* into the SAN (Hoogaars et al. 2004, 2007; Wiese et al. 2009) the morphology of the SAN remains normal, indicating that *Tbx3* is not required per se for the formation of the SAN (Wiese et al. 2009). Mice over-expressing *Tbx3* in the atrial working myocardium display ectopic pacemaking activity in the atria and develop arrhythmias. This would seem to indicate that the major function of Tbx3 is in the regulation of the electrical components of nodal cell. In line with this, ectopic expression of Tbx3 leads to reduced expression of working myocardial genes *Cx40*, *Cx43* and *Scn5a* in the right atrium (Hoogaars et al. 2007).

Hcn4 is initially expressed at the caudal end of the *Nkx2-5* positive heart tube (Garcia-Frigola et al. 2003; Stieber et al. 2003), where its expression will subsequently be inhibited (Mommersteeg et al. 2007). In the SAN domain, the expression of *Hcn4* and *Tbx3* is initiated in the newly added *Tbx18*-positive/*Nkx2-5*-negative cells (Mommersteeg et al. 2007). Disruption of *Nkx2-5* in mouse embryos results in ectopic expression of *Hcn4* and *Tbx3* in heart (Mommersteeg et al. 2007), suggesting the requirement of Nkx2-5 for the repression of these genes and their presence in the *Nkx2-5*-negative SAN. However, the expression of *Tbx3* overlaps that of *Nkx2-5* in many areas including the AV conduction system, which indicates that Nkx2-5 does not work alone to repress *Tbx3*. Other factors, involved in the repression of *Tbx3,* through acting as down-stream targets of Nkx2-5 and/or through interaction with Nkx2-5 are most likely essential to this end. In this respect, Nkx2-5 is also required for the activation of working myocardial genes in the atrium (Stennard and Harvey 2005). The molecular pathways required for maintaining a gene expression border between the SAN and the atrium are depicted in Fig. 8.10a.

Formation of the AVN and the AV Junction

The atrioventricular junction, which includes the AVN, has properties that could allow one to call it a secondary pacemaker of the heart. The phenotype of the AVN cells is similar to that of the SAN, resembling primary myocardium. Different studies support the hypothesis that the AVC will derive the AVN, the AV ring bundles, the myocardium that supports the AV valves and the lower rim of the atrium (Davis et al. 2001; Virágh and Challice 1982; Wessels et al. 1996). Arguably one of the most convincing lines of evidence to this fact made use of a *Tbx2*^{Cre}-mediated lineage tracing experiment. In this study the *Tbx2*$^{+}$ cell population was followed during development and showed that *Tbx2*$^{+}$ primary myocardial cells could be detected in

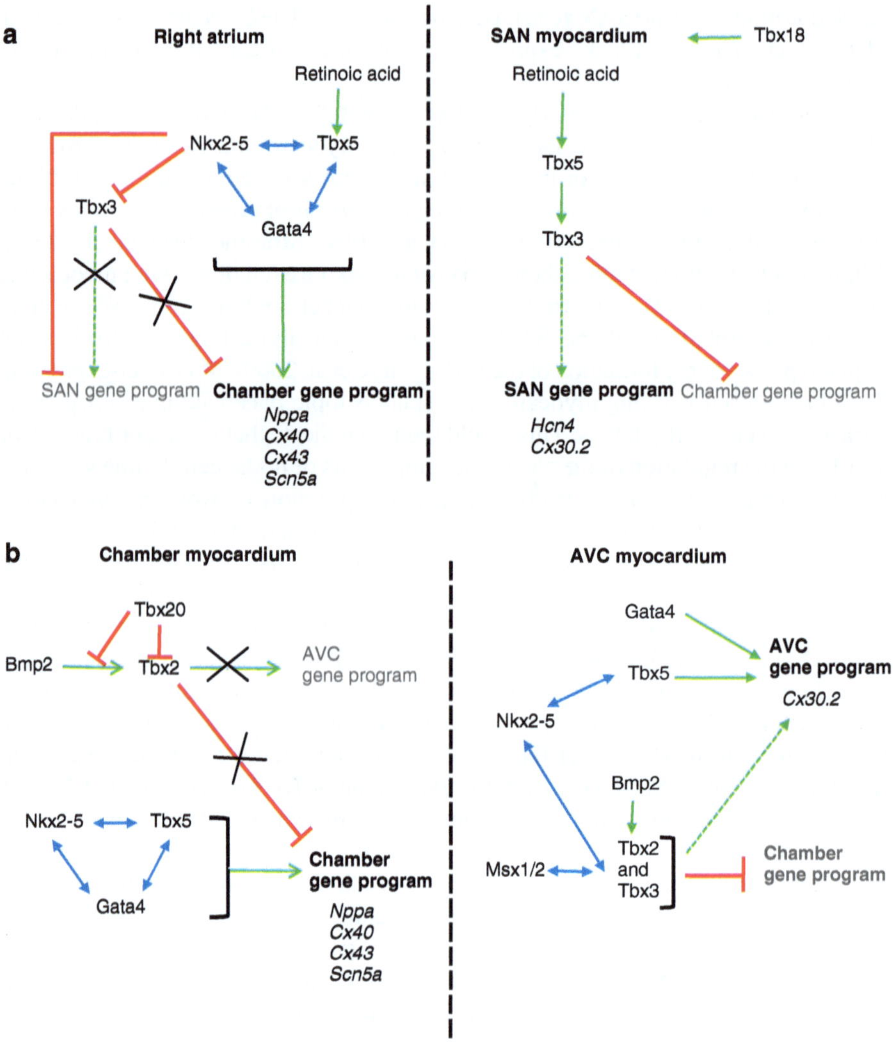

Fig. 8.10 (**a**) Model depicting genetic interactions of factors involved in the formation of the sinuatrial node (SAN) and working myocardium. The *thick dashed line* in the middle represents the border between the right atrium and the SAN. The scheme on the *left side* depicts the molecular pathway regulating the chamber formation of the right atrium. The scheme on the *right side* represents the molecular pathway underlying the formation of the SAN. The sinus venosus myocardium, which will derive the SAN, is formed by recruitment of *Nkx2-5* negative/*Tbx18*-positive mesenchymal precursor cells from the periphery of the inflow tract myocardium. Although not dealt with here, transcription factors regulating development are themselves regulated by a complex interaction of signal molecules and their downstream receptors. In the atria and sinus venosus, *Tbx5* expression is regulated by the retinoic acid signaling pathway (Niederreither et al. 2001). (**b**) A model displaying the network of transcription factors required to form the atrioventricular canal (AVC) and retain a border between the chambers and the AVC. The *thick dashed line* in the middle represents the border between the chamber and AVC myocardium. Note that the atrioventricular node is developed within the AVC myocardium. Although not dealt in the text, Bmp2 and Tbx20,

the cardiac crescent as early as E7.5, which at E9 could be traced to the myocardium of the AVC (Aanhaanen et al. 2009). As development progressed to E17.5, the *Tbx2*+ cell population was detected in the AVN and AV ring bundles, suggesting that these structures are derived from the *Tbx2*+ AVC myocardium (Aanhaanen et al. 2009). Interestingly, the *Tbx2*+ derived cell population could not be detected in the AVB and BBs, indicating that these structures are not derived from the *Tbx2*+ AVC (Aanhaanen et al. 2009). The origin of the AVB and BBs will be discussed further in section "Formation of the Atrioventricular Bundle and Proximal Bundle Branches."

The Molecular Pathways of AVC Formation

As the AVC myocardium gives rise to the AVN, insight into the molecular pathways underlying the specification of the AVC is necessary in order to understand how the AVN is formed. Activation of the AVC gene program and repression of the chamber gene program are required for the specification and morphological patterning of the AVC. A molecular network of transcription factors, in which T-box transcription factors have an important role, is required in order to retain a boundary between the fast-conducting chambers and the slow-conducting AVC. Tbx2 and Tbx3 have both been shown to have a similar functional capacity in their ability to repress the chamber-specific gene program in the AVC (Habets et al. 2002; Hoogaars et al. 2004). Tbx2 is expressed in all primary myocardium (Chapman et al. 1996; Christoffels et al. 2004b; Habets et al. 2002; Yamada et al. 2000), a feature shared, although to a more restrictive degree, with Tbx3. In mouse, *Tbx2* is expressed in the inflow tract, AVC and outflow tract (Chapman et al. 1996; Christoffels et al. 2004b). Expression of *Tbx3* initially overlaps with that of *Tbx2* particularly in the inflow tract and AVC (Christoffels et al. 2004b; Hoogaars et al. 2004), although unlike *Tbx2*, whose expression fades and disappears as the heart develops, *Tbx3* expression, although becoming restricted to the nodal tissues and the proximal AVB, persists into adulthood. The probable redundancy that one may expect between Tbx2 and Tbx3, based on their function and expression, is demonstrated by the limited AVC defect observed in *Tbx2* knockout mice (Dupays et al. 2009). These mice display ectopic expression of chamber-specific genes only in the left-side of AVC, the right-side remaining unaffected (Aanhaanen et al. 2009). This is in line with the observation that *Tbx3* expression is higher in the right side of the AVC (Harrelson et al. 2004; Hoogaars et al. 2004), suggesting that the higher levels of *Tbx3* compensate for loss of *Tbx2*. Further, *Tbx3* deficiency in the heart does not affect AV-conduction velocity, also suggesting a level of Tbx3 redundancy in the AVC (Bakker et al. 2008), a redun-

Fig. 8.10 (continued) both play key roles in chamber formation, regulate *Tbx2* expression (Cai et al. 2005; Ma et al. 2005; Singh et al. 2009; Yamada et al. 2000). For a review on these aspects we refer the reader to Van Wijk et al. (2007). *Cx30.2* has been demonstrated to be activated by Gata4 and Tbx5 (Munshi et al. 2009). *Green arrows* depict "activation"; *red lines*, "repression"; *blue double-arrows*, protein–protein interaction; *dashed arrow*, "regulation might not be direct"

dancy which may go some way to explaining that, to date, no congenital heart disease patients have been identified with a single allelic *Tbx2* mutation.

In contrast to the chamber myocardial transcriptional repressors Tbx2 and Tbx3, Tbx5 is an important activator of chamber-specific genes and thereby essential for working myocardial differentiation (Bruneau et al. 2001; Munshi et al. 2009). Interestingly, its expression also overlaps that of Tbx2 and Tbx3 in the AVC and Tbx5 haploinsufficiency has been shown to underlie Holt-Oram syndrome, these patients often displaying AV conduction defects (Basson et al. 1997; Li et al. 1997). Interestingly, mice with one functional copy of the *Tbx5* allele show morphological and functional defects in the AV conduction system (Moskowitz et al. 2004), a feature not observed in *Tbx2* or *Tbx3* heterozygous knockout mice. Clearly then, Tbx5 is required, alongside Tbx2 and Tbx3, for the formation and normal function of the AV conduction system.

Tbx2, Tbx3 and Tbx5 have been shown, in vitro, to recognize a similar T-box-binding element DNA consensus (Kispert and Herrmann 1993; Smith 1999). As Tbx2 and Tbx3 are co-expressed with Tbx5 in the AVC, it is very plausible that Tbx2 and Tbx3 must compete with Tbx5 for DNA-binding site occupation (Habets et al. 2002) (Fig. 8.7a). Although the exact mechanism that governs this seemingly competitive process has not been elucidated yet, stoichiometry may play a chief role, suggested in part by the redundancy of Tbx2 and Tbx3 and the dose dependence of Tbx5. In fact, the dosage of Tbx5 also seems itself to directly influence the expression of *Tbx3* in the AVC region. Mouse embryos with only one functional allelic copy of Tbx5 show a loss of *Tbx3* expression (Munshi et al. 2009), consistent with the observed defects in the AV conduction system (Moskowitz et al. 2004).

A level of functional similarity between Tbx2, Tbx3 and Tbx5 is demonstrated by their interaction with and apparent dependence on the presence of Nkx2-5. Nkx2-5 interacts with Tbx5 to activate chamber-specific genes (Bruneau et al. 2001; Hiroi et al. 2001), while also interacting with Tbx2 and Tbx3 to repress these genes in the AVC (Habets et al. 2002). Similarly, Tbx2, 3 and 5 have been shown to complex with several other transcription factors, for example, Gata4 (Garg et al. 2003) and the muscle segment homeobox proteins Msx1 and Msx2 (Boogerd et al. 2008), that perhaps in some way either recruit or direct T-box proteins to specific sets of gene targets thereby playing a role in securing the correct terminal differentiation status. The discovery of functional interacting protein partners of T-box factors thus reveals that co-operation of T-box factors with other factors is also a key factor involved in proper activation or repression of gene expression. De novo discovery of such factors will thus assist in unraveling mechanistic issues surrounding the transcriptional regulation of AVC patterning. Figure 8.10b shows the summary of T-box function during AVC development.

Formation of the Atrioventricular Bundle and Proximal Bundle Branches

The AVB and BBs are fast-conducting components of the cardiac conduction system that rapidly transfer the electrical signals from the AVN to the apex and the remainder of the ventricle. In mouse, the *Cx43*-negative/*Tbx3*-positive AVB originates from cells located in the top of the ventricular septum where the onset of AVB specification is first observed at E10-11 (Virágh 1977). As the ventricular septum further develops, its subendocardial myocytes give rise to the *Cx43*-negative/*Tbx3*-positive BBs (Chuck and Watanabe 1997). *Tbx3* expression eventually becomes restricted to the nodal tissues and proximal AVB, *Cx43* expression remaining absent in these regions (Bakker et al. 2008). *Cx40* is absent in the *Tbx3*-positive AVB until E14.5 after which time its expression becomes upregulated (Bakker et al. 2008; Delorme et al. 1995). In *Tbx3* knock-out embryos, *Cx40* is up-regulated in the AVB at earlier stages of development, suggesting that Tbx3 is, at least in part, responsible for the down-regulation of *Cx40* in the AVB and that Tbx3 plays an important role in the specification of the AVB (Bakker et al. 2008). From E14.5 onward, *Cx40* expression increases in the AVB, interestingly, overlapping with *Tbx3* in this region, a feature not observed in nodal tissues (Bakker et al. 2008). This suggests that a mechanism must exist to specifically reduce the repression of some chamber myocardial genes in the presence of Tbx3, something which could in part, perhaps, be brought about by the presence of Tbx5. The observation that Tbx5 exerts dominant effects on the activation of *Cx40* in the AVB suggests a direct role for Tbx5 in the specification of the fast-conducting component of the AV conduction system (Munshi et al. 2009). As discussed in the previous section, the balance of the levels of Tbx3 and Tbx5 in the AV conduction system might shift in such a way that Tbx5 can enforce the activation of *Cx40*.

In summary, the atrioventricular conduction axis comprises multiple components of the cardiac conduction system, including the AVN and the AVB, which have different origins; the AVN is derived from the AVC myocardium, whereas the AVB is from the ventricular septum. Although these components have distinctive phenotypic features, both require Tbx3 for their specification. Differentiation of the AVN and AVB requires repression of the chamber gene program by Tbx3, a factor which also seems required for the persistence of these structures during adulthood. To what extent Tbx3 plays a role in the exact specification of a nodal gene program remains to be resolved but clearly in addition to Tbx3, a complex molecular network involving multiple factors, including Tbx5 and Nkx2-5, is required for the formation of the AV conduction system.

Summary and Perspectives

Animal life starts with fertilization. In a short but key period of development, the germ layers are formed by the fundamental process of gastrulation that lays down the blue print defining an organism's body plan. The heart, derived from the mesodermal germ layer, is the first organ to form and function. Heart development involves a series of evolutionarily conserved events, including the recruitment and differentiation of cardiac progenitor cells, events that are subject to an orchestrated control by transcriptional networks that will eventually define the terminally differentiated function of each cell (Moorman and Christoffels 2003; Olson 2006). Cardiogenesis starts with the specification of heart-forming regions, which, due to embryonic folding, fuse at the midline of the embryonic disc to give rise to the early heart tube. During the subsequent stage of looping, chambers will balloon out from this tube at localized sites of myocardial differentiation. At the same time, large regions of this early tube will remain undifferentiated, retaining a so-called primary myocardial phenotype that is extremely comparable to the nodal tissues of the cardiac conduction system. Key transcriptional factors involved in both the processes of chamber differentiation and chamber repression have been identified. The non-chamber area of the heart tube will give rise to nonworking cardiac structures, including the cardiac conduction system. A well-defined conduction system is responsible for proper, coordinated, function of the heart as a pump. Tbx2 and Tbx3, which are expressed in the sinus venosus and AVC, regions that will give rise to the nodal tissues of the heart, repress the chamber formation in these regions, a process that is of tremendous importance for the correct specification of the nodal tissues. The fact that *Tbx2* expression fades during development while *Tbx3* expression persists into adulthood may indicate that Tbx3 plays a more prominent role than does Tbx2 in the development of the cardiac conduction system.

Congenital heart disease, the most common of all human birth defects, occurs in nearly 1% of live births (Hoffman 1995). It is frequently caused by heterozygous mutations in genes that regulate heart development during the embryonic period. Further, many mutations that affect heart development may also affect other developmental processes not only directly via impaired gene function but also by insufficient early heart function. These aspects of development are only just beginning to be understood. With the advent of modern high-throughput techniques such as deep sequencing and microarrays, we will soon be able to identify all genes involved not only in heart development, but also other developmental processes. Although this will mark the end of a key discovery phase, only by understanding how all these factors work together to drive development will one then be able to truly tackle the mechanisms underlying what happens when a component of this system is faulty. A future goal of medical science is to finally integrate such knowledge to permit not only mechanical surgical intervention, but also make use of molecular techniques such as cellular repair therapy along with molecularly tailored drugs to ease and perhaps even repair diseased and damaged organs.

References

Aanhaanen WT, Brons JF, Dominguez JN, Rana MS, Norden J, Airik R et al (2009) The Tbx2+ primary myocardium of the atrioventricular canal forms the atrioventricular node and the base of the left ventricle. Circ Res 104:1267

Aanhaanen WT, Mommersteeg MT, Norden J, Wakker V, de Gier-de VC, Anderson RH et al (2010) Developmental origin, growth, and three-dimensional architecture of the atrioventricular conduction axis of the mouse heart. Circ Res 107:728–736

Akazawa H, Komuro I (2005) Cardiac transcription factor Csx/Nkx2-5: its role in cardiac development and diseases. Pharmacol Ther 107:252–268

Anderson RH, Wilkinson JL, Becker AE (1978) The bulbus cordis–a misunderstood region of the developing human heart: its significance to the classification of congenital cardiac malformations. Birth Defects Orig Artic Ser 14(7):1–28

Bakker ML, Boukens BJ, Mommersteeg MTM, Brons JF, Wakker V, Moorman AFM et al (2008) Transcription factor Tbx3 is required for the specification of the atrioventricular conduction system. Circ Res 102:1340–1349

Bamshad M, Lin RC, Law DJ, Watkins WS, Krakowiak PA, Moore ME et al (1997) Mutations in human TBX3 alter limb, apocrine and genital development in ulnar-mammary syndrome. Nat Genet 16:311–315

Bamshad M, Le T, Watkins WS, Dixon ME, Kramer BE, Roeder AD et al (1999) The spectrum of mutations in TBX3: Genotype/phenotype relationship in ulnar-mammary syndrome. Am J Hum Genet 64:1550–1562

Basson CT, Bachinsky DR, Lin RC, Levi T, Elkins JA, Soults J et al (1997) Mutations in human TBX5 (corrected) cause limb and cardiac malformation in Holt-Oram syndrome. Nat Genet 15:30–35

Basson CT, Huang T, Lin RC, Bachinsky DR, Weremowicz S, Vaglio A et al (1999) Different *TBX5* interactions in heart and limb defined by Holt-Oram syndrome mutations. Proc Natl Acad Sci USA 96:2919–2924

Benson DW, Silberbach GM, Kavanaugh-McHugh A, Cottrill C, Zhang Y, Riggs S et al (1999) Mutations in the cardiac transcription factor NKX2.5 affect diverse cardiac developmental pathways. J Clin Invest 104:1567–1573

Bodmer R (1993) The gene tinman is required for specification of the heart and visceral muscles in Drosophila. Development 118:719–729

Boogerd KJ, Wong LYE, Christoffels VM, Klarenbeek M, Ruijter JM, Moorman AFM et al (2008) Msx1 and Msx2 are functional interacting partners of T-box factors in the regulation of connexin 43. Cardiovasc Res 78:485–493

Boogerd CJ, Moorman AF, Barnett P (2009) Protein interactions at the heart of cardiac chamber formation. Ann Anat 191:505–517

Boogerd CJ, Dooijes D, Ilgun A, Mathijssen IB, Hordijk R, van de Laar I, Rump P et al (2010) Functional analysis of novel TBX5 T-box mutations associated with Holt-Oram syndrome. Cardiovasc Res 88(1):130–9

Boyett MR, Inada S, Yoo S, Li J, Liu J, Tellez J et al (2006) Connexins in the sinoatrial and atrioventricular nodes. Adv Cardiol 42:175–197

Bruneau BG, Logan M, Davis N, Levi T, Tabin CJ, Seidman JG et al (1999) Chamber-specific cardiac expression of *Tbx5* and heart defects in Holt-Oram syndrome. Dev Biol 211:100–108

Bruneau BG, Nemer G, Schmitt JP, Charron F, Robitaille L, Caron S et al (2001) A murine model of Holt-Oram syndrome defines roles of the T-box transcription factor Tbx5 in cardiogenesis and disease. Cell 106:709–721

Buckingham M, Meilhac S, Zaffran S (2005) Building the mammalian heart from two sources of myocardial cells. Nat Rev Genet 6:826–837

Cai CL, Liang X, Shi Y, Chu PH, Pfaff SL, Chen J et al (2003) Isl1 identifies a cardiac progenitor population that proliferates prior to differentiation and contributes a majority of cells to the heart. Dev Cell 5:877–889

Cai CL, Zhou W, Yang L, Bu L, Qyang Y, Zhang X et al (2005) T-box genes coordinate regional rates of proliferation and regional specification during cardiogenesis. Development 132:2475–2487

Canale ED, Campbell GR, Smolich JJ, Campbell JH (1986) Cardiac muscle. Springer, Berlin

Chapman DL, Garvey N, Hancock S, Alexiou M, Agulnik SI, Gibson-Brown JJ et al (1996) Expression of the T-box family genes, *Tbx1-Tbx5,* during early mouse development. Dev Dyn 206:379–390

Chen JR, Chatterjee B, Meyer R, Yu JC, Borke JL, Isales CM et al (2004) Tbx2 represses expression of connexin43 in osteoblastic-like cells. Calcif Tissue Int 74:561–573

Chen Y, Mao J, Sun Y, Zhang Q, Cheng HB, Yan WH et al (2010) A novel mutation of GATA4 in a familial atrial septal defect. Clin Chim Acta 411:1741–1745

Christoffels VM, Habets PEMH, Franco D, Campione M, de Jong F, Lamers WH et al (2000) Chamber formation and morphogenesis in the developing mammalian heart. Dev Biol 223:266–278

Christoffels VM, Burch JBE, Moorman AFM (2004a) Architectural plan for the heart: early patterning and delineation of the chambers and the nodes. Trends Cardiovasc Med 14:301–307

Christoffels VM, Hoogaars WMH, Tessari A, Clout DEW, Moorman AFM, Campione M (2004b) T-box transcription factor Tbx2 represses differentiation and formation of the cardiac chambers. Dev Dyn 229:763–770

Christoffels VM, Mommersteeg MTM, Trowe MO, Prall OWJ, de Gier-de Vries C, Soufan AT et al (2006) Formation of the venous pole of the heart from an *Nkx2-5*-negative precursor population requires Tbx18. Circ Res 98:1555–1563

Chuck ET, Watanabe M (1997) Differential expression of PSA-NCAM and HNK-1 epitopes in the developing cardiac conduction system of the chick. Dev Dyn 209:182–195

Coppen SR, Gourdie RG, Severs NJ (2001) Connexin45 is the first connexin to be expressed in the central conduction system of the mouse heart. Exp Clin Cardiol 6:17–23

Coppen SR, Kaba RA, Halliday D, Dupont E, Skepper JN, Elneil S et al (2003) Comparison of connexin expression patterns in the developing mouse heart and human foetal heart. Mol Cell Biochem 242:121–127

Davenport TG, Jerome-Majewska LA, Papaioannou VE (2003) Mammary gland, limb and yolk sac defects in mice lacking Tbx3, the gene mutated in human ulnar mammary syndrome. Development 130:2263–2273

Davis DL, Edwards AV, Juraszek AL, Phelps A, Wessels A, Burch JB (2001) A GATA-6 gene heart-region-specific enhancer provides a novel means to mark and probe a discrete component of the mouse cardiac conduction system. Mech Dev 108:105–119

De la Cruz MV, Sánchez-Gómez C, Palomino M (1989) The primitive cardiac regions in the straight tube heart (stage 9) and their anatomical expression in the mature heart: an experimental study in the chick embryo. J Anat 165:121–131

DeHaan RL (1967) Development of form in the embryonic heart. An experimental approach. Circulation 35:821–833

Delorme B, Dahl E, Jarry-Guichard T, Marics I, Briand JP, Willecke K et al (1995) Developmental regulation of connexin40 gene expression in mouse heart correlates with the differentiation of the conduction system. Dev Dyn 204:358–371

Delorme B, Dahl E, Jarry-Guichard T, Briand JP, Willecke K, Gros D et al (1997) Expression pattern of connexin gene products at the early developmental stages of the mouse cardiovascular system. Circ Res 81:423–437

Dupays L, Kotecha S, Mohun TJ (2009) Tbx2 misexpression impairs deployment of second heart field derived progenitor cells to the arterial pole of the embryonic heart. Dev Biol 333(1):121–131

Durocher D, Charron F, Warren R, Schwartz RJ, Nemer M (1997) The cardiac transcription factors Nkx2-5 and GATA-4 are mutual cofactors. EMBO J 18:5687–5696

Fan C, Liu M, Wang Q (2003) Functional analysis of TBX5 missense mutations associated with Holt-Oram syndrome. J Biol Chem 278:8780–8785

Fijnvandraat AC, Lekanne Deprez RH, Christoffels VM, Ruijter JM, Moorman AFM (2003) TBX5 overexpression stimulates differentiation of chamber myocardium in P19C16 embryonic carcinoma cells. J Muscle Res Cell Motil 24:211–218

Garcia-Frigola C, Shi Y, Evans SM (2003) Expression of the hyperpolarization-activated cyclic nucleotide-gated cation channel HCN4 during mouse heart development. Gene Expr Patterns 3:777–783

Garg V, Kathiriya IS, Barnes R, Schluterman MK, King IN, Butler CA et al (2003) GATA4 mutations cause human congenital heart defects and reveal an interaction with TBX5. Nature 424:443–447

Gros D, Theveniau-Ruissy M, Bernard M, Calmels T, Kober F, Sohl G et al (2009) Connexin 30 is expressed in the mouse sino-atrial node, and modulates heart rate. Cardiovasc Res 85(1):45–55

Habets PEMH, Moorman AFM, Clout DEW, van Roon MA, Lingbeek M, Lohuizen M et al (2002) Cooperative action of Tbx2 and Nkx2.5 inhibits ANF expression in the atrioventricular canal: implications for cardiac chamber formation. Genes Dev 16:1234–1246

Harrelson Z, Kelly RG, Goldin SN, Gibson-Brown JJ, Bollag RJ, Silver LM et al (2004) Tbx2 is essential for patterning the atrioventricular canal and for morphogenesis of the outflow tract during heart development. Development 131:5041–5052

He ML, Wen L, Campbell CE, Wu JY, Rao Y (1999) Transcription repression by Xenopus ET and its human ortholog TBX3, a gene involved in ulnar-mammary syndrome. Proc Natl Acad Sci USA 96:10212–10217

Hiroi Y, Kudoh S, Monzen K, Ikeda Y, Yazaki Y, Nagai R et al (2001) Tbx5 associates with Nkx2-5 and synergistically promotes cardiomyocyte differentiation. Nat Genet 28:276–280

Hoffman J (1995) Incidence of congenital heart disease: I. Postnatal incidence. Pediatr Cardiol 16:103–113

Hoogaars WMH, Tessari A, Moorman AFM, de Boer PAJ, Hagoort J, Soufan AT et al (2004) The transcriptional repressor Tbx3 delineates the developing central conduction system of the heart. Cardiovasc Res 62:489–499

Hoogaars WM, Engel A, Brons JF, Verkerk AO, de Lange FJ, Wong LY et al (2007) Tbx3 controls the sinoatrial node gene program and imposes pacemaker function on the atria. Genes Dev 21:1098–1112

Houweling AC, Somi S, van den Hoff MJ, Moorman AFM, Christoffels VM (2002) Developmental pattern of ANF gene expression reveals a strict localization of cardiac chamber formation in chicken. Anat Rec 266:93–102

Houweling AC, van Borren MM, Moorman AFM, Christoffels VM (2005) Expression and regulation of the atrial natriuretic factor encoding gene *Nppa* during development and disease. Cardiovasc Res 67:583–593

Kelly RG, Buckingham ME (2002) The anterior heart-forming field: voyage to the arterial pole of the heart. Trends Genet 18:210–216

Kispert A, Herrmann BG (1993) The Brachyury gene encodes a novel DNA binding protein. EMBO J 12:3211–3220

Kleber AG, Janse MJ, Fast VG (2001) Normal and abnormal conduction in the heart. In: Page E, Fozzard HA, Solaro RJ (eds) Handbook of physiology. Oxford University Press, Chicago, Illinois, pp 455–530

Komuro I, Izumo S (1993) Csx: a murine homeobox-containing gene specifically expressed in the developing heart. Proc Natl Acad Sci USA 90:8145–8149

Kreuzberg MM, Sohl G, Kim JS, Verselis VK, Willecke K, Bukauskas FF (2005) Functional properties of mouse connexin30.2 expressed in the conduction system of the heart. Circ Res 96:1169–1177

Kreuzberg MM, Schrickel JW, Ghanem A, Kim JS, Degen J, Janssen-Bienhold U et al (2006a) Connexin30.2 containing gap junction channels decelerate impulse propagation through the atrioventricular node1. Proc Natl Acad Sci USA 103:5959–5964

Kreuzberg MM, Willecke K, Bukauskas FF (2006b) Connexin-mediated cardiac impulse propagation: connexin 30.2 slows atrioventricular conduction in mouse heart. Trends Cardiovasc Med 16:266–272

Kuo CT, Morrisey EE, Anandappa R, Sigrist K, Lu MM, Parmacek MS et al (1997) GATA4 transcription factor is required for ventral morphogenesis and heart tube formation. Genes Dev 11:1048–1060

Laverriere AC, MacNeill C, Mueller C, Poelmann RE, Burch JBE, Evans T (1994) GATA-4/5/6, a subfamily of three transcription factors transcribed in developing heart and gut. J Biol Chem 269:23177–23184

Lee Y, Shioi T, Kasahara H, Jobe SM, Wiese RJ, Markham BE et al (1998) The cardiac tissue-restricted homeobox protein Csx/Nkx2.5 physically associates with the zinc finger protein GATA4 and cooperatively activates atrial natriuretic factor gene expression. Mol Cell Biol 18:3120–3129

Li QY, Newbury-Ecob RA, Terrett JA, Wilson DI, Curtis AR, Yi CH et al (1997) Holt-Oram syndrome is caused by mutations in TBX5, a member of the Brachyury (T) gene family. Nat Genet 15:21–29

Linden H, Williams R, King J, Blair E, Kini U (2009) Ulnar mammary syndrome and TBX3: expanding the phenotype. Am J Med Genet A 149A:2809–2812

Lints TJ, Parsons LM, Hartley L, Lyons I, Harvey RP (1993) *Nkx-2.5:* a novel murine homeobox gene expressed in early heart progenitor cells and their myogenic descendants. Development 119:419–431

Lyons I, Parsons LM, Hartley L, Li R, Andrews JE, Robb L et al (1995) Myogenic and morphogenetic defects in the heart tubes of murine embryos lacking the homeo box gene Nkx2-5. Gene Dev 9:1654–1666

Ma L, Lu MF, Schwartz RJ, Martin JF (2005) Bmp2 is essential for cardiac cushion epithelial-mesenchymal transition and myocardial patterning. Development 132:5601–5611

Mangoni ME, Nargeot J (2008) Genesis and regulation of the heart automaticity. Physiol Rev 88:919–982

Mjaatvedt CH, Nakaoka T, Moreno-Rodriguez R, Norris RA, Kern MJ, Eisenberg CA et al (2001) The outflow tract of the heart is recruited from a novel heart-forming field. Dev Biol 238:97–109

Mommersteeg MTM, Hoogaars WMH, Prall OWJ, de Gier-de Vries C, Wiese C, Clout DEW et al (2007) Molecular pathway for the localized formation of the sinoatrial node. Circ Res 100:354–362

Moorman AFM, Christoffels VM (2003) Cardiac chamber formation: development, genes and evolution. Physiol Rev 83:1223–1267

Moorman AFM, Lamers WH (1994) Molecular anatomy of the developing heart. Trends Cardiovasc Med 4:257–264

Moorman AF, Schumacher CA, de Boer PA, Hagoort J, Bezstarosti K, van den Hoff MJ et al (2000) Presence of functional sarcoplasmic reticulum in the developing heart and its confinement to chamber myocardium. Dev Biol 223:279–290

Moorman AFM, Webb S, Brown NA, Lamers WH, Anderson RH (2003) Development of the heart: 1. Formation of the cardiac chambers and arterial trunks. Heart 89:806–814

Moorman AFM, Soufan AT, Hagoort J, de Boer PAJ, Christoffels VM (2004) Development of the building plan of the heart. Ann NY Acad Sci 1015:171–181

Moretti A, Caron L, Nakano A, Lam JT, Bernshausen A, Chen Y et al (2006) Multipotent embryonic isl1+ progenitor cells lead to cardiac, smooth muscle, and endothelial cell diversification. Cell 127:1151–1165

Mori AD, Bruneau BG (2004) TBX5 mutations and congenital heart disease: Holt-Oram syndrome revealed. Curr Opin Cardiol 19:211–215

Moskowitz IPG, Pizard A, Patel VV, Bruneau BG, Kim JB, Kupershmidt S et al (2004) The T-Box transcription factor Tbx5 is required for the patterning and maturation of the murine cardiac conduction system. Development 131:4107–4116

Munshi NV, McAnally J, Bezprozvannaya S, Berry JM, Richardson JA, Hill JA et al (2009) Cx30.2 enhancer analysis identifies Gata4 as a novel regulator of atrioventricular delay. Development 136:2665–2674

Niederreither K, Vermot J, Messaddeq N, Schuhbaur B, Chambon P, Dolle P (2001) Embryonic retinoic acid synthesis is essential for heart morphogenesis in the mouse. Development 128:1019–1031

Nowotschin S, Liao J, Gage PJ, Epstein JA, Campione M, Morrow BE (2006) Tbx1 affects asymmetric cardiac morphogenesis by regulating Pitx2 in the secondary heart field. Development 133:1565–1573

Okubo A, Miyoshi O, Baba K, Takagi M, Tsukamoto K, Kinoshita A et al (2004) A novel GATA4 mutation completely segregated with atrial septal defect in a large Japanese family. J Med Genet 41:e97

Olson EN (2006) Gene regulatory networks in the evolution and development of the heart. Science 313:1922–1927

Paff GH, Boucek RJ (1962) Simultaneous electrocardiograms and myograms of the isolated atrium, ventricle and conus of the embryonic chick heart. Anat Rec 142:73–80

Paff GH, Boucek RJ, Harrell TC (1968) Observations on the development of the electrocardiogram. Anat Rec 160:575–582

Park PJ (2009) ChIP-seq: advantages and challenges of a maturing technology. Nat Rev Genet 10:669–680

Patten BM, Kramer TC (1933) The initiation of contraction in the embryonic chicken heart. Am J Anat 53:349–375

Paxton C, Zhao H, Chin Y, Langner K, Reecy J (2002) Murine Tbx2 contains domains that activate and repress gene transcription. Gene 283:117–124

Pepke S, Wold B, Mortazavi A (2009) Computation for ChIP-seq and RNA-seq studies. Nat Methods 6:S22–S32

Plageman TF Jr, Yutzey KE (2004) Differential expression and function of Tbx5 and Tbx20 in cardiac development. J Biol Chem 279:19026–19034

Postma AV, van de Meerakker JBA, Mathijssen IB, Barnett P, Christoffels VM, Ilgun A et al (2008) A gain-of-function TBX5 mutation is associated with atypical Holt-Oram syndrome and paroxysmal atrial fibrillation. Circ Res 102:1433–1442

Rajagopal SK, Ma Q, Obler D, Shen J, Manichaikul A, Tomita-Mitchell A et al (2007) Spectrum of heart disease associated with murine and human GATA4 mutation. J Mol Cell Cardiol 43:677–685

Ramsay G (1998) DNA chips: state-of-the art. Nat Biotechnol 16:40–44

Reamon-Buettner SM, Borlak J (2006) HEY2 mutations in malformed hearts. Hum Mutat 27:118

Rochais F, Mesbah K, Kelly RG (2009) Signaling pathways controlling second heart field development. Circ Res 104:933–942

Satin J, Fujii S, de Haan RL (1988) Development of cardiac heartbeat in early chick embryos is regulated by regional cues. Dev Biol 129:103–113

Schott J-J, Benson DW, Basson CT, Pease W, Silberbach GM, Moak JP et al (1998) Congenital heart disease caused by mutations in the transcription factor NKX2-5. Science 281:108–111

Seidl W, Schulze M, Steding G, Kluth D (1981) A few remarks on the physiology of the chick embryo heart (*Gallus gallus*). Folia Morphol 29:237–242

Sepulveda JL, Belaguli N, Nigam V, Chen CY, Nemer M, Schwartz RJ (1998) GATA-4 and Nkx-2.5 coactivate Nkx-2 DNA binding targets. Role for regulating early cardiac gene expression. Mol Cell Biol 18:3405–3415

Singh R, Horsthuis T, Farin HF, Grieskamp T, Norden J, Petry M et al (2009) Tbx20 interacts with smads to confine tbx2 expression to the atrioventricular canal. Circ Res 105:442–452

Sissman J (1966) Cell multiplication rates during development of the primitive cardiac tube in the chick embryo. Nature 210:504–507

Sizarov A, Ya J, de Boer BA, Lamers WH, Christoffels VM, Moorman AFM (2011) Formation of the building plan of the human heart: morphogenesis, growth and differentiation. Circulation 123(10):1125–35

Small EM, Krieg PA (2003) Transgenic analysis of the atrialnatriuretic factor (ANF) promoter: Nkx2-5 and GATA-4 binding sites are required for atrial specific expression of ANF. Dev Biol 261:116–131

Smith J (1999) T-box genes: what they do and how they do it. Trends Genet 15:154–158

Soufan AT, van den Berg G, Ruijter JM, de Boer PAJ, van den Hoff MJB, Moorman AFM (2006) Regionalized sequence of myocardial cell growth and proliferation characterizes early chamber formation. Circ Res 99:545–552

Stalsberg H (1969) The origin of heart asymmetry: right and left contributions to the early chick embryo heart. Dev Biol 19:109–127

Stennard FA, Harvey RP (2005) T-box transcription factors and their roles in regulatory hierarchies in the developing heart. Development 132:4897–4910

Stieber J, Herrmann S, Feil S, Loster J, Feil R, Biel M et al (2003) The hyperpolarization-activated channel HCN4 is required for the generation of pacemaker action potentials in the embryonic heart. Proc Natl Acad Sci USA 100:15235–15240

Sun Y, Liang X, Najafi N, Cass M, Lin L, Cai C et al (2006) Islet 1 is expressed in distinct cardiovascular lineages, including pacemaker and coronary vascular cells. Dev Biol 304(1):286–96

Tanaka M, Chen Z, Bartunkova S, Yamasaki N, Izumo S (1999) The cardiac homeobox gene *Csx/Nkx2.5* lies genetically upstream of multiple genes essential for heart development. Development 126:1269–1280

Thompson RP, Lindroth JR, Alles AJ, Fazel AR (1990) Cell differentiation birthdates in the embryonic rat heart. Ann NY Acad Sci 588:446–448

van den Berg G, Moorman AF (2009) Concepts of cardiac development in retrospect. Pediatr Cardiol 30:580–587

van den Berg G, Abu-Issa R, de Boer BA, Hutson MR, de Boer PA, Soufan AT et al (2009) A caudal proliferating growth center contributes to both poles of the forming heart tube. Circ Res 104:179–188

van Kempen MJA, Vermeulen JLM, Moorman AFM, Gros DB, Paul DL, Lamers WH (1996) Developmental changes of connexin40 and connexin43 mRNA distribution patterns in the rat heart. Cardiovasc Res 32:886–900

van Mierop LHS (1967) Localization of pacemaker in chick embryo heart at the time of initiation of heartbeat. Am J Physiol 212:407–415

van Mierop LHS, Gessner IH (1970) The morphologic development of the sinoatrial node in the mouse. Am J Cardiol 25:204–212

van Wijk B, Moorman AFM, van den Hoff MJB (2007) Role of bone morphogenetic proteins in cardiac differentiation. Cardiovasc Res 74:244–255

Virágh Sz, Challice CE (1977) The development of the conduction system in the mouse embryo heart. II. Histogenesis of the atrioventricular node and bundle. Dev Biol 56:397–411

Virágh Sz, Challice CE (1980) The development of the conduction system in the mouse embryo heart. III. The development of sinus muscle and sinoatrial node. Dev Biol 80:28–45

Virágh Sz, Challice CE (1982) The development of the conduction system in the mouse embryo heart. IV. Differentiation of the atrioventricular conduction system. Dev Biol 89:25–40

Waldo KL, Kumiski DH, Wallis KT, Stadt HA, Hutson MR, Platt DH et al (2001) Conotruncal myocardium arises from a secondary heart field. Development 128:3179–3188

Wang Z, Gerstein M, Snyder M (2009) RNA-Seq: a revolutionary tool for transcriptomics. Nat Rev Genet 10:57–63

Wessels A, Markman MWM, Vermeulen JLM, Anderson RH, Moorman AFM, Lamers WH (1996) The development of the atrioventricular junction in the human heart. Circ Res 78:110–117

Wiese C, Grieskamp T, Airik R, Mommersteeg MT, Gardiwal A, de Gier-de VC et al (2009) Formation of the sinus node head and differentiation of sinus node myocardium are independently regulated by tbx18 and tbx3. Circ Res 104:388–397

Xu H, Morishima M, Wylie JN, Schwartz RJ, Bruneau BG, Lindsay EA et al (2004) Tbx1 has a dual role in the morphogenesis of the cardiac outflow tract. Development 131:3217–3227

Yamada M, Revelli JP, Eichele G, Barron M, Schwartz RJ (2000) Expression of chick Tbx-2, Tbx-3, and Tbx-5 genes during early heart development: evidence for BMP2 induction of Tbx2. Dev Biol 228:95–105

Zeisberg EM, Ma Q, Juraszek AL, Moses K, Schwartz RJ, Izumo S et al (2005) Morphogenesis of the right ventricle requires myocardial expression of Gata4. J Clin Invest 115:1522–1531

Chapter 9
The Functional Significance of the Reptilian Heart: New Insights into an Old Question

James W. Hicks and Tobias Wang

Introduction

The Phylogeny of Amniotes and the Evolution of a Divided Ventricle

The vertebrate heart has undergone an extensive evolutionary remodeling from a "piscine cardiac design" with a single atrium and a single ventricle to a double circulatory system in air-breathing vertebrates with two separate atria and a partial or complete division of the ventricle. The anatomical division of the ventricle and the complete separation of the systemic and pulmonary circulations seem to have evolved independently on at least two occasions within terrestrial air-breathing vertebrates. This independent evolution is evident from Fig. 9.1, which depicts the phylogenetic relationship amongst amniotes, where a divided ventricle appeared within the synapsid lineage that gave rise to mammals, and within archosaurs that gave rise to crocodilians and birds. It is also possible that the four-chambered heart arose separately within extant crocodilians and birds, respectively, but this scenario would require that the ventricular division had occurred at least three times independently.

The divided ventricle provides the ability to increase systemic arterial blood pressures, while keeping low blood pressures in the pulmonary circulation. This low pulmonary pressure allows for a thinner air–blood barrier within the lungs, which contributes to a larger diffusion capacity and hence more efficient gas exchange. A division of the ventricle also avoids cardiac shunts, i.e. the mixture of oxygen-rich

J.W. Hicks (✉)
Department of Ecology and Evolutionary Biology, University of California, Irvine, CA, USA
e-mail: jhicks@uci.edu

T. Wang
Department of Biological Sciences, Zoophysiology, University of Aarhus, Denmark

D. Sedmera and T. Wang (eds.), *Ontogeny and Phylogeny of the Vertebrate Heart*, DOI 10.1007/978-1-4614-3387-3_9, © Springer Science+Business Media, LLC 2012

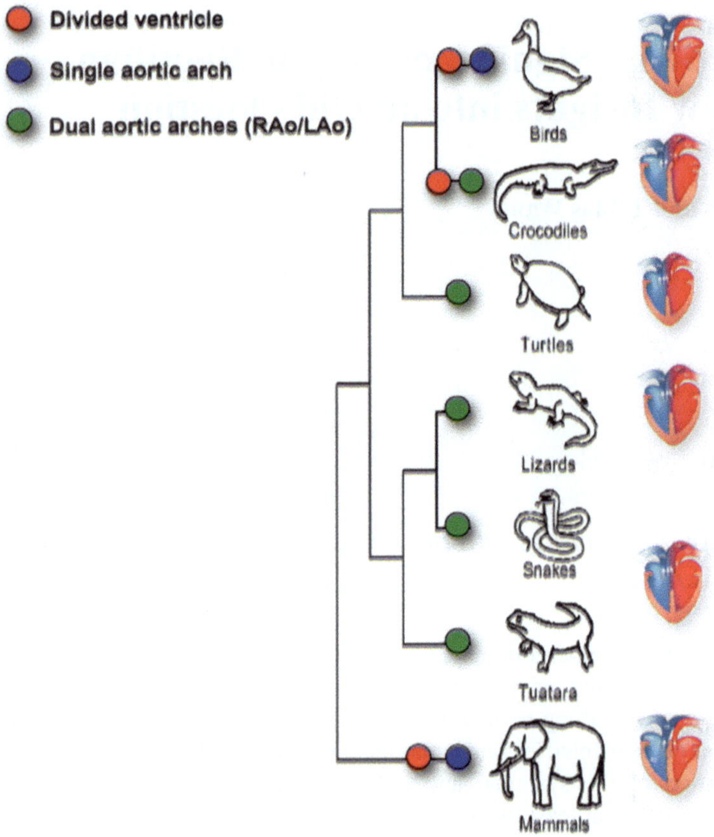

Fig. 9.1 The evolution of the heart in terrestrial vertebrates. Plotted on the phylogeny (modified from Meyer and Zardoya 2003) are the occurrences of the anatomically divided ventricle (*red circles*), single aortic arch (*blue circles*) and dual aortic arches (RAo/LAo; *green circles*). Also shown, for each group, is a schematic representation of heart morphology (modified from Summers 2005). The relationships suggest the lost of the dual aortic arch system twice independently (mammals and birds) and the independent evolution of the four chambered heart, at least twice, (mammals and birds) and possibly three times (mammals, birds and crocodilians)

and oxygen-poor blood within the heart. Both traits are important for effective oxygen transport, particularly with elevated oxygen demands, and it seems likely that the evolution of the divided ventricle was associated with the independent evolution of endothermy and the resulting rise in both resting and maximal metabolic rates within birds and mammals. In this context, it is interesting that crocodilians, in spite of their anatomical division of the ventricle, have retained the ability for R–L cardiac shunts and typical ectothermic rates of metabolism. Recently, Seymour et al. (2004) suggested that the cardiovascular system of extant crocodilians is retained from an ancestral endothermic state. According to this scenario, the divided ventricle did evolve in response to elevated metabolic demands associated with endothermy, while the re-appearance of the capacity to shunt blood in extant crocodilians coincided with the reversion to an ectothermic metabolic strategy.

Cardiac shunting is a functional characteristic of all nonavian reptiles, and is often considered an adaptive trait, i.e. a trait that conveys unique physiological advantages that improve fitness in this group of vertebrates. The goals of this review are to re-examine the functional significance of cardiac shunting in nonavian reptiles and to challenge the notion that cardiac shunts are an adaptive trait.

Comparative Heart Morphologies in Nonavian Reptiles

Testudines and Squamates

The cardiac anatomy of nonavian reptiles has been extensively reviewed (see Hicks 1998; Farrell et al. 1998). Briefly, the heart of testudines (turtles) and squamates (lizards and snakes) contains two anatomically separate atria, but the single ventricle is divided into three subchambers: cavum pulmonale (CP), cavum arteriosum (CA) and cavum venosum (CV). CP can be considered homologues to the right ventricle in mammals because it receives oxygen-poor venous blood returning from the body and contains the outlet to the single pulmonary artery (PA). The blood entering the CP from the right atrium, however, has to pass through the CV, whereas the oxygen-rich blood from the lungs enters directly into the CA from the left atrium. The two aortae supplying the systemic circulation (i.e. the right and left aortae; RAo and LAo, respectively) arise from the CV. The CV is positioned between the CA and CP. This chamber, therefore, serves as a conduit for oxygen-poor blood when the ventricle is filled during diastole, while it conducts oxygen-rich blood when the ventricle empties during systole. A distinctive feature of the ventricle is the muscular ridge (MR) or Muskelleiste. This partial ventricular septum runs from apex to base and may separate the CP from the CV/CA during ventricular systole. There is considerable variation in the morphology of the ventricles of chelonian and squamates. Thus, the MR is particularly large in pythons and varanid lizards, where it provides pronounced separation of the CP from CV/CA during ventricular systole allowing for much higher systemic pressures than pulmonary arterial pressures (see Jensen et al. 2010b, c). While the MR may also appear rather large in several other snakes, such as rattlesnakes and anaconda, there is no evidence of ventricular pressure separation in other snakes than pythons (Jensen et al. 2010a). In testudines, the MR is often smaller than in squamates and the volume of the CV is larger.

Crocodilians

In crocodilians, the ventricle is anatomically divided by a complete ventricular septum. However, in contrast to birds and mammals where a single aorta arises from the left ventricle, crocodilians retain the ancient reptilian phenotype of two aortic arches.

The right aorta (RAo) emerges the left ventricle and conveys the majority of systemic output. In contrast, the left aorta (LAo) emerges from the right ventricle, along with the pulmonary artery, and provides a potential pathway for systemic venous blood to bypass the pulmonary circulation. As an additional feature, the RAo and LAo communicate at two distinct locations. The first is a small opening, foramen of Panizza, where the LAo and RAo share a common wall in the outflow tract immediately anterior to the heart. The second point of communication is via an arterial anastomosis in the abdomen after both arches have looped caudally from the heart. Beyond this anastomosis, the RAo continues as the dorsal aorta, while the LAo becomes the celiac artery, giving rise to smaller arteries that supply most of the splanchnic circulation.

Cardiac Shunts in Nonavian Reptiles

The cardiovascular anatomy described above for crocodilians, squamates and testudines result in the possibility of admixture between oxygen-rich and oxygen-poor blood streams. The potential flow patterns of oxygen-rich and oxygen-poor blood within the reptilian ventricle are summarized in Fig. 9.2. Cardiac shunts are typically defined, either as right-to-left (R–L; a fraction of systemic venous blood that bypasses the pulmonary circulation) or left-to-right (L–R; a fraction of

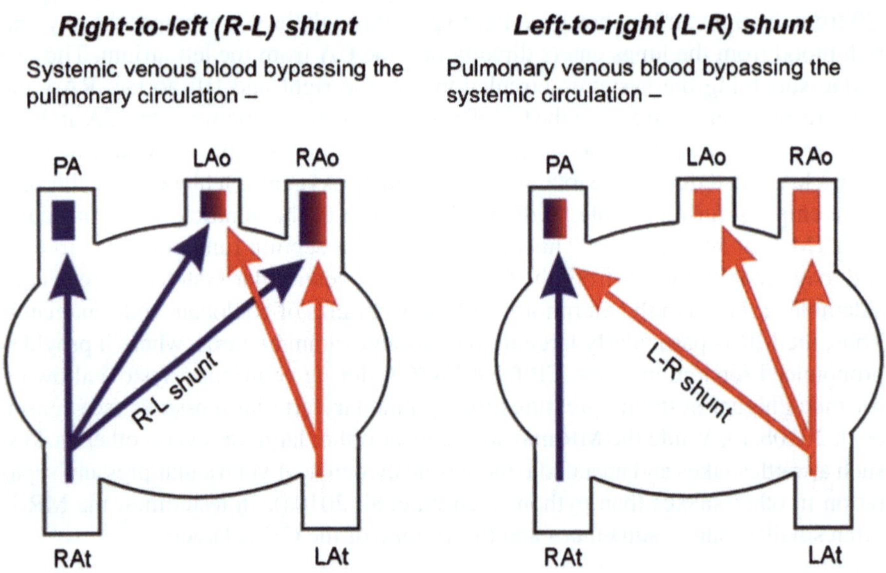

Fig. 9.2 Schematic representation of blood flow patterns associated with right-to-left (R–L) and left-to-right (L–R) cardiac shunting in vertebrates. R–L shunt is represented by a fraction of systemic venous return (RAt; *blue arrow*) bypassing the pulmonary circulation (PA) and entering the left aortic (LAo) and right aortic (RAo) arches. An L–R shunt is represented by a fraction of pulmonary venous return (LAt; *red arrow*), bypassing the systemic circulation (RAo/LAo) and entering the PA. Modified from Hicks (1998)

pulmonary venous blood that recirculates to the lungs). Occasionally the term "pulmonary by-pass shunt" to designate R–L and "systemic bypass shunt" to describe L–R shunts are used. While these phrases may be considered more descriptive, they are not generally accepted in the field of cardiovascular physiology.

During R–L shunts, the fraction of blood that bypasses the lungs has no chance of being oxygenated and therefore, as the magnitude of the R–L shunt increases, the oxygen concentration of systemic arterial blood decreases. The effects of R–L shunt on arterial oxygen can be quantitatively described by the R–L shunt equation:

$$[O_2]a = \frac{\text{Qpul}\,[O_2]\text{LAt} + \text{QRL}\,[O_2]\text{RAt}}{\text{Qpul} + \text{QRL}}. \tag{9.1}$$

Where $[O_2]$ = oxygen content of blood, LAt = left atrium and RAt = right atrium.

This standard shunt equation (9.1) provides a means to evaluate the effects of R–L shunt on oxygen transport, but can also, upon proper re-arrangement, be used to calculate the R–L shunt if oxygen concentrations are measured in both atria and the aorta. Nevertheless, because RAo of squamates and turtles normally carries blood with higher oxygen concentrations than the LAo, a precise description of the influence of cardiac shunts on arterial oxygen concentrations requires a more complicated set of equations that takes into account the differential distribution of systemic venous blood into the two aortae (Ishimatsu et al. 1988).

During L–R shunt blood returning from the lungs (pulmonary venous blood) is recirculated back into the pulmonary circulation, and is quantitatively described by the following equation:

$$[O_2]\text{pa} = \frac{\text{Qpul}\,[O_2]\text{RAt} + \text{QLR}\,[O_2]\text{LAt}}{\text{Qpul} + \text{QLR}}. \tag{9.2}$$

Thus, as L–R shunt increases, the oxygen concentration, and thus PO_2, of pulmonary arterial blood increases. In crocodilians, the existence of a complete ventricular septum precludes the development of an L–R shunt.

Historical View of Cardiac Shunting in Vertebrates

The undivided ventricle in testudines and squamates appears to represent an intermediate stage between the fish heart and the fully divided ventricle in mammals and birds, and consequently it is not surprising that earlier studies interpreted cardiac function in these ectothermic tetrapods as a functionally inefficient phylogenetic step in comparison to the "perfect" circulation of birds and mammals; For example:

> This arrangement is highly impractical (Hyman 1942)
>
> Considered only in the light of oxygen transport the left systemic arch appears to be not merely useless, but positively deleterious (Ewer 1950)
>
> [Reptiles] are still in the process of evolution and hence imperfect. This is particularly so in regard to the circulatory organs (Kashyap 1959)

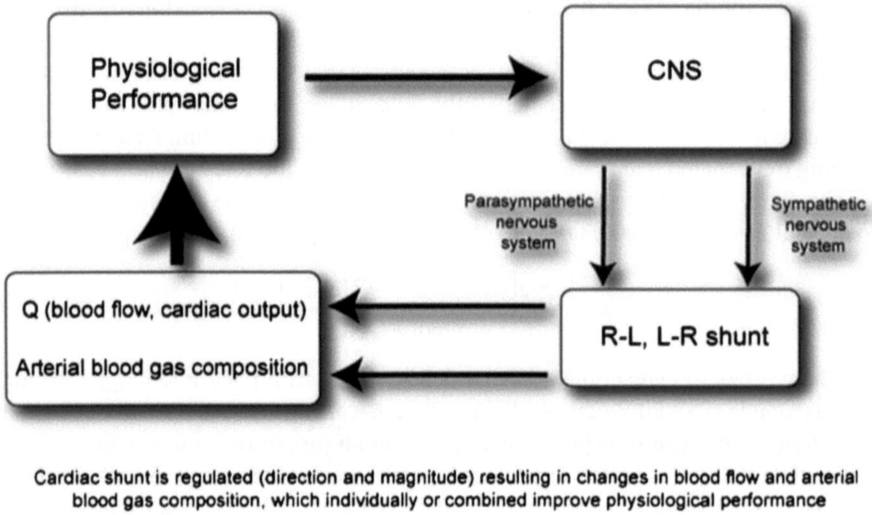

Fig. 9.3 Schematic representation showing the hypothesis that cardiac shunts are a regulated variable. Briefly, the central nervous system (CNS) alters the autonomic outflow to the heart and to the pulmonary and systemic circulations, altering R–L and/or L–R shunt, which then influences cardiac output (Q) and arterial blood gas composition. This in turn influences some physiological performance, which is monitored, ultimately by the CNS

These disdainful views of cardiovascular function in terrestrial ectothermic vertebrates became antiquated in the latter half of the twentieth century when a growing number of studies revealed that the magnitude of the cardiac shunts varies consistently with physiological state and that the degree of admixture is under autonomic regulation (see Hicks 1998, 2002 for review of relevant literature). Thus, rather than placing focus on the negative physiological consequences of cardiac shunts, akin to the devastating symptoms in human congenital heart disease, cardiac shunting and its exquisite control were now suggested to offer unique functional advantages in nonavian reptiles with their low metabolic rates and intermittent breathing patterns. This view is encapsulated in the following quote from Burggren and Warburton (1994):

> An unfortunate tendency of earlier studies on the heart of reptiles was to take a mammalicentric view that the ventricle suffered an intraventricular defect. In fact, the squamate heart is a highly derived condition that provides reptiles with the ability to regulate the intracardiac shunt to their advantage, depending on respiratory needs.

Thus, shunts in turtles, lizards, snakes and crocodilians were viewed as a cardiac variable that could be controlled and regulated to improve the overall physiological performance (Fig. 9.3). While there has been no shortage on proposals and ideas about the physiological benefits of cardiac shunts in nonavian reptiles (Table 9.1), the experimental evidence, however, remains sparse. The lack of experimental support for the professed benefits of cardiac shunting has, nevertheless, not deterred

Table 9.1 Hypothesized physiological functions of cardiac shunting in nonavian reptiles

Physiological function	Type of shunt	Reference
Reduces CO_2 flux into lung	R–L	White et al. (1989)
Saves cardiac energy	R–L	Burggren (1987)
"Meters" lung O_2 stores during apnea	R–L	Burggren and Shelton (1979), Burggren et al. (1989)
Reduces plasma filtration into the lungs	R–L	Burggren and Johansen (1982)
Facilitates warming	R–L	Tucker (1966), Baker and White (1970)
Triggers hypometabolism	R–L	Hicks and Wang (1999), Platzack and Hicks (2001)
Involved in acid secretion in stomach	R–L	Jones and Shelton (1993), Farmer et al. (2008)
Facilitates CO_2 elimination into the lung	L–R	White (1989), Ackerman and White (1979)
Minimizes V/Q mismatching	L–R	Wood (1984), West et al. (1992)
Improves systemic O_2 transport	L–R	Hicks (1994)
Important for myocardial oxygenation	L–R	Farmer and Hicks (2002)

much of the literature within comparative physiology to perpetuate the perception that cardiac shunts represent an adaptive phenotypic trait.

Cardiac Shunts: An Adaptive Feature of Nonavian Reptiles?

Approaches Used to Study Adaptive Features of Shunts

In the comparative physiological literature, interpretation of many physiological traits is predisposed by the underlying assumption that all observed characters in living animals are adaptive. Clearly, natural selection acts to improve fitness, and animals are adapted to their environments. Consequently, the argument goes that if cardiac shunts were not adaptive, why would this unique cardiac morphology and functional capacity in nonavian reptiles be maintained? Nevertheless, despite more than a century of anatomical and physiological studies, it remains a mystery whether or not cardiac shunts represent an adaptive phenotypic trait in reptiles.

For any physiological or morphological trait to be considered adaptive, it is important to provide convincing evidence that the absence of the trait reduces physiological performance and/or reproductive fitness. However, measuring fitness in a large representative sample of the reptilian clade is nearly impossible. The "adaptiveness" of cardiac shunts has been inferred from describing the morphological features of the heart, understanding the physiological factors that alter the direction and size of cardiac shunt, measuring the patterns of cardiac shunts during various physiological states and/or environmental challenges and then analyzing this information within the context of the natural history and phylogenetic relationships between various species.

Physiological Factors that Alter Cardiovascular Shunts

The factors that alter the direction and magnitude of cardiac shunting have been studied in a number of species and there is now a reasonable understanding of the mechanisms dictating the shunt patterns in monitor lizards (*Varanus exanthematicus*) (Webb et al. 1971; Burggren and Johansen 1982; Ishimatsu et al. 1988), pythons (*Python regius* and *Python molurus*) (Jensen et al. 2010b, c), freshwater turtles (*Trachemys scripta*) (Burggren and Shelton 1979; Burggren 1985; Comeau and Hicks 1994; Hicks and Comeau 1994; Hicks 1994; Wang and Hicks 1996) and crocodilians (*Alligator mississipiensis* and *Crocodylus porusus*) (Jones and Shelton 1993; Axelsson et al. 1996; Axelsson and Franklin 2001; Malvin et al. 1995; Webb 1979).

Apart from specific aspects of the heart anatomy, which differ considerably within the groups, the cardiac shunt is influenced by factors that affect heart rate, myocardial contractility and most importantly the vascular resistance in the pulmonary and systemic circulations. For example, the small MR of freshwater turtles cannot effectively separate CP from CV and CA during ventricular contraction and the pulmonary and systemic circulations therefore remain in continuity throughout the cardiac cycle. Consequently, the direction and magnitude of cardiac shunts are determined by the vascular resistance of the pulmonary and systemic circulations (Rp/Rs) (Fig. 9.4). Thus as Rp/Rs increases, the pulmonary circulation becomes the least favored route for blood flow and the size of R–L shunt increases (evidenced by an decrease in the ratio of pulmonary to systemic blood flow, or Qp/Qs) (Fig. 9.4). The opposite occurs as the Rp/Rs is reduced and an L–R shunt develops (evidenced by an increase in Qp/Qs) (Fig. 9.4). A similar relationship between Rp/Rs and Qp/Qs has been observed in rattlesnakes (Galli et al. 2004, 2007).

Regulation of pulmonary and systemic vascular resistances results from changes in both cholinergic and adrenergic tone, as well as release of neurohumoral factors that may influence vascular resistance. Changes in autonomic tone are partially generated within the central nervous system, but stimulation of pulmonary stretch receptors and vascular chemoreceptors may also be involved (Wang et al. 1997). During conditions with high parasympathetic drive (i.e. rest), the pulmonary vascular resistance is elevated and the magnitude of the R–L shunt typically increases, and there will be a concomitant reduction in L–R shunt. The opposite occurs under conditions, such as exercise, that promote a greater sympathetic drive. The relationship between the balance of parasympathetic and sympathetic tone, the directionality of cardiac shunts and the resulting influence on arterial oxygen content is illustrated in Fig. 9.5. It should be noted again that crocodilians cannot develop a L–R shunt because the ventricle is anatomically divided, but that the same general principle applies. Thus, resting undisturbed crocodilians often exhibit sizable R–L shunts, but this shunt is diminished or even eliminated when sympathetic drive is increased during disturbance or during spontaneous exercise (Owerkowicz et al. (2010).

In some squamates (lizards and snakes), the influence of the autonomic nervous system on cardiac shunting may be diminished. For example, in monitor lizards and pythons, the ventricle is nearly fully divided by a large horizontal septum and a

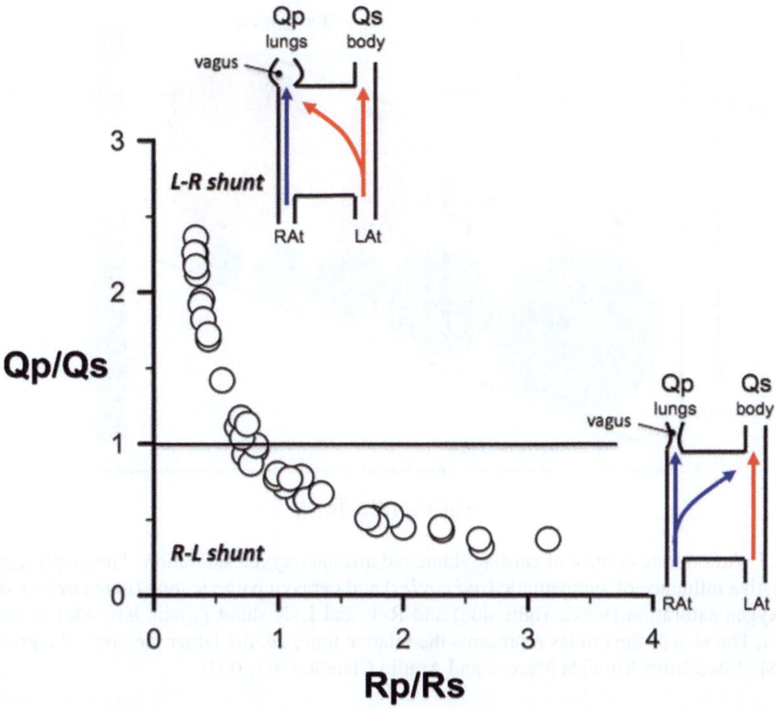

Fig. 9.4 The effects of changing vascular resistances in the pulmonary and systemic circulations (Rp/Rs) on the pulmonary blood flow (Qp) to systemic blood flow (Qs) ratio (Qp/Qs). Qp/Qs greater than 1 represents L–R shunt and Qp/Qs less than 1 represents an R–L shunt. Each axis has a schematic representation of the heart with the direction of blood flow. The lung vasculature is innervated by the vagus nerve as indicated in figure. *RAt* right atrium, *LAt* left atrium. Data from Crossley et al. (1998)

well-developed MR. Because the MR abuts against the ventricular wall, during systole, the ventricle can separate high systemic blood pressures from low pulmonary pressures. The magnitude of the cardiac shunting in these animals is rather small, but deoxygenated and oxygenated blood may mix during primarily during diastole (Jensen et al. 2011). In these species, ventricular pressure separation during systole can approach the values measured in equivalent size birds and mammals. Consequently, the size and direction of the cardiac shunt are influenced more by those factors that influence ventricular volumes and contractility and less by changes in pulmonary and systemic vascular resistance (Hicks 1998).

Patterns of Cardiac Shunting in Nonavian Reptiles

The patterns of cardiac shunting associated with environmental and/or physiological challenges have now been investigated in representatives from each major group of nonavian reptiles, i.e. testudines, squamates and crocodilians. Measuring shunt

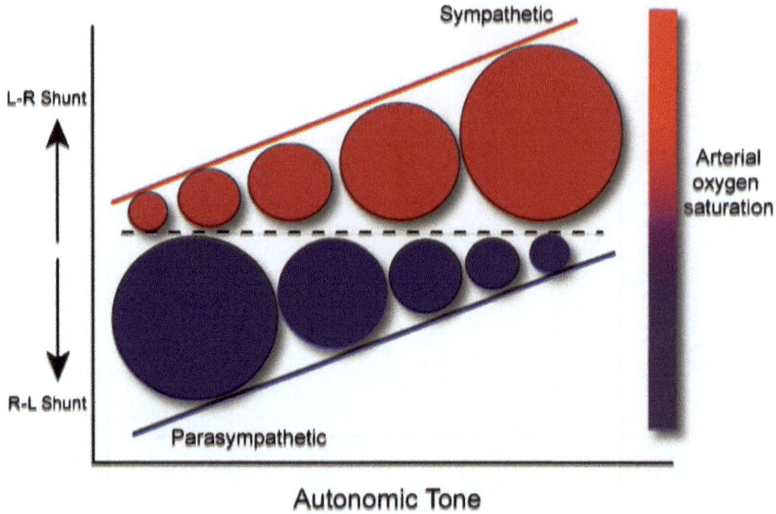

Fig. 9.5 Autonomic control of cardiac shunt and arterial oxygen saturation. The graph represents the relative influence of sympathetic (*red circles*) and parasympathetic tone (*blue circles*), on arterial oxygen saturation (*y*-axis right side) and R–L and L–R shunt (*y*-axis left side) in nonavian reptiles. The size of the circles represents the relative tone, i.e. the larger the circle the greater the tone. Modified from Almeida Marcos and Araújo Claudio Gil (2003)

patterns typically requires chronic implantation of blood pressure and/or blood flow probes, which can be difficult. Therefore, most studies are limited to relatively large species.

Many nonavian reptiles are intermittent lung breathers, where brief ventilatory periods are interspersed with periods of breath holding (apnea), lasting from less than 1 min to over an hour. Consequently, the patterns of cardiac shunting, associated with ventilatory state have garnered much attention (Burggren and Shelton 1979; Wang and Hicks 1996). Briefly, during apnea there is an increase in parasympathetic tone resulting in a bradycardia and an increased pulmonary vascular resistance. These cardiovascular changes result in a reduction in pulmonary blood flow and the development of an R–L shunt (e.g. Hicks 1998). In contrast, lung ventilation is characterized by tachycardia, decreased pulmonary vascular resistance, increase in pulmonary blood flow, and reduction of R–L shunt and development of an L–R shunt.

Krosniunas and Hicks (2003) measured HR, Qpul, Qsys in chronically instrumented freshwater turtles (*Trachemys scripta*) that were free to move within an environment that allowed for a variety of natural behaviors, including thermoregulation, diving, exploration, foraging and swimming (Krosniunas and Hicks 2003).

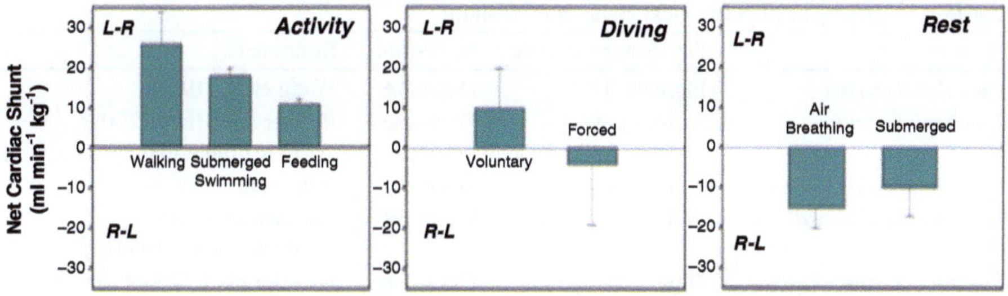

Fig. 9.6 The net cardiac shunt flow, defined as the difference between total pulmonary blood flow and systemic blood flow, during different behavioral states (activity, diving, rest) in the red-eared slider, *Trachemys scripta*. Activity included walking (terrestrial), scratching, digging and clawing; Swimming, defined as submerged swimming only, and diving, defined as moving into water from the tank's platform above the water surface. Feeding included the act of capturing and feeding on goldfish. Rest was defined by the continuous absence of movement and includes rest in air, with the turtle completely outside water, and rest totally submerged in water (not including events during dives). Net cardiac shunt flows with a negative value indicate a net R–L shunt and a positive value indicates a net L–R shunt. Except for feeding, all data and behavioral definitions were derived from Krosniunas and Hicks (2003). Net cardiac shunts during feeding are derived from unpublished data obtained during the study

Voluntary behaviors were correlated with cardiovascular measurements and indicated that the cardiovascular responses were more variable than previously appreciated. For example, any behavior associated with activity and hence increased metabolic demands (walking, swimming, foraging) were associated with an L–R shunt. In contrast, R–L shunt dominated during quiescent behaviors, such as quiet breathing or submersion (Fig. 9.6). Finally, diving behavior and subsequent apnea did not always induce an R–L shunt. In voluntarily submerged turtles, usually associated with underwater swimming and foraging, an L–R shunt developed. However, forced submergence and "fright" dives resulted in an R–L shunt (Fig. 9.6).

R–L Shunt and Organismal Oxygen Demands

It is evident from the standard shunt equation (9.1) that increased R–L shunt reduces arterial oxygen levels and hence systemic oxygen delivery. From the standpoint of oxygen transport, any physiological or environmental challenge or behavior that increases tissue oxygen consumption would benefit from reductions in R–L shunt (Wang and Hicks 2002).

A variety of studies indicate that R–L cardiac shunt decreases during periods of elevated oxygen demands (Table 9.2). For example, in the South American rattlesnake

Table 9.2 Effects of physiological state on R–L shunt

Species	Physiological State	R–L shunt	References
Crotalus durissus	Increase Tb	Decrease	Wang et al. (1998)
Trachemys scripta	Swimming	Decrease	Farmer and Hicks (2002), Krosniunas and Hicks (2003)
Chelonia mydas	Swimming	Decrease	West et al. (1992)
Varanus exanthematicus	Walking/running	Decrease	Mitchell et al. (1981), Butler et al. (1984)
Varanus exanthematicus	Digestion	Decrease	Hartzler et al. (2006)
Alligator mississippiensis	Walking	Decrease	Owerkowicz et al. (2010)
Alligator mississippiensis	Digestion	Reduced	Busk et al. (2000)
Python morulus	Digestion	Reduced	Overgaard et al. (1999), Secor et al. (2000)

(*Crotalus durissus*), R–L shunt is reduced with elevation in body temperature (Wang et al. 1998). In addition, R–L shunt is diminished and L–R shunt is enhanced during swimming activity in aquatic turtles (Farmer and Hicks 2002; Krosniunas and Hicks 2003; West et al. 1992). Preliminary studies in the American alligator indicate a prevailing R–L shunt during resting periods, which is eliminated during activity. This is evidenced by the reduction in flow in the LAo during activity (Owerkowicz and Hicks unpublished).

Interestingly, digestion is a physiological challenge that may best illustrate the connections between increased metabolism and reduced R–L shunt. In many vertebrates, digestion is a physiological process commonly associated with high parasympathetic activity, and should, based on the relationships illustrated in Fig. 9.5, be a physiological state characterized by increased R–L shunt. However, in many carnivorous reptiles, digestion is associated with large increases in metabolism, termed specific dynamic action (SDA), where oxygen consumption may increase several-fold above standard metabolism. Thus, in terms of providing adequate oxygen transport to support the SDA response, it would be predicted that the level of R–L shunt should be decreased. Consistent with this view, arterial PO_2 and HbO_2 saturation remain high or even increases during digestion in pythons, rattlesnakes, varanid lizards and alligators (Overgaard et al. 1999; Busk et al. 2000; Arvedsen et al. 2005; Hartzler et al. 2006), although a small increase in the calculated R–L was recently reported for digesting pythons (Jensen et al. 2011).

Phenotypic Manipulation to Test Adaptive Significance of Shunt

Many studies have correlated cardiac shunting patterns with different physiological states and much is now known about the physiological mechanisms and anatomical features that determine cardiac shunts. However, understanding the adaptive significance of this unique organismal trait remains elusive.

Determining whether or not a specific trait is adaptive can be difficult, and in the case of cardiac shunts, particularly challenging. One experimental approach to test adaptive significance is to manipulate the specific phenotypic trait, and then measure the effects of this manipulation on some organismal performance variable (Sinervo and Basolo 1996).

In reptiles, two approaches to investigating cardiac shunts have been used. The first is pharmacological blockade or denervation to obstruct the autonomic neural regulation of the shunt patterns. Administration of the cholinergic antagonist, atropine, for example, abolishes the capacity to increase pulmonary vascular resistance and hence the ability to develop an R–L shunt. The second approach involves surgical manipulation of the nerves innervating the heart or surgical manipulation of the great vessels emerging from the ventricle. In either case, the strategy is to reduce or eliminate the capacity to develop cardiac shunts. Below we briefly summarize recent studies that have used either pharmacological or surgical manipulations to investigate some of the hypotheses that focus on adaptive significance of R–L shunt (Table 9.1).

Pharmacological Manipulation: Role of R–L Shunts During Heating and Cooling

Regulation of body temperature in reptiles results from the coordinated interaction of both behavioral and physiological responses. In 1970, Baker and White hypothesized that the redistribution of blood flow, resulting from the development of R–L shunt, would play an important role in thermoegulation, specifically in lizards, and possibly in other reptiles. Briefly, to maximize heating rates, the R–L shunt would redistribute blood away from the lungs thus reducing heat loss across the pulmonary vascular bed, and "optimize" the warming of the body. Preliminary experiments in *T. scripta* (Hicks unpublished) indicated that an R–L shunt did indeed develop during heating and was consistent with previous results in Iguanas (Baker and White 1970). Interestingly, when the turtles were treated with atropine (a cholinergic antagonist), the R–L continued to develop during periods of warming. This result suggested that peripheral vasodilation was important during heating, and the reduction in systemic vascular resistance was the driving factor behind the development of the R–L shunt. Therefore, in contrast to the suggestion of Baker and White, R–L shunt was not leading the cardiovascular response, but was passively tracking the changes in vascular resistances. This notion was affirmed by measuring heating and cooling rates in *T. scripta*, in both control and atropinized animals (Galli et al. 2004). This study convincingly showed that heating rates were the same in untreated and atropinized animals and the authors concluded, "shunt patterns contribute little to heat exchange."

Surgical Manipulations: Role of R–L Shunts During Digestion and Growth

As described above, it has been proposed that increased R–L shunt serves to provide CO_2 and rich blood to the parietal cells and thus plays an important role in promoting gastric acid secretion during digestion (Jones and Shelton 1993; Farmer et al. 2008; Gardner et al. 2011).

In all carnivorous vertebrates, gastric H^+ secretion increases as an immediate response to ingestion of a meal. The secretion of H^+ to the stomach lumen by the parietal results from hydration of CO_2, and causes an accumulation of HCO_3^- in body fluids (the alkaline tide):

$$CO_2 + H_2O \longleftrightarrow H_2CO_3 \longleftrightarrow H^+ + HCO_3^- \quad (9.3)$$

If CO_2 is limiting for gastric secretion, then the development of a large R–L shunt could be an important source of CO_2 rich blood and thus facilitate the production of H^+ within the parietal cell as described in (9.3). The importance of R–L shunting to digestion is underscored by Farmer et al. (2008): "We hypothesize…that the right-to-left shunt serves to provide carbon dioxide to gastrointestinal organs besides the stomach, such as the pancreas, spleen, upper small intestine, and liver."

This hypothesis could explain why crocodilians have retained the capacity for R–L shunts. However, actual measurements of R–L shunt during digestion have not been reported in alligators or other crocodilians. Farmer et al. (2008) provided measurements of increased LAo flow during digestion in the American alligator, but it was not shown that this rise in flow originated from the right ventricle. The general value of increased R–L shunt also remains unclear with respect to non-crocodilian reptiles where the splanchnic circulation arises from the dorsal aorta after the merger of the LAo and the RAo.

If the development of an R–L shunt is critical for digestion and assimilation of energy in carnivorous reptiles, then an alternative approach would be to eliminate the capacity for R–L shunt and measure growth rates. Growth is often considered an integrated measure of physiological function and thus has been used as a proxy for fitness, e.g. insects (Lampert and Trubetskova 1996), reptiles (Sinervo and Adolph 1989), mammals (Kraus et al. 2005) and birds (Richner 1989).

In crocodilians, surgical occlussion the left aorta (LAo) upstream and downstream of the foramen of Panizza results in an animal with an obligatory, avian/mammalian central circulation. Recently Eme et al. (2010) took this approach to investigate the effects of removing R–L shunt on growth in the American alligator. In this study, the R–L cardiac shunt was surgically eliminated in age-matched, female American alligators (*Alligator mississippiensis*; 5–7 months of age) and indices of growth, i.e. snout-vent length, total length and body mass, were measured every 2 weeks for an additional 17 months (Fig. 9.7). This indicates that elimination of the capacity for R–L cardiac shunt does not greatly reduce animal growth, despite a chronic ventricular enlargement in surgically altered juvenile alligators. If R–L cardiac shunt conferred a strong adaptive advantage to crocodilians, one would

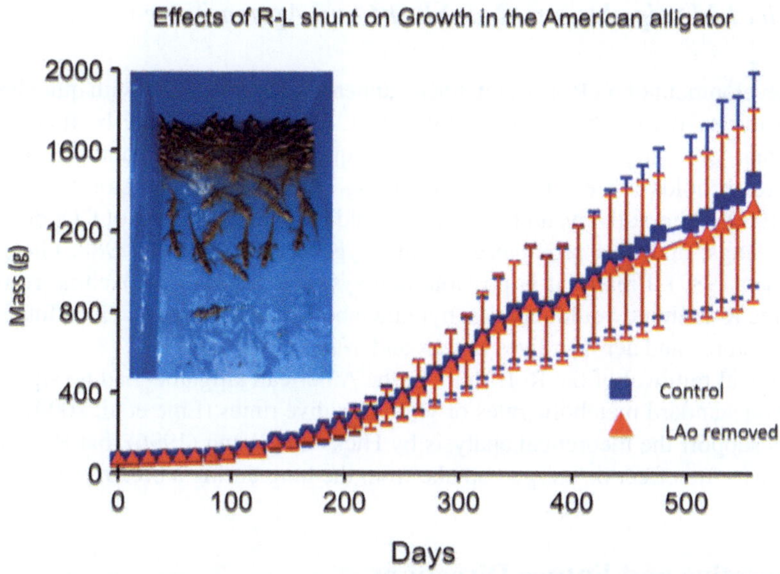

Fig. 9.7 Longitudinal growth study in the American alligator, *Alligator mississippiensis* with (*blue squares*) and without (*red triangles*) the ability to generate an R–L cardiac shunt. Graph from Eme et al. (2010)

expect a large and consistent decrement in growth to be apparent in all animals with their LAo occluded. No matter how data in this study are analyzed, this is not the case. This study suggests that crocodilian R–L cardiac shunt does not provide an adaptive advantage for juvenile alligator growth and supports the logic that cardiac shunts persist in crocodilians because they have not been selected against (Eme et al. 2010).

The importance of R–L shunting in digestion and assimilation has been recently tested in the South American rattlesnake, *Crotalus durissus* (Leite 2011). As in all reptiles, the cardiovascular system in this species is controlled by both subdivisions of the autonomic nervous system, with heart rate, pulmonary blood flow and cardiac shunt patterns primarily controlled by the parasympathetic nervous system. However, in these snakes, Taylor et al. (2009) reveal that whereas both the left and right vagi influence heart rate, it is only the left vagus that influences pulmonary vascular resistance, and thus shunts. In a subsequent series of experiments, snakes underwent a bilateral vagotomy resulting in marked increases in heart rate and pulmonary blood flow, and the inability to develop an R–L shunt. Body mass was measured over a 12-month period in a control group, a group of snakes with a left vagotomy and a group with a right vagotomy. The measurements revealed no significant difference in growth rates between the groups. Together with the work by Eme et al. (2010) these results do not support the hypothesis that R–L shunt is an adaptive feature required for digestion and growth in carnivorous reptiles.

Surgical Manipulation: R–L Shunt and Apnea Time

The development of an R–L shunt during apnea, either associated with quiet breathing or during diving, has suggested that R–L cardiac shunts may be functionally important for diving (Hicks 2002). One consequence of the R–L shunt is that voluntary breath holds are extended by "metering" the lung oxygen stores (Burggren et al. 1989). Alternatively, an R–L shunt would reduce the amount of CO_2 delivered to the lungs, and thus promote efficient oxygen uptake (White 1988; Grigg and Johansen 1987). Finally, it is possible that systemic arterial hypoxemia, resulting from the R–L shunt would induce a hypometabolism in the tissues, extending blood oxygen stores and aerobic dives (Hicks and Wang 1996, 2004).

Surgical removal of the R–L shunt in the American alligators had no significant effect on standard metabolic rates or voluntary dive times (Eme et al. 2009). These results support the theoretical analysis by Hicks and Wang (1996) that R–L shunts per se have no effect on oxygen uptake from the lung during a breath hold.

Perspective and Future Directions

Experimental studies on cardiovascular shunting in nonavian reptiles have shown that the direction and magnitude of the shunt are primarily controlled through changes in the vascular resistances of the pulmonary and systemic circulations. These resistances are in turn primarily controlled by both subdivision of the autonomic nervous system, such that under conditions primarily dominated by parasympathetic tone, an R–L shunt persists, with desaturation of arterial blood. Conversely, under conditions associated primarily by sympathetic tone, an L–R shunt (if possible) persists (Fig. 9.4).

In contrast to earlier views, cardiac shunts per se are not a regulated variable (Fig. 9.3). Rather, we suggest that a combination of physiological state and environmental stressors, acting together, influence the regulatory centers of the CNS changing the balance of autonomic outflow and subsequently altering cardiac shunts, i.e. shunt direction and magnitude are a consequence of factors that alter autonomic tone and cardiac shunts are not a regulated variable (Fig. 9.8).

Based on this scenario and a review of the literature, any condition that is associated with low metabolic rates in combination with dominant parasympathetic tone will be associated with R–L shunt. In contrast, any physiological state or environmental stress that elevates tissue demands for oxygen in combination with a dominant sympathetic tone will reduce R–L shunt and in some cases (e.g. noncrocodilian reptiles), result in the development of an L–R shunt (Fig. 9.9).

Experimental evidence supporting the notion that cardiac shunts are a derived state, or provide an adaptive advantage does not exist. More likely, the unique cardiovascular system of reptiles simply represents an ancestral condition and/or an embryonic character trait that combined with low metabolic demands persist because there has been no selection pressure against the presence of cardiac shunts.

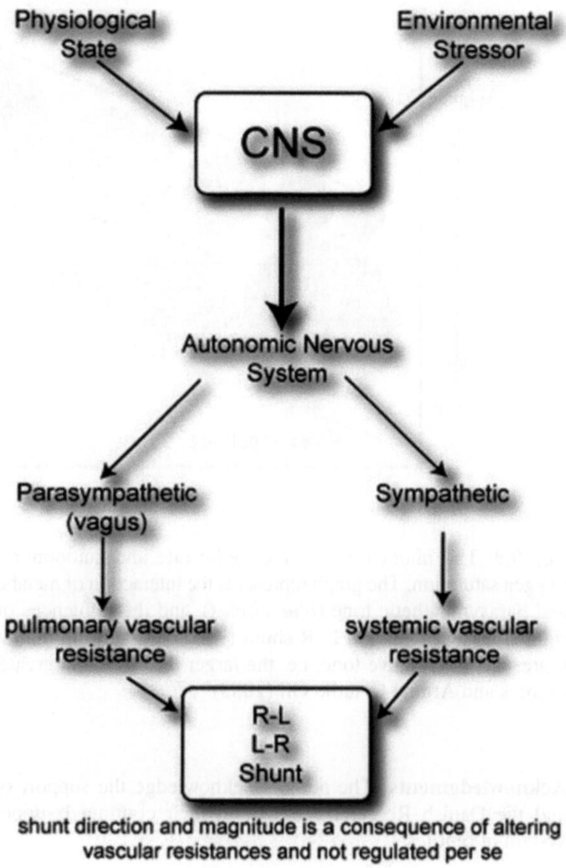

Fig. 9.8 Flow chart representation of the influences of physiological state and environmental stressors on R–L and L–R cardiac shunts. Both physiological states and environmental stresses are processed through the central nervous system (CNS), activating both components of the autonomic nervous system (ANS). Differential activation of both subdivisions of the ANS alters vascular resistances in the pulmonary and systemic circulations, altering the level of R–L and/or L–R cardiac shunting

There are still many mysteries surrounding the evolution of the vertebrate heart. The developmental genetics of the reptilian heart and the correlation to the developmental events that occur in the mammalian and avian heart still need to be resolved. Recent work has pointed to the importance of the expression pattern of the Tbx transcriptions factors (Koshiba-Takeuchi et al. 2009) in determining ventricular morphology in reptiles. The capacity to generate genomic information from a variety of species will provide the foundation to investigate, within a phylogenetic context, the genetic networks that determine the diversity of cardiac morphologies in vertebrates. Such information will ultimately provide stronger inferences for the functional significance associated with the anatomical variations observed in the vertebrate heart.

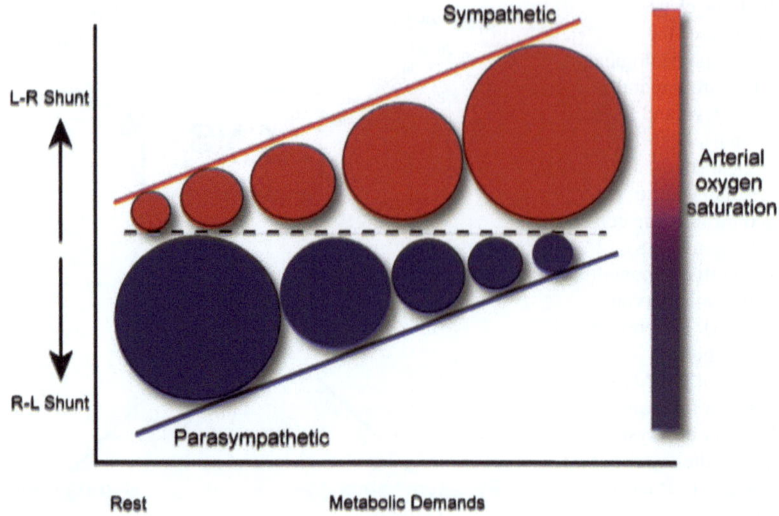

Fig. 9.9 The interactions of metabolic rate and autonomic tone on cardiac shunt and arterial oxygen saturation. The graph represents the interaction of metabolic rates on sympathetic (*red circles*) and parasympathetic tone (*blue circles*), and the influences on arterial oxygen saturation (*y*-axis right side) and R–L and L–R shunt (*y*-axis left side) in nonavian reptiles. The size of the circles represents the relative tone, i.e. the larger the circle the greater the tone. Modified from Almeida Marcos and Araújo Claudio Gil (2003)

Acknowledgments The authors acknowledge the support of the National Science Foundation and the Danish Research Council in their continued support investigating the cardiovascular system of reptiles.

References

Ackerman RA, White FN (1979) Cyclic carbon dioxide exchange in the turtle *Pseudemys scripta*. Physiol Zool 52:378–389

Almeida Marcos B, Araújo Claudio Gil S (2003) Effects of aerobic training on heart rate. Rev Bras Med Esporte 9(2):113–120. http://www.scielo.br/scielo.php?script=sci_arttext&pid=S1517-86922003000200006&lng=en. http://dx.doi.org/10.1590/S1517-86922003000200006. Accessed 02 Sept 2011

Arvedsen SK, Andersen JB, Zaar M, Andrade D, Abe AS, Wang T (2005) Arterial acid-base status during digestion and following vascular infusion of HCO3Na and HCl in the South American rattlesnake *Crotalus durissus*. Comp Biochem Physiol 142A:495–502

Axelsson M, Franklin CE (2001) The calibre of the foramen of Panizza in *Crocodylus porosus* is variable and under andrenergic control. J Comp Physiol B Biochem Syst Environ Physiol 171:341–346

Axelsson M, Franklin CE, Lofman CO, Nilsson S, Grigg GC (1996) Dynamic anatomical study of cardiac shunting in crocodiles using high resolution angioscopy. J Exp Biol 199:359–365

Baker LA, White FN (1970) Redistribution of cardiac output in response to heating in *Iguana iguana*. Comp Biochem Physiol 35:253–262

Burggren WW (1985) Hemodynamics and regulation of central cardiovascular shunts in reptiles. In: Johansen K, Burggren WW (eds) Cardiovascular shunts: phylogenetic, ontogenetic and clinical aspects. Munksgaard, Copenhagen, pp 121–142

Burggren WW (1987) Form and function in reptilian circulations. Am Zool 27:5–19

Burggren WW, Johansen K (1982) Ventricular hemodynamics in the monitor lizard *Varanus exanthematicus*: pulmonary and systemic pressure separation. J Exp Biol 96:343–354

Burggren WW, Shelton G (1979) Gas exchange and transport during intermittent breathing in chelonian reptiles. J Exp Biol 8:75–92

Burggren WW, Warburton SJ (1994) Patterns of form and function in developing hearts: contributions from non-mammalian vertebrates. Cardioscience 5:183–191

Burggren W, Smits A, Evans B (1989) Arterial O_2 homeostasis during diving in the turtle *Chelodina longicollis*. Physiol Zool 62:668–686

Busk M, Overgaard J, Hicks JW, Bennett AF, Wang T (2000) Effects of feeding on arterial blood gases in the American alligator *Alligator mississippiensis*. J Exp Biol 203:3117–3124

Butler, Milsom PJ, Woakes AJ (1984) Respiratory, cardiovascular and metabolic adjustments during steady state swimming in the green turtle, *Chelonia mydas*. J Comp Physiol B 154:167–174

Comeau SG, Hicks JW (1994) Regulation of central vascular blood flow in the turtle. Am J Physiol 267:R569–R578

Crossley D, Altimiras J, Wang T (1998) Hypoxia elicits an increase in pulmonary vascular resistance of anaesthetised turtles (*Trachemys scripta*). J Exp Biol 201:3367–3375

Eme J, Gwalthney J, Blank JM, Owerkowicz T, Barron G, Hicks JW (2009) Surgical removal of right-to-left cardiac shunt in the American alligator (*Alligator mississippiensis*) causes ventricular enlargement but does not alter apnoea or metabolism during diving. J Exp Biol 212:3553–3563

Eme J, Gwalthney J, Owerkowicz T, Blank JM, Hicks JW (2010) Turning Crocodilian hearts into bird hearts: growth rates are similar for alligators with and without cardiac shunt. J Exp Biol 213:2673–2680

Ewer RF (1950) Haemodynamic factors in the evolution of the double circulation in the vertebrates. Am Nat 84(816):215–220

Farmer CG, Hicks JW (2002) The intracardiac shunt as a source of myocardial oxygen in a turtle, *Trachemys scripta*. Int Comp Biol 42:208–215

Farmer CG, Uriona TJ, Olsen DB, Steenblick M, Sanders K (2008) The right-to-left shunt of crocodilians serves digestion. Physiol Biochem Zool 81:125–137

Farrell AP, Gamperl AK, Francis TB (1998) Comparative aspects of heart morphology. In: Gans C, Gaunt AS (eds) Biology of reptilia, vol 19: Morphology G: Visceral organs. Society for Study of Amphibians and Reptiles Press, Ithaca, New York, pp 375–424

Galli G, Taylor EW, Wang T (2004) The cardiovascular responses of the freshwater turtle *Trachemys scripta* to warming and cooling. J Exp Biol 207:1471–1478

Galli GLJ, Skovgaard N, Abe AS, Taylor EW, Wang T (2007) The adrenergic regulation of the cardiovascular system in the South American rattlesnake, *Crotalus durissus* Comp Biochem Physiology 148A:510–520

Gardner MN, Sterba-Boatwright B, Jones DR (2011) Ligation of the left aorta in alligators affects acid–base balance: a role for the R→L shunt. Resp Physiol Neurobiol 178:315–322

Grigg GC, Johansen K (1987) Cardiovascular dynamics in Crocodylus porosus breathing air and during voluntary aerobic dives. J Comp Physiol B Biochem Syst Environ Physiol 157:381–392

Hartzler LK, Munns SL, Bennett AF, Hicks JW (2006) Metabolic and blood gas dependence on digestive state in the Savannah monitor lizard *Varanus exanthematicus:* an assessment of the alkaline tide. J Exp Biol 209:1052–1057

Hicks JW (1994) Adrenergic and cholinergic regulation of intracardiac shunting. Physiol Zool 67:1325–1346

Hicks JW (1998) Cardiac shunting in reptiles: mechanisms, regulation and physiological function. In: Gans C, Gaunt AS (eds) Biology of reptilia, vol 19: Morphology G: Visceral organs. Society for Study of Amphibians and Reptiles Press, Ithaca, New York, pp 425–483

Hicks JW (2002) The physiological and evolutionary significance of cardiovascular shunting patterns in reptiles. News Physiol Sci 17:241–245

Hicks JW, Comeau SG (1994) Vagal regulation of intracardiac shunting in the turtle *Pseudemys scripta*. J Exp Biol 186:109–126

Hicks JW, Wang T (1996) Functional role of cardiac shunts in reptiles. J Exp Zool 275:204–216

Hicks JW, Wang T (1999) Hypoxic hypometabolism in the anesthetized turtle, *Trachemys scripta*. Am J Physiol 277:R18–R23

Hicks JW, Wang T (2004) Hypometabolism in reptiles: behavioural and physiological mechanisms that reduce aerobic demands. Resp Physiol Neurobiol 141:261–271

Hyman LH (1942) Comparative anatomy of vertebrates, 2nd edn. The University of Chicago, Chicago, p 346

Ishimatsu A, Hicks JW, Heisler N (1988) Analysis of intracardiac shunting in the lizard, Varanus nilodicus: a new model based on blood oxygen levels and microsphere distribution. Respir Physiol 71:83–100

Jensen B, Abe AS, Andrade DV, Nyengaard JR, Wang T (2010a) The heart of the South American rattlesnake, *Crotalus durissus*. J Morph 271:1066–1077

Jensen B, Nyengaard JR, Pedersen M, Wang T (2010b) The anatomy of the python heart. Anat Sci Int 85:194–203

Jensen B, Nielsen JM, Axelsson M, Pedersen M, Löfman C, Wang T (2010c) How the python heart separates pulmonary and systemic blood pressures and blood flows. J Exp Biol 213:1611–1617

Jensen B, Larsen CK, Nielsen JM, Simonsen LS, Wang T (2011) Change of cardiac function, but not form, in postprandial pythons. Comp Biochem Physiol 160:35–42

Jones DR, Shelton G (1993) The physiology of the alligator heart-left aortic flow patterns and right-to-left shunts. J Exp Biol 176:247–269

Kashyap HV (1959) The reptilian heart. Proc Natl Inst Sci India 26B:234–254

Koshiba-Takeuchi K, Mori AD, Kaynak BL, Cebra-Thomas J, Sukonnik T, Georges RO, Latham S, Beck L, Beck L, Henkelman RM, Black BL, Olson EN, Wade J, Takeuchi JK, Nemer M, Gilbert SF, Bruneau BG (2009) Reptilian heart development and the molecular basis of cardiac chamber evolution. Nature 461(7260):95–98

Kraus C, Trillmich F, Kunkele J (2005) Reproduction and growth in a precocial small mammal *Cavia magna*. J Mammol 86:763–772

Krosniunas EH, Hicks JW (2003) Cardiac output and shunt during voluntary activity at different temperatures in the turtle, *Trachemys scripta*. Physiol Biochem Zool 76:679–694

Lampert W, Trubetskova I (1996) Juvenile growth rate as a measure of fitness in *Daphnia*. Funct Ecol 10:631–635

Leite C (2011) Haemodynamic alterations due to ablation of control of the cardiac shunt in the South American rattlesnake, *Crotalus durissus*. In: Society for Experimental Biology Meeting, Glasgow, [A6.64]

Malving M, Hicks JW, Greenee R (1995) Central vascular flow patterns in the alligator *Alligator niississipiensis*. American J Physiol 38:R1133–1139

Meyer A, Zardoya R (2003) Recent advances in the (molecular) phylogeny of vertebrates. Annu Rev Ecol Evol Syst 34:311–338

Mitchell GS, Gleeson TT, Bennett AF (1981) Pulmonary oxygen transport during activity in lizards. Respir Physiol 43:365–375

Overgaard J, Busk M, Hicks JW, Jensen FB, Wang T (1999) Respiratory consequences of feeding in the snake *Python molorus*. Comp Biochem Physiol 124A:359–365

Owerkowicz T, Eme J, Gwalthney J, Blank JM, Hicks JW (2010) Cardiac shunting does not constrain aerobic capacity of the American alligator. Abstract P2.78, SICB Meeting, http://www.sicb.org/meetings/2010/schedule/abstractdetails.php3?id=1228

Platzack BR, Hicks JW (2001) Reductions in systemic oxygen delivery induce a hypometabolic state in the turtle *Trachemys scripta*. Am J Physiol 281:R1295–R1301

Richner H (1989) Habitat-specific growth and fitness in Carrion crows (*Corvus corone corone*). J Anim Ecol 58:427–440

Secor SM, Hicks JW, Bennett AF (2000) Ventilatory and cardiovascular responses of a python (*Python molurus*) to exercise and digestion. J Exp Biol 203:2447–2454

Seymour RS, Bennett-Stamper CL, Johnston SD, Carrier DR, Grigg GC (2004) Evidence for endothermic ancestors of crocodiles at the stem of archosaur evolution. Physiol Biochem Zool 77(6):1051–1067

Sinervo B, Adolph SC (1989) Thermal sensitivity of growth rate in hatchling Sceloporus lizards: environmental, behavioral and genetic aspects. Oecologica 78:411–419

Sinervo B, Basolo AL (1996) Testing adaptation with phenotypic manipulations. In: Rose MR, Lauder GV (eds) Evolutionary biology of adaptation. Academic, New York, pp 149–185

Summers AP (2005) Evolution: warm-hearted crocs. Nature 434:834–844

Taylor EW, Andrade DV, Abe AS, Leite CAC, Wang T (2009) The unequal influence of the left and the right vagi on the control of the heart and pulmonary artery in the rattlesnake, Crotalus durissus. J Exp Biol 212:145–151

Tucker VA (1966) Oxygen transport by the circulatory system of the green iguana (*Iguana iguana*) at different body temperatures. J Exp Biol 44:77–92

Wang T, Hicks JW (1996) Cardiorespiratory synchrony in turtles. J Exp Biol 199:1791–1800

Wang T, Hicks JW (2002) An integrative model to predict maximum oxygen uptake of animals with central vascular shunts. Zoology 105:45–53

Wang T, Krosniunas EH, Hicks JW (1997) The role of cardiac shunts in the regulation of arterial blood gases. Am Zool 37:12–22

Wang T, Abe AS, Glass ML (1998) Effects of temperature on lung and blood gases in the South American rattlesnake *Crotalus durissus terrificus*. Comp Biochem Physiol A Physiol 121:7–11

Webb GJW (1979) Comparative cardiac anatomy of the Reptilian. III. The heart of crocodilians and a hypothesis on the completion of the interventricular septum of crocodilians and birds. J Morphol 161:221–240

Webb G, Heatwole H, Bavay J (1971) Comparative cardiac anatomy of the Reptilia. I. The chambers and septa of the varanid ventricle. J Morphol 134:335–350

West NH, Butler PJ, Bevan RM (1992) Pulmonary blood flow at rest and during swimming in the green turtle, Chelonia mydas. Physiol Zool 65:287–310

White FN (1988) Carbon dioxide homeostasis Wood SC (ed.), Comparatice Pulmonary Physiology: Current Concepts, Marcel Dekker, New York (1988), pp. 439–466

White FN (1989) Carbon dioxide homeostasis. In: Wood SC (ed) Comparative pulmonary physiology: current concepts. Marcel Dekker, New York, pp 439–462

White FN, Hicks JW, Ishimatsu A (1989) Respiratory states and intracardiac shunts in turtles. Am J Physiol 256:R240–R247

Wood SC (1984) Cardiovascular shunts and oxygen transport in lower vertebrates. Am J Physiol Regul Integr Comp Physiol 247:R240–R247

Index

A
Adaptation, 23–27, 39, 46, 55, 84, 155, 158–159, 162–166
Adrenergic stimulation, 126, 128–134, 136, 165
Air breathing fish, 85, 92–95, 97
Allometry, 56–58
Amphibians, 95, 96, 108, 117, 125, 134, 135, 141, 142, 156
Anatomy, 1, 11, 13, 27, 55, 64, 78, 80, 81, 155, 209, 210, 214
Atrioventricular canal (AVC), 66, 83, 150, 152, 153, 155–157, 181, 182, 186–188, 190, 192, 195–200
Atrioventricular region, 47–49, 56
Atrium, 35, 37, 41, 47–50, 81, 95, 109, 125, 156, 181, 182, 184, 186, 191, 194–196, 207, 209, 216
AVC. *See* Atrioventricular canal (AVC)

B
Blood pressure, 42, 46, 77, 81, 83, 84, 89, 90, 95–97, 109, 159, 207, 215, 216
Bulbus arteriosus, 35–37, 39, 81, 84, 90

C
Calcium (Ca^{2+}), 61, 69, 103, 125, 161
Calcium activation, 127, 129, 136, 138, 140
Calsequestrin (CASQ), 105, 116, 119
Cardiac contractility, 129, 138–141, 149
Cardiac growth, 56–58, 60, 63, 64, 66, 77, 79, 96, 182–183, 194, 195
Cardiac metabolism, 150–152, 166
Cardiac output, 11–13, 16–18, 46, 77, 84, 89, 90, 119, 126, 141, 154, 155, 157–159, 165, 212
Cardiac oxygen supply, 75, 84, 97
Cardiac regeneration, 64–67
Cardiac work, 77, 80, 83, 84, 88–90, 165
Cardial differentiation, 199
Chick (embryo), 66, 150–152, 185
Circulation, 18, 20, 23, 25, 26, 47, 75–98, 147, 149, 150, 153, 155, 158, 159, 191, 207, 209–212, 214, 216, 220, 222, 223
Collagen, 37–39, 44–49, 61, 64
Compacta, 36, 38, 39, 41–48, 58, 60, 61, 63, 64
Compact myocardium, 39, 41, 47, 59, 66, 76–85, 87–90, 92–98, 150, 155
Control mechanisms, 12, 13, 126
Conus arteriosus, 35, 36, 39, 42, 47, 48, 81, 84
Coronaries, 42, 43, 76, 85
Coronary arteries, 58, 66, 75, 76, 80, 84–88, 90–95, 97, 155
Coronary blood flow, 75, 77, 78, 83, 84, 89, 90
Cyclostomes, 87, 88

D
Dihydropyridine receptor, 104

E
Eel, 38, 40, 44, 46, 60, 61, 63, 64, 67–71, 93
Elasmobranchs, 10, 11, 82, 84, 87–90, 92, 93, 96, 97
Embryogenesis, 48, 177, 185, 188, 192, 193

D. Sedmera and T. Wang (eds.), *Ontogeny and Phylogeny of the Vertebrate Heart*, DOI 10.1007/978-1-4614-3387-3, © Springer Science+Business Media, LLC 2012

Endocardium, 37, 38, 47, 56, 150
Endothelium, 47
Epicardium, 37, 56, 65, 155, 156
Evolution, 1–28, 39, 49, 55, 56, 60, 70, 75–98, 119, 125–142, 207–209, 211, 223
Evolution of endothermy, 125, 138–142, 208
Excitation-contraction coupling (E-C coupling), 103–106, 108, 109, 112, 117, 120
Exercise, 61, 93, 153, 214

F
Fish, 18, 25, 35–40, 42, 44, 46, 48, 49, 55–71, 75–98, 105, 108, 109, 112, 113, 115–119, 134–136, 141, 156, 211
Functional morphology, 36, 46, 55, 56
Functional plasticity, 61

G
Growth factors, 66

H
Hypoxia, 16, 18, 76, 80, 88–90, 93, 96, 120, 157–158, 163–166

I
Invertebrate, 2, 4, 6–12, 18, 19, 23, 26–28

L
L-type Ca^{2+} channel, 69, 104–106, 109, 114, 115, 117, 118, 136, 154

M
Mouse (embryo), 66, 154, 157, 181, 192, 195, 198
Muscle contraction, 136
Myocardial hyperplasia, 61
Myocardial hypertrophy, 155
Myocardium, 7, 9, 11, 14, 15, 39–41, 44, 47–49, 56, 59, 60, 65, 66, 76–85, 88–98, 105, 134, 150–157, 160–166, 180–187, 190–197, 199
Myocyte, 1, 43, 58, 77, 104, 125, 155, 182
Myosin binding protein C (MyBP-C), 126, 130, 134–136

N
Neuro-hormonal, 13–16
Nitric oxide, 38
Nitric oxide synthase (NOS), 67, 69–71

O
OFT. *See* Outflow tract (OFT)
Ontogeny, 2, 12, 19–22, 76, 88, 98, 149
Outflow tract (OFT), 36–40, 56, 66, 76, 81, 84, 90–92, 95, 97, 98, 153, 154, 181, 183, 184, 186, 188–190, 192, 197, 210
Oxygen partial pressure, 77, 81, 88, 96, 97

P
Phospholamban, 69, 70, 130, 136
Phylogenetic analysis, 126
Phylogeny, 2, 6–8, 26, 71, 76, 86–88, 95, 98, 109, 127, 207–209
Physiology, 2, 13, 35, 39, 40, 89, 166, 211, 213
Plasticity, 60, 61, 79, 119
Protein evolution, 125–142
Protein phosphorylation, 126, 139

R
Remodeling, 60–64, 70, 71, 156–158, 165, 207
Reptiles, 95, 96, 105, 107–109, 117, 134, 141, 142, 209–224
Ryanodine receptor (RyR), 69, 104, 106, 115–117, 119, 154
RyR. *See* Ryanodine receptor (RyR)

S
SERCA (SR Ca^{2+} ATPase), 104, 105, 112–114, 117–119, 130, 132
Spongiosa, 41–46, 48, 58, 60, 61, 63, 64
Spongious, 59
Spongy myocardium, 76–78, 81–83, 87, 88, 91, 95, 96

T
Teleost, 35–50, 56, 58, 61, 66, 81, 84, 85, 89–93, 96–98, 109, 116, 125, 126, 134, 142
Temperature, 14, 18, 39, 95, 105, 106, 109, 111–113, 117–119, 125, 127, 134, 139, 141, 142, 159, 218, 219
Thin filament regulation, 129
Troponin complex, 130, 131, 133, 138, 141
Troponin I, 126, 129, 130, 133, 135

V

Venous, 18, 24, 36, 58, 59, 75, 77, 81, 83, 88, 89, 92, 93, 95–97, 180–183, 185, 194, 209–211
Ventricle, 4, 35, 56, 75, 105, 125, 150, 181, 207

X

Xenopus, 52, 53, 74, 174, 203

Z

Zebrafish, 36, 47, 50, 55, 65–67, 71, 115

MIX
Papier aus verantwortungsvollen Quellen
Paper from responsible sources
FSC® C105338

If you have any concerns about our products,
you can contact us on
ProductSafety@springernature.com

In case Publisher is established outside the EU,
the EU authorized representative is:
Springer Nature Customer Service Center GmbH
Europaplatz 3, 69115 Heidelberg, Germany

Printed by Libri Plureos GmbH
in Hamburg, Germany